TRADITION AND INNOVATION

# THE UNIVERSITY OF WESTERN ONTARIO SERIES IN PHILOSOPHY OF SCIENCE

A SERIES OF BOOKS
IN PHILOSOPHY OF SCIENCE, METHODOLOGY, EPISTEMOLOGY,
LOGIC, HISTORY OF SCIENCE, AND RELATED FIELDS



VOLUME 56

J. E. McGUIRE

*Department of History and Philosophy of Science,*
*University of Pittsburgh, U.S.A.*

# TRADITION AND INNOVATION
## *NEWTON'S METAPHYSICS OF NATURE*

KLUWER ACADEMIC PUBLISHERS
DORDRECHT / BOSTON / LONDON

**Library of Congress Cataloging-in-Publication Data**

McGuire, J. E.
Tradition and innovation : Newton's metaphysics of nature / by J.E. McGuire.
p. cm. -- (The University of Western Ontario series in philosophy of science ; v. 56)
Includes bibliographical references and index.
ISBN 0-7923-3617-8 (alk. paper)
1. Newton, Isaac, Sir, 1642-1727--Contributions in philosophy of nature. 2. Philosophy of nature--History--17th century. 3. Philosophy of nature--History--18th century. I. Title. II. Series.
B1299.N34M37 1996
113'.092--dc20 95-30104



ISBN 0-7923-3617-8

---

Published by Kluwer Academic Publishers,
P.O. Box 17, 3300 AA Dordrecht, The Netherlands.

Kluwer Academic Publishers incorporates
the publishing programmes of
D. Reidel, Martinus Nijhoff, Dr W. Junk and MTP Press.

Sold and distributed in the U.S.A. and Canada
by Kluwer Academic Publishers,
101 Philip Drive, Norwell, MA 02061, U.S.A.

In all other countries, sold and distributed
by Kluwer Academic Publishers Group,
P.O. Box 322, 3300 AH Dordrecht, The Netherlands.

*Printed on acid-free paper*



Printed in the Netherlands

In memory of Betty Jo Teeter Dobbs: she dwelt in the temple of the Colossus of Cambridge.

TABLE OF CONTENTS

# PREFACE

In my early years I was constituted in the exacting imperatives of philosophical analysis. That stern face is present in the composition of the Newton essays chosen here for republication. It is my hope that potential readers will be patient with the old Adam of analysis, and seize the portrait of Newton's intellectual world presented in these essays. It is gratifying for me to acknowledge the encouragement of Robert Butts and John Nicholas of the University of Western Ontario, intellectual comrades in arms. It was at Western that I began my intellectual journey, and many of the present members of the Philosophy Department remain my friends and mentors. I thank also Marta Spranzi-Zuber who long ago believed in the merit of my Newton scholarship. But most important to me is the sustaining encouragement of Professor Barbara Tuchanska, who shares my vision of the historicity of scientific thought. It is a pleasure to express my gratitude for membership, over twenty years, in the Department of History and Philosophy of Science at the University of Pittsburgh. It is the mecca for one who seeks to understand.

J. E. McGuire, *Pittsburgh*

# ACKNOWLEDGEMENTS

I wish to thank the following publishers, institutions, and journals for permission to reproduce the essays included in this volume: The Society for the Study of Alchemy and Early Science who publish *Ambix*; the Reidel Publishing Company for permission to reproduce my essay in *Nature Mathematized*: Taylor & Francis Publishers who publish *Annals of Science*; Springer-Verlag Publishing Company who publish *Archive for the History of Exact Sciences*; The History of Science Department, University of Aarhus, and Munksgaard International Publishers Ltd., who publish *Centaurus*; and Elsevier Science Ltd., who publish *Studies in the History and Philosophy of Science.*

## BIBLIOGRAPHICAL DETAILS IN THE ORDER OF THE TABLE OF CONTENTS

1. "Existence, Actuality and Necessity: Newton on Space and Time," *Annals of Science* **35** (1978), 463–508.
2. "Atoms and the 'Analogy of Nature': Newton's Third Rule of Philosophizing," *Studies in the History and Philosophy of Science* **I** (1970), 3–58.
3. "Body and Void and Newton's *De Mundi Systemate*: Some New Sources," *Archive for the History of Exact Science* **3** (1966), 206–248.
4. "Space, Geometrical Objects and Infinity: Newton and Descartes on Extension": *Nature Mathematized*, ed. William R. Shea, 69–112 (Reidel, Dordrecht: Holland, 1983).
5. "Force, Active Principles, and Newton's Invisible Realm," *Ambix* **15** (1968), 154–208.
6. "The Origin of Newton's Doctrine of Essential Qualities," *Centaurus* **12** (1968), 233–260.
7. "Transmutation and Immutability: Newton's Doctrine of Physical Qualities," *Ambix* **14** (1967), 69–95.

# INTRODUCTION

Written some years ago the essays selected for this volume have made a modest way in the world. For this reason I have decided to bring them together. I shall not try the reader's forbearance with candid avowals of their imperfection: these are evident enough. Yet it is doubtful that I would now reach conclusions differing greatly from those they express. It is not my intention, in saying this, to mitigate the value of subsequent scholarship from which I have learned much. Rather, my hope is that these interpretations still have a useful role to play in understanding Newton.

Let me say something about these essays. First, they have a unified subject matter. As a group they focus on the central concepts of Newton's metaphysics of nature: God, infinity, space, time, matter, force, and activity. Moreover, they are textually oriented. They deal with texts published by Newton in his lifetime, and with manuscript sources pertinent to the central issues of his natural philosophy. Most of these manuscript sources appear only in these essays.

But the essays manifest a deeper unity: they project a particular vision of Newton's thought. In 1966, with Piyo Rattansi, I published "Newton and the Pipes of Pan."[1] This article is based on manuscript material Newton prepared for the opening propositions of the "System of the World" in Book III of the *Principia*. In these manuscripts Newton analyzes ancient thought in philosophy, mythology, and theology. He intended to append this material to the opening propositions of Book III, in an unpublished second edition of his master-work prepared during the early 1690s. These propositions embody the argument that all planetary bodies, their interstitial particles included, mutually gravitate toward each other in empty space. Newton's purpose in these "classical scholia" to Book III, is to argue that the ancients had foreshadowed a central doctrine of the *Principia*: that all bodies gravitate mutually in direct proportion to their masses and inversely as the square of the distance between them.

For Newton, the incorporation of ancient wisdom into his vision of nature is more than a ritualistic deference to tradition: it constitutes an active appropriation of tradition into the structure of his understanding of nature. As Newton construes it, understanding comprises more than sets of propositions linked together into chains of argument, or the active comprehension of the content either of propositions or occurrent mental states. For him, it is an event dynamically poised at the cognitive interface between historical patterns constitutive of particular cultural embodiments and individual perspectives emerging from those embodiments. Indeed, for Newton, tradition is a cultural appropriation that both enables and limits innovative thought. It does not exist passively in an objectified past, but actively in the very interstices of intellectual life. Thus, the transformation of *renovatio* into *innovatio* involves an active interrogation of a living past by a mind at liberty to think. Newton's position well illustrates Pelikan's dictum: "Traditionalism is the dead faith of the living: Tradition is the living faith of the dead".

Interpretation, then, is a central Newtonian concern. Newton left no source unturned in furthering the recovery of original wisdom. Scripture, history, the chronicles of the ancient kingdoms, religion, mythopoeic thought, the hexaemeral tradition, hermeticism, astronomy and astrology, philosophy, the wisdom of Judaeo, and alchemy, are marshalled in his quest for the origin of truth and reality. As Dobbs has recently reaffirmed,[2] central to this quest is insight into how divine activity fashioned the intricacies of material world. In this regard, the Arian Christ – in Newton's view a Christian demiurge enacting the bidding of the high God – constitutes the paradigm of creative mediation in the natural world. Newton's models of aethereal action – whether material, elastic, alchemical, or electrical – are further attempts to comprehend the active mediation that exists between the high God and nature.

The notion of active mediation is important. Newton is concerned not only with the symmetry of the world, but also with the agency that connects and forms that symmetry. In Newton's view, truth is indivisible and expressed in different discourses. Therefore, the discourses he scrutinizes are not for him isolated vocabularies, but alternative representations of one and the same reality – the truth of the created world. Consequently, his quest seeks for an understanding of the concealed wisdom present in the texts and traditions at his disposal, and the unity embodied in their intertextual and cultural vision. In this regard, the discourse of the *Principia* – the mathematical language of the book of nature – is itself a part of the grand intertextual discourse representing that unified vision. Newton does not, however, view the work of his predecessors as a standard to be slavishly copied. Quite the contrary, their writings must be interpretatively transformed into holistic hypertexts that will help in the restoration of the true philosophy. Moreover, these hypertexts in turn constitute an event of understanding in an on-going process of interpretation. Thus, for Newton interpretation is not the mere reduction of documents to preferred textual protocols; it mandates the construction of hypertexts that historically relate to their subtexts and enter the emerging continuum of human understanding.

But understanding involves more than interpretation and textuality. It is an event embedded always and already in discourse, the historico-social, and the real. For Newton the lost perennial philosophy must be restored. He is not content, however, simply to show that the ancient atomists anticipated his theory of matter, Aristotle his conception of inertial motion, the Stoics his vision of divine action in the world, and the "Platonizing" philosophers his mathematical representation of gravitational action on passive matter. Every expression of knowledge, ancient as well as modern, must satisfy the tribunal of communal experience; must be embodied effectively in human practices; and must have a purchase on the real. In Newton's view, if a scriptural text forecasts an event's occurrence, its advent must be ascertainable through the practices of science, or the techniques of historical reconstruction, or the methodologies of textual criticism. In fact, for Newton, a consilience of practices toward an end serves to authenticate the events in question. Given his cognitive orientation, the priority of the real over the social or the discursive over the historical is never in question. They are parts of a seamless whole comprising the emerging totality of human experience. The pressing task is to recast human practices and representations to better reveal the perennial philosophy of nature. In this endeavor, the distance of time is vital to the recovery of what is lost.

Newton is a hermeneutical thinker, alchemical exegete and mathematician of nature. His unified vision of the true natural philosophy, and the conception of understanding it embodies, is the central concern of these essays. Many deal explicitly with Newton's use of ancient sources, and all with a central issue in his thought – the relation of divinity to the natural world. In this regard, Newton commits himself to sustained reflection on the religious, moral, historical, social, and cosmological reality of the universe. This sensibility he expresses eloquently in the *Opticks*:

> And if natural Philosophy in all its Parts – shall at length be perfected, the Bounds of Moral Philosophy will be also enlarged. For so far as we can know by natural Philosophy what is the first Cause, what Power he has over us, and what Benefits we receive from him, so far our Duty towards him, as well as that towards one another, will appear to us by the Light of Nature. And no doubt, if the Worship of false Gods had not blinded the Heathen, their moral Philosophy would have gone farther than the four Cardinal Virtues; and instead of teaching the Transmigration of Souls, and to worship the Sun and the Moon, and dead Heroes, they would have taught us to worship our true Author and Benefactor, as their Ancestors did under the Government of *Noah* and his Sons before they corrupted themselves.[3]

This closing passage of his great scientific treatise, links together the pristine wisdom of nature and the primitive theology. Clearly, Newton believes that the moral and social orders are rooted in the natural order. Thus, true natural philosophy leads to a surer knowledge of God, and hence to a firmly-grounded moral philosophy. The reference to Noah and his sons is explained on the assumption that, if the true religion and morality follow from the true natural philosophy, then the latter served as the foundation for the former in the pristine age before the corruption of Noah and his sons.

Central to how Newton understands God's relation to creation are the themes of anthropocentrism, geocentrism, and theocentrism current in Renaissance and early modern cultures. A brief consideration of these notions challenges the belief that clear historical connections link an anthropocentric view of human nature to the geocentric cosmology of Aristotle. Despite the fact that his cosmology is geocentric, Aristotle believes that anthropology is subordinate to cosmology. After all, the lesser gods dwell in the perfection of the heavenly bodies, the distant majesty of which is hardly subservient to human needs and purposes. Of the ancients, only the Stoics hold that God and humankind are citizens of one cosmic *polis*, with God, man, and the cosmos engaged in a drama of mutual fulfillment. These views differ from those that predominate in Christianity. Standardly, in Christian thought, the cosmos is alien to the redemptive needs of human nature. Moreover, it yields to the biblical injunction that mandates divine not human dominion. Ultimately, this leads to the destruction of the cosmos in the final conflagration. In Christianity, therefore, a tension is present between the cosmology of the created world and the redemptive imperatives of human nature. This view is clearly theocentric, and indicates that there is no dominant anthropocentric interpretation of geocentrism.

If humanity's removal from the center of the world was culturally traumatic, that was not because Copernicus, by putting the earth into motion, prised apart anthropocentrism and geocentrism. For Copernicus, we are at the center of the world, not in virtue of our place in the hierarchy of being, or because of the privilege of geocentric position, but in terms of our ability to understand the world as a harmonious object of reflective thought. Accordingly, we are a cognitive microcosm able to interiorize the macrocosm in terms of geometrical representations of its outer structure. In this way, in our capacity to construct representations, we are the mirror in which the cosmos is cognitively reflected. It is this privilege that puts us at the center of the scheme of things.

Copernicus' cognitive anthropocentrism, with its denial of geocentrism as a condition of human dignity, is shared by both Kepler and Newton. Like Copernicus, they believe that human reason can unravel the mystery of the cosmos through experience, and on this basis construct an understanding of how God stands to creation. Indeed, for Kepler, reason grasps the geometrical harmonies reflecting the divine archetypes of creation. Kepler is the Copernicus of physical astronomy, who breaks with the circularity of planetary motion, and with the physics of the sphere in favor of the ellipse. Moreover, for Kepler the sun is the physical center of the planetary system and the force that drives the planets. Causality is directed from the center, and its effects are expressed precisely in geometrical laws relating changes in planetary speeds and positions along determinate elliptical paths. Kepler's program for a dynamic astronomy, based on physical causes that emanate from a sun-centered world, is the true foundation of Copernicanism. It is the program brilliantly brought to completion with the publication of Newton's *Principia* in 1687.

But the optimistic anthropocentrism of Copernicus, Kepler, and Newton is tempered by the theology of God's *potentia absoluta* and *potentia ordinata*.

This theology, constitutive of the reformed cultures of Europe, affected the interplay of reason and experience in 17th century natural philosophy. The idea that God had created the world by an arbitrary fiat originated in the thought of Origin, Paul, and Augustine. This view was developed further in the medieval thought of Buridan, Scotus and Ockham. This voluntarist view of God's sovereign dominion was later reaffirmed in the theology of Luther and Calvin. According to this tradition, the divine attribute the human mind comprehends is divine omnipotence expressed through the natural order. Emphasizing the absoluteness of divine liberty, voluntarism stresses the impossibility of knowing the purposes of God's *potentia absoluta* implicit in the final causes of creation. Thus, reason cannot grasp the inner workings of the divine mind, and cannot know the ultimate causes that underlie the structure of creation. This theological dialectic provided cultural support for an increasing appeal to experience and experimental practice in the face of the diminished role of reason. If reason is unable to fathom the archetypal causes of creation, it can accumulate descriptive knowledge of experience and thereby explore the ways in which God's *potentia ordinata* manifests itself through efficient causes. As Anneliese Maier points out, voluntarist conceptions of God in medieval culture entrenched the methodological distinction between theological thought and natural philosophy.[4] The latter enterprise works with experience and studies God's ordinary power which it conceptualizes through the operation of efficient causes. Final causes have no immediate role to play in natural philosophy as so conceived.

As my article "Force, Active Principles and Newton's Invisible Realm" shows, Newton accepted this distinction and the voluntarist tradition behind it. He stresses the power and freedom of God's will, arguing that God abides with the lawful constitution of nature as he has in fact ordained it. Nevertheless, in virtue of his absolute power, God could have created a world other than he did, and can now alter nature's present constitution by an act of unconstrained volition. Newton's voluntarism did not diminish his adherence to cognitive anthropocentrism. He never wavers in his belief (one shared with Copernicus and Kepler) that the human mind can approach final causes anagogically through experience. But Newton insists that the design of nature must be established first before we can argue to the presence of the hidden designer. Even so, an ascent to higher principles will not reveal the inner essence of God's absolute will. Newton is firmly committed to the view that divinity manifests and communicates itself in what it creates. His God is the pantokrator, or the lord of the universe, the supreme being who exercises direct dominion over the constitution of creation. It is God's actions, emanating from a free and omnipresent will, that are manifest in the natural world, not the transcendent perfection of divine nature. Newton insists that God is best known from the design of creation, a structure that reveals the power and wisdom of divine action.

Despite the endemic skepticism of his culture, Newton never doubted that the wisdom of creation is recoverable. He views the traditions he studies as records of the complex human endeavor to understand creation. Although these human products lack consistency with one another, Newton believes that a deeper interpretive synthesis will transcend these discrepancies, and reveal the fundamental unity of divine truth. Interestingly enough, Newton's notion of experience goes beyond the deliverance of the senses and the protocols of experimental practice. Natural, moral, and divine knowledge comprise a seamless whole and conspire to reveal the harmonious truth of God's creation. Moreover, these traditions are constitutive of the emerging character of human experience and, when correctly interpreted, mutually reinforce and correct one another.

I close with a few remarks on the writing of these articles. They are exercises in neither the social nor the institutional history of science. My approach could well be called historico-philosophical. It attempts to consider Newton's views within their context, and to assess critically the traditions that he affirms and rejects. Moreover, it treats Newton's work as the product of a mind in action. To this end, my purpose is to establish the motives and intentions that inform his reasoning; to unearth the unstated assumptions and presuppositions that drive his arguments; and to assess the implications of the positions he develops. Consequently, the unit of analysis is not the disembodied "idea" or "proposition" detached from embodiment in Newton's mental activity. Rather, I focus on how Newton organizes his experience in light of the cultural resources available to him, and how he appropriates texts expressing certain beliefs and representing certain traditions. In sum, Newton's discourse is contextualize to show the concrete historicity of his achievement.

## NOTES

[1] McGuire, J. E. and Rattansi, P. M., "Newton and The Pipes of Pan", *Notes and Records of The Royal Society of London* **21** (1966).

[2] Dobbs, B. J. T., *The Janus Faces of Genius: The Role of Alchemy in Newton's Thought* (Cambridge, University Press, 1991).

[3] Maier, Annaliese, *Studien zur Naturphilosophie der Spätscholastik* (Rome: Ed. di Storia e letteratura, 1949).

CHAPTER 1

# EXISTENCE, ACTUALITY AND NECESSITY: NEWTON ON SPACE AND TIME

## 1. INTRODUCTION

In his learned work *The True Intellectual System of the Universe* (1678), the Cambridge Platonist Ralph Cudworth (1617–1688) criticizes Pierre Gassendi's view that space is neither a substance nor an affection of substance. Cudworth notes that Frenchman allows space to be an 'incorporeal thing'. Nevertheless, he rightly observes that for Gassendi it neither follows 'of necessity from thence, that it is an incorporeal substance or affection thereof; nor yet that it is an accident existing alone by itself, without a substance; because this space is really neither accident, nor substance, but a certain middle nature or essence betwixt both'. Cudworth speaks for a majority of thinkers in the seventeenth century when he argues that whatsoever exists 'or hath any kind of entity, doth either subsist by itself, or else in an attribute, affection, or mode of something, that doth subsist by itself'.[1] These contrasting opinions divide the thought not only of Cudowrth and Gassendi but also of Henry More and Isaac Newton. Like Gassendi, Newton holds that space (and also time) is strictly neither a substance nor an affection of substance. Henry More, on the other hand, agrees with the opinion that Cudworth defends, which is that all existents are in either the category of corporeal or incorporeal substance or else dependent attributes and modes. For these reasons alone it is worth investigating the possibility that Newton was indebted to Gassendi. In addition, the question of More's influence on Newton's thought will be re-examined.

These questions are pursued in the context of considering Newton's views concerning the existence of space and time in relation to that of God. This analysis leads to special issues which require separate comment. Chief amongst them are: the nature and necessity of Divine existence; Newton's special ontology of space and time; and the question of what it is to say that something exists. The evidence on which my analysis rests is drawn from three main sources in Newton's published writings: *De gravitatione et aequipondio fluidorum* (c. 1666–70); the Scholium on space and time of the *Principia* (1713).[2] In addition, evidence will be cited from a manuscript which I entitle 'Tempus et Locus'. This document can be dated from the early 1690s, and I analyse it in another paper.[3] Moreover, it was written during the period in which Newton was busy revising the first edition of his *Principia*. Other sources are cited when necessary.

As I argue in the above mentioned paper, the general metaphysical position that Newton adopts in *De gravitatione* provides the ongoing framework in which he develops his views on space, time, motion and related issues.[4] In the present essay I shall assume the validity of this claim in dealing with the questions to be considered. The basic conclusion to emerge is this: Newton is largely a Gassendist on the problems of space, time and existence. This is not to deny that Henry More's Neo-Platonism contributed to his intellectual background. However, Newton's indebtedness to More is more ramified than has previously been appreciated.[5] Lastly, I shall contend throughout that Newton's views on space and time are best understood with respect to the relevant philosophical discussions of his period.

## 2. SPACE AND TIME

Newton's most succinct statement of how place and time relate to existing things occurs in the manuscript 'Tempus et Locus' (c. 1692–93). It begins: 'Time and Place are common affections of all things without which nothing whatsoever can exist. All things are in time as regards duration of existence, and in place as regards amplitude of presence. And what is never and nowhere, it is not in the nature of things [*rerum natura*]'.[6]

Newton answers the question: what is it for anything to exist in nature? It is to exist in a place and at a time. As the text implies, existing in place and time is what counts as actually existing; in contrast, for example, to existing in the manner of an abstract entity or as a number. This contention is supported by Newton's use of the phrase 'rerum natura'. In the early 1690s Newton read Lucretius's *De rerum natura*, in which an account of the nature of real existence is put forward.[7] Lucretius holds that things exist actually just in case they can be said to endure and to have a place in space.[8] In general terms Lucretius's position agrees with the view which Newton expresses. This is further borne out by the fact that Newton also cites Lucretius's doctrine of atoms moving in void space as an anticipation of his own account of the nature of matter.[9]

Unremarkable though Newton's position on actual existence seems, it rests on a particular view of predication and its appropriate objects. In order to develop this, let us begin by considering Newton's disarmingly simple statement. It characterizes time and place as 'common affections' of all existing things. In the seventeenth century there was no general agreement as to what counts as a common affection. Lists of such notions are as varied as the writers who discuss them.[10] Broadly construed, however, an affection is a general characteristic of things, which pertains to either their nature or to their existence. A *common* affection is one which applies universally to all things, and is not restricted to particular kinds. Moreover, affections can be conceived as conceptual characterizations through which we understand something's nature or existence, or can be said to be ontological features pertaining to things. From the context of the passage it is clear that Newton conceives time and place to be common affections in an ontological sense. In his view, real things require them so that they can be said to endure in their individual actuality.

We can now state more fully what Newton intends to convey in calling place and time 'common affections'. As the passage indicates, they are not properties which inhere in things, but rather characteristics pertaining to their existence. For the weight of interpretation falls on the phrases 'duration of existence' and 'amplitude of presence'. As employed in the passage they indicate that time and place specify anything's actual existence, rather than constituting a part of its defining nature. So considered it is probable that Newton has in mind a traditional distinction among affections. Those which characterize things in virtue of their nature, and those that wholly pertain to their sheer existence. In Newton's view, time and place are in the latter category. Individual things not only endure in time; they are present in the same or different places through time. Because place and time universally pertain to the duration and presence of things, they are common affections of their existence.

But what commitments lie behind Newton's view? The key phrases in the passage express the view that 'all things are in time as regards duration of existence, and in place as regards amplitude of presence'. These same phrases are found in another important passage from Newton's earlier *De gravitatione*. It unmistakably relates to the first, not only with respect to its general conceptual orientation, but also in its style of language. Accordingly, it can be considered as expressing at least some of the basic commitments implied by the cryptic statements of the former. Moreover, it is a paradigmatic passage on which to base a summary of Newton's conceptions of place and time. It sets a framework for what is to come, and is thus worth quoting in full:

> Space is an affection of being in so far as it is being. No being exists, or can exist, which is not related to space in some way. God is everywhere, created minds are somewhere, and body is in the space that it occupies; and what is neither everywhere nor anywhere does not exist. Hence it follows that space is an emanative effect of the primarily existing being, because when any being is posited, space is posited. The same may be affirmed of duration: both are affections of being or attributes according to which the quantity of existence of each individual [being] is denominated as regards its amplitude of presence [*amplitudinem praesentiae*] and its perseverance in existence. So the quantity of existence of God is eternal in relation to duration, and infinite in relation to the space in which he is present; and the quantity of the existence of a created thing is as great, in relation to duration, as the duration since the beginning of its existence, and in relation to its amplitude of presence as great as the space in which it is.[11]

In the first instance, Newton states that space and duration are general affections of all being. However, this statement is qualified significantly. His considered opinion is that space and duration are affections of things which pertain to the fact of their existence. In other words, space and time are said to characterize the *existence* of all being. This is the thrust of the notion that everything has a 'quantity of existence'. Moreover, Newton relates this notion to the concepts of 'amplitude of presence' and 'duration of existence', whereas in the 1690s passage the relation is merely implied. So the duration of anything in time and its presence in space – the defining characteristic of which is extension – are said to characterize its actual 'quantity of existence'. Thus in saying that space and duration are 'affections of being', Newton is not saying that being *per se* entails extension and duration (that is, that to be is to be extended and to endure, though anything that is *in* space is extended). Rather he invokes the distinction between affections that characterize natures *qua* natures, and those that pertain to the *fact* that an individual nature exists. Of the latter kind are space and time, including the relational properties among existents that this involves; while the former kind specify the essential, appropriate (*propria*) or accidental features of a given nature. As the passage states, space and time are also 'common affections' of God's 'quantity of existence'. The main burden of the following analysis will be to concentrate on the clarification of this claim.

Nevertheless, as it stands, its meaning is straightforward enough. Contingent things have a finite quantity of existence in relation to their individual times and places, while God's is eternal with respect to infinite time, and omnipresent in relation to infinite space. Newton also indicates that these features of existing things provide a means of understanding the measure of time and quantity in so far as they apply to notions of change and permanence. Moreover, to say of something that it endures is not to ascribe a real property to it. For in reality something's duration is indistinguishable from its continuing existence. So to claim that something endures is merely to refer to its disposition with respect to space and time.[12]

Place, space and time function at two distinct levels, both of which relate to the conceptions of duration and 'amplitude of presence'. First, they have an ontological status. To say of anything that it endures, is to say that it exists and is localized in place and time. This is what it is to be in existence or to have individual actuality. Moreover, in Newton's view existing things are really distinct from space and time: 'although we can possibly image that there is nothing in space, yet we cannot think that duration lacks existence, even though it might be possible to suppose [*fingere*] that nothing whatsoever endures'. Of space he says: '. . . we can clearly conceive extension existing without any subject, as when we imagine spaces outside the world or places empty of body'.[13] The same point is put more succinctly in the Scholiumm on space and time of the *Principia*: 'absolute space, in its own nature, without relation to anything external, remains always similar and immovable'.[14] Time and space are therefore actual existents. By nature they are independent of all finite things which exist in them, whereas finite things depend on space and time for existence. Moreover, spaces and times empty of solid matter are not only comprehensible, but are required for real motion: 'Absolute motion is the translation of a body from one absolute place into another', as it endures through a determinate stretch of absolute time.[15] Time and space *per se*, however, are infinite consequences of Divine existence. For God is eternal. His life is infinitely extended both forward and backward in time. Moreover, he is substantially omnipresent at all times whatsoever.[16] So in contrast to the existence of finite things, God's existence is unending in duration and uncircumscribed by place. And there can be no truly empty times or places, since God is actually present with respect to all times and places whatsoever.

Space, place and time also have an epistemological status. The mind comprehends the duration and presence of things through the measures of time and quantity.[17] This means that the duration of sensible things is conceived in terms of temporal successiveness, whereas their individual extension, along with the places they occupy, is conceived through the category of permanent quantity. All sensible things are therefore perceived, according to time, as existing over a definite period and, according to quantity, as existing with a determinate extension (presence), and in a determinate place. God's quantity of existence, on the other hand, is said to be without limitation. For it is understood through the conception of infinity with respect to extension, and through that of eternity with respect to time.[18]

Newton is able in terms of this epistemological framework to distinguish his conceptions of absolute and relative place, space and time. These he discusses in the *Principia*.[19] Relative place, space and time are quantitative measures through which we conceive the independent existence of absolute space and time. The existence of the latter is not an object of sensory experience. It must be conceptualized according to relative measures which refer to phenomena affecting the senses.[20] These relative measures can be refined, however. Moreover, Newton holds that they progressively approximate to the true nature of actual space and time.[21] While in order of being space and time are prior to their conceptualization (absolute is prior to relative), in order of knowing they are not (relative is prior to absolute). Accordingly, absolute space and time are prior to, and independent of, human cognition.[22] They constitute the foundation for the intelligible order existing beyond the varieties of direct experience. Nevertheless, according to Newton, the mind is able to understand the infinite characteristics of absolute space and time, even though it may never have a completely adequate grasp of all that is intelligible in this respect.[23] In any event, relative measures of space and time should not be confused with their real natures. This would be to 'confound real quantities with their relations and sensible measures'.[24]

As we might expect, Newton's talk of the 'quantity of existence' of things, ultimately rests on distinguishing it from what counts as a real attribute, and on settling what sorts of characteristics can be directly predicated of individual things. So, too, does his conception of space and time as affections of existence. The *De gravitatione* passage is eclectic in character, and clearly reflects these concerns. Moreover, it manifests the influence of both Henry More and

Gassendi, the latter certainly known to Newton through his reading of Walter Charleton's *Physiologia*.[25] Indeed it was through his reading of Charleton's presentation of Gassendi's philosophy that Newton probably got his first acquaintance with the views on space and time of the ancient atomists. But the question of the proximate influences on Newton (More, Charleton and Gassendi), its intrinsic merits apart, provides a vehicle for developing an analysis of Newton's conception of predication. This in turn reveals further important aspects of his ontology and epistemology.

In its Latin text the first sentence of the *De gravitatione* passage reads: 'Spatium est entis quatenus ens affectio'. Here Newton speaks as a metaphysician. As such, he begins with what he takes to be the general characteristics of being *qua* being. Space is one of them. In his *Enchiridion metaphysicum* Henry More also begins by considering the fundamental attributes of 'Ens quatenus Ens'.[26] Among them is the doctrine that extension is a necessary attribute of all being whatsoever. The same point is made by More in his *Divine dialogues*: '*Extension or Amplitude* is an intrinsicall or essential Property of *Ens quatenus Ens*'.[27] Thus the claim that all being is extended is for More an integral part of his conception of the general and disjunctive features which pertain to being as such. In fact he states that 'ens quatenus ens' is 'an object of logic not of metaphysics'.[28] Moreover being, *per se*, not substance, is the *summum genus*. For being is predicated univocally of all individuals and their dependent modes.[29] God is necessary being; all else is contingent in nature.

Although other sources cannot be ruled out, it is reasonable to assume that Newton was at least familiar with More's *Divine dialogues*. While we know that he read an earlier treatise, *The Immortality of the Soul*,[30] in that work More does not express his doctrine that all being is extended by means of the phrase *ens quatenus ens*. Nor does he speak of extended space as the 'amplitude of presence' in any given thing. However, these phrases are used by Newton in *De gravitatione*, the latter being employed in the 1692 manuscript. In any event, there are striking similarities in language and concept between More and Newton, especially as More also characterizes duration as an affection of all being.[31] More also holds that if anything is said to be never and nowhere (that is, if it is said to be without extension and unrelated to the temporal order), it cannot exist, for only something extended or related to the extended, can be somewhere.[32] This conception is shared by Newton in *De gravitatione*. Indeed, it grounds Newton's objections to Descartes's conception of Divine existence.[33]

The case for More's influence can be strengthened. Newton uses the curious phrase 'effectus emanativus' three times in *De gravitatione*.[34] In each case he is characterizing the relation of space to God's being and existence. More also uses it in the context of saying how dependent things stand to their causes. In particular, he is concerned to characterize the way in which secondary substances relate to the essence of Divine nature.[35] He defines 'an emanative effect' as 'coexistent with the very substance of that which is said to be the cause thereof'. 'An emanative cause' is understood as 'such a cause as merely be Being, no other activity or causality interposed, produces an Effect'.[36] Clearly More's distinctions are relevant to Newton's thinking. It will therefore be important to decide whether they are pertinent to Newton's conception of how space and time relate to the existence of God.

However, the question of the sources from which Newton drew is more complex than this. The 'Quantity of existence' of things, as it relates to their actuality, does not appear to function as an attribute if that term is understood as pertaining to a given characteristic of their nature. Moreover, existing *per se* with respect to space and time is said to be a prerequisite for speaking of any individual thing's quantity of existence. This should lead us to suspect that Newton does not consider time, space and existence to function according to the logic of substance and attribute. The suspicion is well-founded. In Newton's view the ontology of substance and attribute is indeed inadequate as a complete explanation of existing things. But according to More it is a perfectly adequate ontology. For him all attributes depend on either corporeal or incorporeal substance.[37] Beyond this nothing whatsoever can be said to exist.

Following Charleton, and through him Gassendi himself, Newton denies that extended space is to be defined 'as substance or accident or else nothing at all'.[38] In an important passage from *De gravitatione*, which sets the scene for his discussion of the properties of extended space, Newton makes the following observation:

> [extension] has a mode of existing proper to itself that pertains neither to substance nor to accidents. It is not substance both because it does not subsist absolutely *per se*, but is, as it were, an emanative effect of God, [*sed tanquam Dei effectus emanativus*] and an affection of all being, and because it does not underlie [*substat*] proper affections of the kind that denominate substance, namely actions such as thoughts in the mind, and motions in body.[39]

In the course of his analysis, Newton applies the same reasoning to the nature of absolute time and duration. We could wish, perhaps, for a fuller discussion of the notions of substance, attribute and accident. There is no doubt, however, that Newton denies to space and time the logic which he takes to govern these general concepts. In fact, he denies the coherence of the notion of substance itself. It refers to 'a certain unintelligible reality that they call substance', in which all the qualities of bodies 'are said to be inherent'. However, in Newton's view, how is the notion of an unobservable substratum to be understood, especially as it is uncharacterizable independently of its role as a basis of instantiation; and what is meant by saying that qualities inhere in it? In contrast to this conception, Newton claims without argument that 'we have an exceptionally clear idea of extension' and of the general notion of extended space.[40] For More, on the other hand, God's eternal presence and infinite space are characterized by the same set of formal attributes, attributes that presuppose an incorporeal substance.[41] Duration is also seen by More as a real attribute of all things that exist, God included.[42] But here, as in the passage from the 1690s manuscript, Newton simply characterizes extension – the defining characteristics of space – as an affection of all being as such. What does this mean?

It is clear from the context that the term 'affection' connotes a general condition that pertains to any being whatsoever. This is distinct from Newton's characterization of the defining attribute of substance as its intrinsic capacity for action. After all, his examples of the relation that attributes bear to substance are thoughts *in* mind and motions *in* body. Thus, for Newton, attributes can define the nature of a substance, or at least can be seen as part of its essential definition. Space and time, however, cannot be said to contribute to the determination of a sort of thing. Nor are they substances *per se*. That category refers only to entities that are able to act or to be acted upon. By nature, space and time are devoid of causal properties and capable only of the passive disposition of receptivity.[43] Moreover, while a substance can be said to exist *per se*, as a complete and independent being, accidents, on the other hand 'are inherent in some subject'.[44] And they would cease to exist were all substance annihilated. Space and time, however, would continue to exist. If God, for example, were to destroy the defining properties of matter, say impenetrability and individual extension, there would remain 'only the uniform and unlimited stretching out of space in length, breadth and depth', as it coexists with infinite duration.[45]

So space and time, *qua* affections, simply characterize all being, whether it is substance, attribute or accident and are really distinct from existing things and independent of their natures. Their mode of predication is therefore external, as they are not real properties which attach to the nature of things in virtue of inherence. Space and time are neither *in* a substance in the manner of its extension and shape, nor are they *in* things as parts are in wholes and as a whole contain parts. Furthermore, they are not accidents or modes of existing things; for this allows that they exist only when finite things exist. In Magirus's *Physiologia peripatelica*, from which the young Newton made lengthy notes, the basis of these distinctions is put thus: space and time are said to be external affections in which things are contained, whereas properties are said to be internal affections of their nature.[46] It is clear that this remains the broad framework of Newton's mature position. In the end, however, he is unable to give an entirely satisfactory characterization of the ontology of space and time. Nevertheless, he does say that space (and by implication time) 'is more thing-like than an accident, and more closely approaches the nature of substance'.[47] This indicates that Newton has in mind a sense in which space is a substance. While it is not a substance in the sense of existing in itself, or in the sense of being the source of change in other things, it is nevertheless a distinct individual of which many things can be said. After all, in essence, space is fully actual and infinite extension to which many definite properties are truly ascribable. Given this sense of the term substance, Newton attempts to convey a conception of space as constituting an ominpresent 'receptable' or 'field' within which all things separately endure. However, those spatial notions do not pertain straightforwardly to the nature of God's existence. But this is to anticipate.

Thus actual or real existence requires reference to the independent existence of space and time. A similar claim is found in the writings of Charleton and Gassendi. For instance, Gassendi says that 'there is no substance and no accident for which it is not appropriate to say that it exists somewhere, or in some place, and exists sometime, or at some moment, and in such a way that even if the substance of the accident would perish, the place would continue, nonetheless, to abide and the time would continue to flow'.[48] Like More, Gassendi begins his discussion by considering the fundamental features of the *Ens quatenus ens*.[49] However, there is an important difference in their conclusions. More holds that the *summum genus* is the category of extended being under which the subaltern genera, corporeal and incorporeal substance, are subsumed.[50] Although he could not have known More's expression of it, the view that being is the *summum genus* is unacceptable to Gassendi. He regards substance as the basic category of individual things; and attributes are construed according to the Aristotelian conception of subject-predicate statements.[51] However, Gassendi also holds that place, space and time do not fit that model of predication.[52] They have their own individual mode of being. Moreover, their relation to substances and their properties is different from the relation of properties to a given substance. For substances, along with their defining properties, require the independent existence of space and time for their individual existence.[53] Both Charleton and Newton agree.[54] Indeed, it is in this sense that Newton understands the status of space when he asserts that it is 'an affection of being in so far as it is being'.[55]

## 3. EXISTENCE AND NATURES

In the wake of the Clarke-Leibniz correspondence, Newton wrote a number of draft sheets defining his conception of space, time, substance, attribute and existence. These were written about 1719, and represent a late account of his ontology of space and time. Newton's views on this question certainly underwent change in emphasis since the time of *De gravitatione*. However, the evident continuity in his position is of equal importance, and it is this that I wish to consider. Newton's drafts are versions of an 'Avertissement au lecteur' which Des Maizeaux in part incorporated into the preface of his *Recueil de diverses pièces* (1720).[56] The following passage is from one of the many similar drafts:

The Reader is desired to observe, that wherever in the following papers through unavoidable narrowness of language, infinite space or Immensity & endless duration or Eternity, are spoken of as *Qualities* or *Properties* of the substance wch is Immense or Eternal, the terms *Quality & Property* are not taken in that sense wherein they are vulgarly, by the writers of *Logick & Metaphysicks* applied to *matter*; but in such a sense as only implies them to be modes of existence in all beings, & unbounded *modes* & consequences of the existence of a substance which is really necessarily & substantially Omnipresent & Eternal; Which existence is neither a substance nor a quality, but the existence of a substance with all its attributes properties & qualities, & yet is so modified by place & duration that those modes cannot be rejected without rejecting the existence.[57]

Here Newton expresses the full implications of his position. Though written late in life, the passage can be shown to be germane to Newton's discussion of space and time in *De gravitatione*, as well as to views set forth in the *Principia*. Moreover, it allows us to see both continuity and change in Newton's treatment of the problems under discussion.

Newton makes three explicit claims. First, he claims that existence is not a real attribute or quality of the defining nature of substance. Second, he connects space and time with the concept of actual existence. Lastly, he states that these conceptions also pertain to God's existence.

Before discussing these claims it is well to notice Newton's denial that 'infinite space or Immenisty' and 'endless duration or Eternity' are '*Qualities or Properties*' of God's defining nature. These are not predicated of God in the manner in which qualities or properties are predicated of finite things. Rather, as he puts it in another draft variant, they should be taken 'in such a sense as if the Predicaments of *Ubi & Quando* should be taken 'in such a sense as if the Predicaments of *Ubi & Quando* should be called qualities or properties when applied to the existence of a being which is omnipresent & eternal'.[58] In other words, when God's being is characterized as immense and eternal we are not in fact strictly ascribing properties specific to him. Rather we refer to the manner of his existence. Thus the predicaments *Ubi* and *Quando* answer two general questions. Where does God exist?;' everywhere', in respect to space. When does God exist?; 'always', in respect to time. This is just as we say of finite things that they are 'here' or 'there' and 'now' and 'yesterday'.[59] This line of argument clearly relates to the claim that space and time are common and external affections which denominate the 'quantity of existence' of all things. Newton therefore treats those questions as making direct reference to things in the "parts" of absolute space and time, and to Divine existence in all times and at all places. Putting the point thus is also characteristic of the thought of Gassendi and Charleton. For them, as for Newton, there are no specific categories called "where" and "when" that are distinct from the common affections "place" and "time".[60] Given Newton's account of how space and time relate to finite things and to God, it is clear that he is not guilty of the charge levied by Leibniz: that he makes space and time real affections of God's essential nature. According to Leibniz's ontology of individual substance, all affections are simply attributes which literally and intrinsically inhere in the specific nature of a given substance.

So interpreted, Newton's concepts of space and time would indeed denote properties which inhere in Divine nature. However, it is just an ontology of this sort (of which Leibniz's interpretation is an extreme version) that Newton is concerned to contrast with his own ontology of space and time. Nevertheless, in the light of his doctrine of attributes, it is not surprising that Leibniz accuses Newton of conceiving God's nature as potentially mutable, and also of the absurdity of viewing God as a substance which exists in its own attributes.[61]

We can now turn to Newton's main claims. These clarify the proceeding analysis, and take us to the basis of his position. The explicit claim that existence is neither a quality nor a substance is crucial. Newton clearly denies that existence is a disjunctive property of a substance in the sense in which something is said to be a member of the class of a substance's 'attributes, properties & qualities'.[62] What we can reasonably assume to be implied is this. Existence *per se* is neither manifested as a single property, nor directly exemplified through properties (*propria*) which are appropriate to the nature of a given thing. It is not substance, since anything that is a substance has a nature and an essence which are its appropriate and defining properties respectively. Moreover, a substance *per se* is that in which its properties exist and through which it itself is perceived. Existence does not fit this ontology.[63] It functions in second-level statements which refer to the sheer actuality of individual natures. Such statements point to the fact that things exist extramentally. As such they exist *qua* actual beings together with their individuating features. Existence is therefore not that which is truly predicated in the category of property: that is, it is not a separate property or attribute which contributes to the defining nature of a thing. Rather, it is an irreducible "feature" of all individuals in so far as they are actual. But to be actual does not confer a distinct property or attribute. As is evident from the passage, Newton applies this conception to the existence of God as well as to finite substance. Thus God is a being – a single individual possessing negative as well as positive attributes – who in fact exists.

This seems a reasonable analysis of Newton's position as it is implied in the text. We must be cautious, however. As it stands, the text does not allow us to state categorically that Newton either affirms or denies that existence is a predicate. He certainly denies that it is a real attribute, property or quality of anything. However, to say this is perfectly compatible with saying that existence is a transcendental predicate. Or put otherwise, it is a second-level predicate: one which functions differently from attributes of the type that pertain to the defining natures of things. After all, Newton wants to say of individual things and of God that they exist. But this is to treat existence as a predicate of individuals in the sense in which it conveys information about them. Thus interpreted, Newton's conception of existence fits the logic of subject and predicate, but only if existence is construed as a transcendental or second-level predicate, not as a real attribute. This of course is consistent with saying, to use Mackie's phrase, that existence is 'colourless'.[64] For it makes no important difference to the concept of a thing, except in the broad sense of saying that something is a "such-and-such" whose existing is compatible with being in space and time. Nor does it contribute to the specific determination of a sort of thing. Both claims are in fact compatible with the analysis of Newton's position thus far. I shall return to this topic when considering the influence of Gassendi's epistemology on Newton's conception of existence.

Newton's second claim is equally important. As he denies, on the one hand, that space and time are substances *per se*, and on the other, that they are real attributes of God's defining nature, so equally he denies that they are attributes and properties of existence *per se*. Rather, in the first instance they are said to be "modes" of both Divine and finite existence. This is just to say that they are not real predicates of existence. For how could existence *per se* instantiate properties? But the terms 'attribute' and 'property' connote instantiation. Thus Newton employs the term 'mode', in referring to the relation that space and time have to something's existence, as he uses the term 'affection' in the other passages. Now as the term 'existence' functions as a transcendental predicate in reference to the individual actuality of any given thing, so the terms 'space' and 'time' are modes or affections which refer to the fact that a given thing exists in the general order of space and time. If this is not so, it can no longer be said to exist actually. Newton thus extends the traditional conception of individual actuality beyond the notion of that which exists *per se*. He insists that the concept also includes the notion of existing in the independent nexus of space and time. In this sense space and time are modes which 'cannot be rejected without rejecting the existence' of whatsoever is said to have actual existence.

For Newton, then, existing is what individuals do irrespective of what they are doing. By existing they instantiate the concepts which they in fact conform to. In its Latin usage *existere* literally means to stand out or to be visible: a usage more familiar to Newton's than to the modern mind.[65] Hence for Newton an existent is literally something that is *there* wating to be discovered at a particular time and in a particular place. Accordingly, he urges that the basic sense of existence is to occupy a definite position, or successively different positions, in space and time. To say of something that it actually exists is to say it is *there* in the objective order of space and time. Therefore, the 'Predicaments of *Quantitas Ubi & Quando* . . . are not qualities in a proper sense, but denote only places & times of the existence of beings contained in them'.[66]

This brings us to Newton's third claim, which involves saying that God's actual existence is with respect to space and time. Space and time are said to be 'unbounded *modes* & consequences of the existence of a substance which is really, necessarily & substantially Omnipresent & Eternal'. This statement relates to the earlier *De gravitatione* passages in which space and time are said to be emanative effects of Divine existence. However, it also relates to an important passage in the General Scholium to the *Principia* (1713):

He is not eternity and infinity, but eternal and infinite; he is not duration and space, but he endures and is present. He endures forever, and is everywhere present; and by existing always and everywhere, he constitutes duration and space. Since every part of space is *always*, and every indivisible moment of time is *everywhere*, certainly the Maker and Lord of all things cannot be *never and nowhere*.[67]

These passages, in conjunction with the previous analysis, reveal important aspects of Newton's conception of the actuality of Divine existence.

In the first place, in the *Principia* passage Newton distinguishes between the nature of Divine substance, and the space and time of his existence. In other words, he states that space and time are not real attributes of God's nature. On the contrary they themselves are, as Newton rather oddly puts it, *constituted* by God's eternal and omnipresent existence. This way of phrasing the point is odd because it implies that space and time are "ingredients" of God's nature. But Newton explicitly states that God is 'not duration and space'; rather they themselves exist because God *exists*. As such they alone denominate the actuality of his 'quantity of existence'. The Des Maizeaux drafts confirm this view:

. . . when the Hebrews called God *MAKOM* place, the place in wch we live & move & have our being & yet did not mean that space is God in a literal sense. For they used to speak of God by figures & allusions & put space for his omnipresence by a figure. And so space & duration are by the writers of Logicks & Metaphysicks called quantities & with respect to God's ubiquity & omnipresence they have a nearer relation to the Predicaments of *Ubi & Quando* then to that of Quality & therefore where they are called Qualities the Reader is to understand it with a figure.[68]

It is in this sense that Newton wishes to speak of 'the boundless extent of God's existence with respect to his ubiquity & eternity'.[69] Just as infinite space and time are not literally "part" of Divine nature, so God is not literally immense in the sense of being "infinitely extended", or literally identical with eternity itself. These are all 'modes' or 'affections' which uniquely characterize the actuality of God's omnipresent existence. For 'to exist in this manner is proper to him alone'.[70] Notice that Newton is not embracing pantheism, since he wishes to assert the contrary truth that God is not identical with space.

These passages give a clear expression of Newton's general conception of predication. As he denies that God *is* Eternity and *is* Infinity, so he denies that God is identical with Wisdom and Power. In other words, he opposes those theologians who hold that God's attributes are strictly one and the same with Divine essence and existence, as well as identical among themselves. According to Newton, Divine nature is just, wise, eternal and infinite. These characteristics are attributable to God in a perfect manner (in comparison with those which are appropriately ascribable to finite things), but none is identical with his nature. Equally, in characterising God as existing and enduring with respect to space and time, we are not saying that God is identical with infinite space and time.[71] So in denying that strict identity conditions pertain to both the individual nature of God and to his attributes, Newton must answer the objection that his conception of predication makes God divisible and a composite nature of substance and attribute. Equally, he must show that his conception of God's existence with respect to space and time, does not make Divine nature liable to change. This and the following sections will discuss these questions in connexion with Newton's accounts of Divine perfection, infinity and necessity. Not surprisingly, his position is different from those theologians who hold that God's nature is identical with his attributes.

Let us bring out the consequences of Newton's position on "Divine" predication, by contrasting it with More's. More holds that the extension of all things proceeds by emanation from the 'first and primest essence'. For More this is ultimately the spiritual presence of God whose extension or Divine amplitude is infinite. Thus the extended natures of secondary substances are produced 'by gradual emanation' from the primary substance of God, so long as the 'substance produced be in due gradual proportions inferior to that which causes it'. Furthermore, More holds that 'it implyes a contradiction that an Emanative effect should be disjoyned from its originall'.[72] He later argues for the same general conception of God's nature in relation to the existence of created things. In his *Divine dialogues* he says that 'every physical Entity has an intrinsicall Extension of its own, as much as it has an intrinsecall Essence of its own'. Moreover, what is intrinsic and essential to a thing 'is the thing itself'.[73] Furthermore, all spirits, including God himself, must have 'some Amplitude of essence'.[74] Otherwise they are literally nothing at all. In his *Enchiridion metaphysicum* he again states that 'omnis Spiritus est Extensus'.[75] But he also connects the extension of God with the infinite extension of space. Thus the "amplitude" of extension *in* God is the essence of Divine immensity. Space itself becomes the innermost and internal attribute of Divine being.[76] For extension itself 'is really and identically coincident with the Amplitude of the Essence of God'.[77]

It must not be concluded that More's position entails that the defining attributes of God and space are one. More is at pains to assert the contrary. 'This immense *locus internus* or space really distinct from matter . . . is a representation of the Divine essence or essential presence, in so far as it is distinguished from his life and activities. For none of these attributes which we have been recounting [the twenty titles of space which are coincident with those of God] appears to concern the divine life and activity, but simply his bare essence and existence'.[78] Thus space *per se* does not possess any of the defining perfections of Divine nature. It is, however, indistinguishable *in re* from the 'Amplitude' or 'essential presence' of God's extended being. As such it is identical with the 'bare essence' of Divine extension. So considered the infinite extension of space is a real and internal attribute of Divine nature, since it is indistinguishable from God's eternal presence. On the other hand, finitely extended things relate to the amplitude of Divine presence by way of emanative causality, so long as that relation is not conceived as necessary. In reality, however, finite things are not separately distinct from God's extended being. And in the end More's metaphysics is a form of pantheism, combined with an emanationist conception of causality.

In *De gravitatione* it would appear that Newton fully accepts More's emanationist account of how space and time relate to Divine existence. However, consider the first of the three occasions on which Newton uses the phrase 'effectus emanativus'. He says (see the first lengthy quotation from *De gravitatione* in Section 2) that 'space is an emanative effect of the primarily existing being, because when any being is posited, space is posited'. If emanationism necessarily involves the notion of the hierarchical dependence of each successive level of being on that proceeding it, then Newton is not invoking More's doctrine. This is so on either of two interpretations that can be given to Newton's statement. In the first place, we can take the phrase 'primarily existing being' to refer to being as such, as opposed to particular kinds of being. On this interpretation Newton is simply saying that when any being *exists* space exists. This says nothing about hierarchical interconnections in levels of being. Secondly, 'primary being' can be taken as referring to God. In this case, Newton is merely claiming that God's *existence* grounds space's existence and *a fortiori* the existence of any related being. On both interpretations the relation between the existence of being and that of space is not causal, but one of ontic dependence. Newton is defining one condition which must be satisfied so that any being can be said to exist. In short, the phrase 'when any being is posited, space is posited' denotes an ontic relation between the existence of any kind of being and the condition of its existence.

This interpretation is confirmed by another passage in which Newton says that space is neither substance nor accident, 'but is, as it were, an emanative effect of God'.[79] The caveat 'as it were' is there for a purpose. We are not to take literally the emanationist metaphysics implied by the phrase 'emanative effect'. Rather, space is again conceived as the general condition required for the existence of any individual substance including its characteristics. If this interpretation is right, Newtons understand the phrase 'effectus emanativus' and its later surrogate 'unbounded consequences', as characterizing his individual conception of the link between God's eternal existence and that of infinite space and time. In other words, there is continuity between his conception of this relation in *De gravitatione* and his later formulations of it. This claim, however, needs further development.

First, what is Newton's conception? It can be stated quite simply. God's eternal existence is coeval with the existence of infinite space and time. They exist because God exists. In other words, built into Newton's conception of the actuality of Divine existence is the claim that God necessarily possesses a nature that exists omnitemporally and omnipresently with respect to infinite space. For God's 'duration reaches from eternity to eternity' and 'his presence from infinity to infinity'.[80] If God's existence is therefore eternal and uncreated, and if all things including God must exist with respect to space and time to be actual existents, the fact of God's sheer existence 'constitutes' the existence of infinite space and time. They are therefore consequences of God's eternal existence. For if God did not exist, there would be no sufficient reason for their existence. But Newton understands this relation differently from More. For More, infinitely extended space is identical with the 'amplitude' of God's presence. In short, it is an *intrinsic* attribute which defines the essence of Divine omnipresence. For Newton, on the other hand, neither space nor time is an intrinsic attribute of Divine Nature, but both are consequences of God's infinite existence, where existence is not a perfection identical with God's defining nature. Moreover, given Newton's view of Divine existence, the possibility of spaces and times with respect to which God does not exist is absurd. For there cannot be a place or a time, according to Newton's conception, with respect to which God can possibly fail to be present and endure. With respect to infinite time, for example, Newton says: 'Can God be nowhere when the moment of time is everywhere?'.[81] God's existence is therefore such that he exists with respect to all possible places and times.

∴ infinite + place.

The implications of Newton's position can be seen in the following passage: 'Space is of eternal duration and immutable nature, because it is an emanative effect of an eternal and immutable being. If ever space had not been, God at that time would have been nowhere, and hence he either created space later, where he himself was not present or else, which is not less absurd to reason, he created his own ubiquity'.[82] Here again we see that space (and by implication time) are coeternal with the existence of God. On this assumption, infinite space and time cannot be caused by God's existence. For if anything is truly infinite it is uncreated. Moreover, the conception of efficient causation, where a cause is said to be prior to its effects, is incompatible with Newton's claim that God, space and time are coeternal. For it involves the absurdity (as Newton himself points out) of saying that God either existed nowhere before he efficiently created space and time, or that he continuously creates his own ubiquity. The latter consequence must also be rejected, since it conceives God as actively creating his own presence in infinite space and time. This, of course, is inconsistent with Newton's conception that infinite space and time exist coeternally with God. In any event, what would it mean to say that God eternally actualizes them? Newton's use of the phrase 'emanative effect' is therefore misleading in so far as it connotes adherence to any form of emanationist causality. Thus his conception of the relation of space and time to God's existence turns on commitments different in kind from More's.

However, there seem to be difficulties. Consider space in terms of an argument common in the seventeenth century.[83] In general it can be said that infinite space is self-existent from all eternity, or what it is created by God. If the former, either space will not differ from God himself (or rather from the immensity of God), or else it will exist apart from God. However, given that being self-existent is the same as being in possession of every perfection, nothing but God can be necessarily self-existent. Therefore, since infinite space is self-existent and increated, it is the essence of Divine immensity. Consequently, infinite space cannot be necessary and self-existent apart from Divine omnipresence.

This chain of reasoning nearly enough reflects More's position.[84] However, Newton denies two of its assumptions. First, he denies that what is infinite is self-existent and that infinity is an intrinsic perfection of that which is infinite. Second, he rejects the view that God's unique perfections are limited if a truly infinite thing other than God is said to exist. Newton's position turns on the claim that the conceptions of perfection and infinity are not identical, whereas given More's adherence to the ontology of substance and attribute, for him infinity must be an intrinsic perfection of Divine nature.

In Newton's view to say of something that it is infinite *per se*, is not to say that it is perfect *simpliciter*. In *De gravitatione* he observes: 'I see what Descartes feared, namely, that if he should consider space infinite, it would perhaps constitute God because of the perfection of infinity. But by no means, for infinity is not a perfection except in so far as it is ascribed to perfections'. As an example, he states that 'Infinity of extension is so far perfect as that which is extended'.[85] Clearly, then, infinity *per se* is not a perfection. It merely 'completes' any perfections or imperfections that a nature in fact possesses. Thus 'infinity of intellect, power, happiness and so forth is the height [*summa*] of perfection, but infinity of ignorance, importence, wretchedness and so on is the height of imperfection'.[86] Again in the 1690s manuscript he says: 'Things which are neither good nor evil do not by eternity and infinity become either good or evil, but merely perfect in their own kind . . . No thing is by eternity and infinity made better or of more perfect nature, but only of longer duration in its own kind'.[87]

Newton's position is clear. The concept of infinity is not identical with the conception of perfection. Nor is it a real or first-level attribute which "adds" something to the intrinsic nature of a thing. Rather, it is a second-level characterization of a thing's intrinsic perfections or imperfections but not in itself a perfection. Moreover, there are levels of perfection or imperfection with respect to kinds of beings, but also degrees of perfection within each kind. Infinity, then, is the highest degree of a kind, where a nature specifies the kind or class to which a thing belongs. Hence to say of something that it possesses a perfection infinitely, is to state a "measure" of the exemplary manner in which that particular perfection is manifest in its nature. Equally, another nature can be said to instantiate the same property in a non-exemplary way: which is to say, it possesses it in a less perfect manner.

These distinctions enable Newton to distinguish God's infinity from that of space. God, according to Newton, is an *ens realissimum*. Not only does God contain every perfection in an intrinsically highest degree, but his perfections possess an intrinsically greater reality than the perfections of other existing things. 'From his true dominion it follows that the true God is a living, intelligent and powerful being; and from his other perfections that he is supreme, or most perfect . . . he governs all things, and knows all things that are or can be done'.[88] God is therefore infinite in that he is infinitely wise, infinitely powerful, and so on. Accordingly, it is clear on Newton's reasoning why 'it follows that not everything eternal and infinite will be God'.[89] However, on Newton's showing God is not an *ens perfectissimum*. That is, on his theory of predication, Divine attributes are not identical, a conception essential to the view that God is a simple and unitary being without distinction. Also thinking beings are not a *type* that is ontologically inferior to God; rather they own deficiently what God possesses infinitely. So Newton has to find grounds for Divine simplicity other than the claim that in God being and unity are one. This he attempts to provide by arguing that Divine being is causally immune to change.

In contrast to these "great-making" attributes which God possesses in an exemplary manner, space 'by reason of its eternity and infinity' will not be 'wise nor powerful nor alive, but will merely be shown to be increased in duration and magnitude'.[90] The sole perfection of infinite space is to be unending in duration and unbounded in extension. Its intrinsic perfection is thus completed. Moreover, being extended involves being passive and causally inert.[91] As such, space cannot posses the properties of an active agent. Accordingly, the essential nature of space is not one with the unique perfections of God's intelligent nature. Its infinity in no way competes with the infinity of Divine perfections. So in regard to their respective natures God and space are really distinct. The fact that both possess their perfections infinitely makes no difference.

On Newton's theory it makes sense to say that being infinite is not the same as possessing an intrinsic perfection. Rather, according to Newton's conception of predication infinity is a second-level characterization of any attribute which truly inheres in a given nature. In this sense, the intrinsic perfections or imperfections of things are merely completed by infinity. Thus to say of something that it is infinite, is not to say, *contra* More, that it is intrinsically perfect in that respect. Accordingly, on Newton's view both God and space can be said to be infinite and coeternal. Moreover, it is also possible to say that more than one infinite thing in fact exists. On More's view that is impossible since infinity is a perfection which is attributable only to God.

Unfortunately Newton's account of the relation between perfection and infinity does not help to clarify some genuine difficulties. If God's existence, as well as that of infinite space and time are coeternal, and if the latter are uncreated as Newton claims, how precisely is their existence related to God's? Since they are said to be 'unbounded consequences' of Divine existence, and distinct from God himself, on Newton's theory of predication they are not entailed by Divine nature. Nor are they a result of any sort of creative activity on the part of God. On the other hand, given Newton's conception of what it is for God to exist, it makes no sense that infinite space and time could exist without God's existence. But simply to say that infinite space and time exist because of the sheer existence of an eternal God, is not in itself sufficiently explanatory. Nor, for that matter, is it clear what kind of explanation is being invoked. So to claim that space and time are coeternal with God's existence does not in itself prevent someone questioning the nature of their relation to Divine existence.

It must be noted, furthermore, that this conceptual framework does not in itself provide a basis for claiming that God is a necessary being. Nor does it allow Newton to claim that God in fact exists. It is quite another matter, based on different sorts of arguments, whether anything in fact satisfies Newton's conception of Divine existence. It can also be asked whether Newton has a conception of necessity which justifies his claim that God is an eternal being. After all, he does maintain that God exists omnitemporally and omnipresently with respect to infinite space. These difficulties bring us to the question of Gassendi's early influence on Newton's conception of God's individual existence. Moreover, it allows us to understand the basis of Newton's epistemology; an epistemology which squares with his conception that existence is not a first-level attribute of Divine nature.

## 4. KNOWLEDGE AND ESSENCE

In the *Fifth meditation*, Descartes advances his version of the ontological argument. He argues to this conclusion: 'I clearly see that existence can no more be separated from the essence of God than can its having three angles equal to two right angles be separated from the essence of a rectilinear triangle'.[92] Later, in 1644, he repeated the same argument in his widely read *Principia philosophiae*.[93] In his objections to the *Fifth meditation* Gassendi argues that existence is not a first-level attribute of any existing thing.[94] His reasoning can be shown to be similar to Newton's arguments in the Des Maizeaux drafts. Moreover, the assumptions apparent in his objections are also similar to Newton's.

In both his *Meditations* and his *Principia philosophiae* Descartes assumes that 'God Exists' is a proposition on all fours with the certainty of the geometrical proposition. For Descartes it is an undoubted fact that both propositions predicate a property of something. The one is a property of 'God', the other a property of 'the triangle'.[95] It is on this assumption alone that Descartes bases his argument to the effect that God's existence is demonstrable as the properties of the triangle are demonstrated.

If is of more than passing interest that the young Newton made reading notes on Descartes's *Meditations*. Since they are few in number, it is significant that one is an observation on the ontological argument of the *Fifth mediation*. Newton wrote in his *Quaestiones*: 'A Necessary being is ye cause of it selfe or its existence after ye same manner yt a mountaine is ye cause of a valley or a △ ye cause of its 3 angles are = to 2 right ones (wch is not from power or excellency, but ye peculiarity of theire natures)'.[96] Notice that Newton distinguishes between God's existing in virtue of his power alone, and existing in virtue of the fact that his existence is implicit in the defining concept of his nature. It is this latter view that he attributes to Descartes. Thus Newton seems to interpret Descartes as holding that the mere conception of a necessary being ensures its actual existence. It is just this particular view of God's necessary existence that he will later reject. Ironically enough, Descartes in fact holds that God exists *per se* and necessarily because he contains in his nature 'a great and inexhaustable power'. Thus in contrast with finite things, God's being needs 'no assistence in order to exist, and requires none for its preservation, and hence is in a certain way the cause of its own existence'.[97]

In any event, Newton shows more than a cursory interest in Descartes's argument; he even advances an interpretation of its significance. Moreover, *De gravitatione* is further witness to the fact that Newton was deeply familiar with Descartes's *Principia philosophiae*, in both its scientific and philosophical aspects. We also know that he was familiar with Gassendi's intellectual position, at least through the mediation of Charleton. Nor is it unreasonable to assume that Newton was directly familiar with Gassendi's arguments against Descartes's conception of God's necessary existence. They were probably available to him, and are consistent with the implicit interpretation which Newton himself advances in his *Quaestiones* note.

Gassendi's arguments in his *Fifth objections* focus on the parallel Descartes draws in the *Fifth meditation* between a triangle's properties being inherent in its nature, and God's existence being intrinsic to his essence.[98] For Gassendi, 'existence is a perfection neither in God nor in anything else'. On the contrary, he argues that

> that which exists and has various perfections, does not have existence as a particular perfection and as one of the number of its perfections, but as that by means of which the thing itself equally with its perfections is in existence, and without which neither can it be said to possess perfections, nor can perfections be said to be possessed by it. Hence neither is existence said to exist in a thing in the way that perfections do, nor if the thing lacks existence is it said to be imperfect (or deprived of the perfections), so much as to be nothing.[99]

Although a common source cannot be discounted, the resemblance between Gassendi's and Newton's reasoning is striking. Both emphasize that existence is not a real attribute or disjunctive property of existing things. Gassendi is emphatic on the point. In his later and more lengthy anti-Cartesian tract, the *Disquisition metaphysica* (1644) he states: 'Existence is not a property, nor determinate genus among things'.[100] Moreover, like Newton, Gassendi links existence in space and time with the concept of Divine actuality.[101] However, unlike Gassendi, Newton does not imply that God would be non-existent were he deprived of actual existence. In fact, Newton's statement that 'what is never and nowhere, it is not in *rerum natura*', contrasts the conception of actual existence with a notion of possible or non-actual existence. That is: *possibilia*, and imaginary and abstract entities can be said to exist, but they do not exist in the world of space and time.

In his notes on the *Meditations*, Newton gives page references to an edition or editions of Decartes's works.[102] The range of citations could well refer to the original six sets of 'Objections and replies', which include those of Gassendi. However, after 1650, many editions of Descartes's works omit Gassendi's objections.[103] For definitive evidence that Newton knew of these objections, it is important that we ascertain the edition or editions of Descartes's works that he used. Unfortunately, I have not been able to match Newton's page references in any consistent way with editions of Descartes's works available in England during the 1650s and 1660s.[104] Nevertheless, since he actually cites the *Meditations*, he probably read Gassendi's objections. Moreover, the affinities between Newton and Gassendi on the question of God's actual existence do not rest entirely on asking whether existence is a real attribute. It can be shown that Newton's basic epistemology is similar in many ways to Gassendi's.

This point can be developed by first considering the candid reply which Gassendi's objections drew from Descartes:

I do not see what kind of thing [*generis rerum*] you wish experience to be, nor do I see why it may not be said to be a property as well as omnipotence, taking the term property as equivalent to any attribute or anything which an be predicated of a thing . . . for necessary existence in the case of God is also a true property in the strictest sense of the word, because it belongs to Him and forms part of His essence alone.[105]

Descartes's position is unambiguous: existence is a real attribute of God's essence. However, he distinguishes how existence relates to the essence of triangle and the way in which it relates to the essence of God. 'God is His existence, while a triangle is not its own existence'.[106] In the idea of God existence is a necessary and inseparable perfection, whereas in a triangle it is only a possible one.[107] Thus a triangle can exist because its essence is a *verum et reale ens*[108] It has a 'determinate nature, essence or form',[109] the idea of which is intrinsically non-contradictory. Its *esse* in the mind is to exist *qua* idea. The mind has therefore got direct knowledge of the properties necessarily implied by the essence of the triangle. But this in no way implies the property of actual existence extramentally.[110] Though, of course, the possibility of thus existing is involved in the triangle's essence.

However, God's actual existence is a property necessarily entailed by his essence. To suppose God without existence is to suppose something self-contradictory. It violates the conception of a supremely perfect and unique being.[111] Hence, accordingly to Descartes, God's real existence is strictly identical with his essence. In the case of the triangle it is not. In the idea God, existence can no more be separated from essence than the idea of a mountain from the idea of a valley.[112] Thus, as disclosed in the clear and distinct idea which the mind has of him, God's nature exists *in re* as a necessarily existing essence. The mind is thus compelled to recognize 'the necessity which lies in the thing itself'.[113]

More fully accepts Descartes's reasoning: 'The idea of God, or what is the same, of a fully perfect Being, containing in it necessary existency, demonstrates to us that God does exist: this in brief is Cartesius his reasoning'.[114] In fact More's position on this question affirms the view, previously discussed, that God's attributes are identical with God himself:

> . . . when we say that the *Existence* of God is *Necessary*, we are to take notice that Necessity is a *Logicall Terme*, and signifies so firme a connection between the *Subject and Predicate* (as they call them) that it is impossible that they should bee disservered, or should not hold to-gether, and therefore if they bee affirm'd one of the other, that they make an *Axioma Necessarium*.[115]

Thus More holds that the relation between God and existence is one of logical necessity. We have already seen that Newton, in virtue of his view of 'Divine predication', opposes this reasoning. In this next section we will also see that his view of the necessity of Divine existence is not grounded in any conception of logical necessity.

At this point it is clear that Gassendi cannot accept Descartes's (and More's) reasoning, since the French thinker's version of the ontological argument rests on conceiving existence as an intrinsic perfection of Divine essence. This conception is incompatible with his ontological presuppositions. However, this is not the place to discuss the merits of Descartes's ontological proof, nor to consider the merits of Gassendi's objections. Rather, the pertinent task is to show how far Newton's reasoning is similar to Gassendi's. Especially in regard to the conception of what it is for something to have actual existence. But before this can be done, further analysis of the epistemological positions of Gassendi and Descartes is necessary.

Gassendi's epistemological position can be put thus. Existence is not an intrinisic perfection inherent in things. Nor is it a property which can be predicated of them. It is an irreducible feature of their individual actuality, without which they are merely non-being. As a thing cannot be said to exist actually, unless it has a nature, so it cannot be said to have a real nature, unless it exists.[116] Thus the essence and existence of things are inseparably combined in sensory experience, and are only distinguished by reason. To say that something has real existence is to state a fact about it. This also applies to God's existence. In the end, the fact that God exists, can only be established *a posteriori*. Whatever arguments are employed must be based on the perceptible characteristics of the natural world.[117]

For his part, Descartes accepts the view that created things do not contain the ground of their own existence. With respect to them, existence is a transcendental predicate.[118] It succeeds in referring because it presupposes experience of things *qua* existents.[119] Thus, if it can be said of created things that they exist, this is merely a contingent, not a necessary fact about them. It is not logically contradictory to suppose that they can cease to exist.[120] However, in regard to God, existence necessarily is a necessary part of Divine essence.

Now epistemologically, Descartes's position rests on the view that the mind possesses ideas of the real essences of things, including God's eternal essence.[121] In Descartes's causal scheme the objective reality of true ideas possessed by the mind, requires a cause in which the same reality is contained, not just objectively (in the manner in which ideas exist mentally), but either *formaliter* (that is, actual reality) or eminently.[122] Similarly, the mind has an idea of Divine perfections. The objective reality of this idea cannot be caused by our natures either *formaliter* or eminently. That is to say, it is not induced by the mind. Nor does it depend on the existence of the human mind, for the content of the idea of an infinite being has greater objective reality than the mind possesses actual reality. Furthermore, the positive idea of an infinite and necessary being cannot be caused through finite experience. Thus an idea of this sort requires God for its true and proper cause.[123] The mind therefore grasps through the possession of a clear and distinct idea, the eternal perfections of Divine nature; realizes that they cannot have their origin in the mind itself; and recognizes by the natural light that in reality Divine existence is necessarily linked with those perfections.

It is this move that Gassendi rejects. For him 'essences' are only general terms or linguistic universals derived or abstracted from perceived resemblances among individually existing things.[124] Hence the truth of the generality "man is an animal" does not refer to an eternal essence which exists *per se*. Rather its truth depends on prior experience of individually existing things, 'this man', 'that man', and so on. The common concept "animal" is then established by reference to recurring patterns of features manifested by individuals called men.[125] Accordingly, the link between the general terms 'man' and 'animal' is forged by the mind alone, and apart from the mind it lacks reality.

Following his nominalistic reasoning, Gassendi denies that the mind has an innate idea of God's immutable essence. He also denies that God's actual existence can logically be demonstrated from such an idea:

Is not that just what has to be proved, that it pertains to his nature to exist? Will you perhaps say that it remains proven by the description you make? But is to describe how you conceive of a thing to prove that it exists? And do you prove it by enumerating all God's perfections, eternity, immensity, supreme power, wisdom, existence, and so forth? But what else are you doing but putting existence among his perfections so that you may prove that existence exists, which is proving a thing by itself.[126]

He therefore accuses Descartes of passing in thought alone 'ex statu ideali in statum rerum'.[127] If existence is put among God's perfections, this is merely an attempt to prove the existence of God in terms of his defining nature. For to say that the conception of God involves his real existence as a necessary perfection, is to assume what must be demonstrated on independent grounds, namely, that God in fact exists:

Do you prove it on the grounds that no other existence but possible and contingent existence belongs to other things; but actual and necessary existence belongs to God himself? But is not this also the subject of dispute, namely, whether existence, whatever kind it may be, belongs to God? For if someone denies it in general [existence], he also denies actual and necessary existence.[128]

Gassendi evokes some important principles. Whatever a definition or concept includes, it can always be asked whether something exists to satisfy the definition or to instantiate the concept. It makes no difference that something is conceived as a necessary existent. Its existence can still be rejected. This applies to God's existence as well. Even if the conception of God implicitly includes the notion of an essence that self-exists, it can still coherently be denied that such an individual in fact exists. For the denial amounts to saying that 'no individual called God exists'. This does not involve contradiction, since the reference of both the subject and the predicate terms of the statement 'God exists' are in fact rejected.[129] Thus Gassendi is not merely denying that the conception of God's existence entails its self-realization. Rather, he rejects the claim that in the act of knowing God, the mind is innately aware of an eternal essence that exists *in intellectu*, and which includes in its nature real existence. For according to Gassendi, an independent causal basis for the idea of an infinitely perfect being can be established only with reference to direct sensory experience. This being so, the mind *per se* cannot be said to grasp an innate idea of an individual essence, the mental representation of which is directly caused by a being that actually exists. It is these principles which constitute Gassendi's conceptual framework. And it is on this basis that he claims that existence is not first-level attribute of God's nature.

There is much here that agrees with Newton's reasoning. In particular, these principles are compatible with the views expressed in the Des Maizeaux passages. While Newton does not explicitly mention them, his position depends on their implicit recognition. How clearly Newton saw this, I hesitate to say; but at some level they are surely what he is aware of. Newton also agrees with Gassendi that all knowledge has sensory experience as its ultimate source. To talk of 'essences' is to talk of those perceptible features of things which are invariably taken to characterize them over time. An uncompromisingly skeptical attitude towards essence is evident in Newton's thinking as early as *De gravitatione*. There the Cartesian concepts of essence and substance are vigorously criticized.[130] Descartes is also the target in the General Scholium. Newton states that 'what the real substance of anything is we know not'. For the 'inward substances' of things 'are not to be known either by our senses or by an reflex act of our mind'. Newton is thus sceptical of our ability to know the real essence of sensory things: 'much less, then, have we any *idea* of the substance of God'.[131] In his view, then, there are no eternal and immutable essences which the mind can grasp independently or sensory experience.The primary properties of things, and the fact that they exist, can only be established on the evidential basis open to the senses. In fact Newton's notion of an individual thing having an individual 'essence', means only that it is relatable to a particular sort or kind, itself established through sensory experience. Similarly, it is on sensory foundations alone, that the nature and existence of God can be argued for. Thus the presuppositions of Descartes's version of the ontological argument have no role in Newton's epistemology. The necessity of Divine existence, as well as our knowledge of God's attributes, can be demonstrated only from established design in the works of creation. Thus the necessary truth of Divine existence is not *a priori* truth. 'We know him only by his most wise and excellent contrivances of things; we admire him for his perfections; but we reverence and adore him on account of his Dominion'.[132]

So Gassendi and Newton, for similar reasons, agree that God's actual existence cannot logically be demonstrated from a direct and intuitive awareness of the nature of his eternal essence. On the basis of their epistemological principles, there can be no argument which shows that God's existence is directly grounded in his eternal nature. It may be the case that this is so, but the human mind is incapable of knowing that it is so. Nevertheless, they both hold that God exists necessarily. For he exists always and everywhere. On this showing some account has to be given of how Divine nature relates to the modality of necessity, and also to the claim that God is an actual existent. On neither score does Gassendi provide an answer. Newton does.

## 5. NECESSITY AND ACTUALITY

We are now in a position to return to the difficulties raised at the end of Section 3. They all turn on the one central issue: how does Newton's conception of God's actual existence with respect to space and time relate to the claim that he is a necessary being?

Let us first consider the question with respect to time. There are three distinct conceptions of how God relates to time. The first is that he exists supra-temporally. Second, that he exists in a *nunc stans* (literally, a 'standing now'), or a "frozen" unchanging present. Lastly, the view that Divine existence is omnitemporal. Each of these conceptions involves a different view of God's eternity. The first maintains that temporal concepts do not apply to

Divine existence. His eternal existence is, so to speak, 'above', 'beyond' or 'outside' time. The second view holds that God exists in an unchanging, indivisible and ever-present eternity. God's eternity is indivisible because eternity *is* God's Divine and indivisible Substance. And the last view, which characterizes Newton's position, maintains that Divine nature endures through all possible times. In other words, God exists in unending time, or sempiternally. He has always existed, is existing and always will exist.

Newton's conception of God's necessary being can be usefully highlighted by contrasting it with the first two positions. Spinoza holds that God's existence is timeless. He also rejects the view that Divine existence endures in unending time. While there is no direct historical connection between the views of Spinoza and Newton, Newton's conception of Divine duration is similar to the one that Spinoza criticizes. In fact, Spinoza advanced his criticisms of Divine omnitemporality as a means of attacking an entrenched theological opinion current in the Low Countries during the seventeenth century.[133] Furthermore, Spinoza's pantheistic conception of God and Nature was widely rejected and criticized by Restoration thinkers in England, especially by the Cambridge Platonists. Moreover, Newton himself was probably aware of Spinoza, and in company with Clarke sought to distinguish his views from those of the Dutch philosopher.[134]

In Spinoza's view God's eternity *is* his actually infinite existence.[135] This conception is based on two related principles. That God's existence is entailed by his essence; and secondly that 'duration is an affection of the existence of things, and not truly of their essence'.[136] The first principle means that in God's nature existence and essence are inseparable, but in the nature of finite things existence is not entailed by essence, so that the two can be considered as separate. Thus the essence of created things is not affected, when duration is ascribed to their existence.[137] To God's nature, however, 'we cannot attribute a future existence – for the very existence he would then have, must even now be actually attributed to him', since 'an actually infinite existence belongs to God just as an actually infinite intellect does'. On the other hand, 'all created things, while having present existence, are entirely devoid of future existence, since obviously, future existence must be given to them continually, yet nothing similar can be said of their essence'.[138] Thus from the time at which they begin to exist, the quantity of duration (Spinoza's conception) of created things increases. But this conception cannot be ascribed to the immutable eternity of God himself.[139]

It is the second principle which in part embraces Spinoza's reason for this last claim. Partly it turns on the notion that the existence and essence of created things can be considered separately; but equally it rests on Spinoza's view that though, *qua* existents, created things are finite, their essences are immutable.[140] So conceived the essence of a created thing is unchanging, even though its existence and duration can be said to increase or diminish with respect to time.[141] However, if the existence of God is involved necessarily in his essence, to ascribe duration to his existence is to make Divine essence potentially mutable.[142] For this is like saying that the essence of a circle or of a triangle 'has endured longer now than it did in the time of Adam',[143] which involves the absurdity that the essence of the triangle is presently different just because it has endured longer. But such a description cannot pertain to the triangle's essence *qua* eternal truth. So it is with God's essence. Duration cannot pertain to his existence; for this is simply identical with his essential nature. However, on the assumptions which Spinoza is attacking, to say that God endures is to say that God's existence can increase with time as does the existence of finite things. But this makes God's eternity a 'species of duration'. In other words, God becomes a being whose 'duration is unlimited in the past or in the future'.[144] This, however, contradicts Spinoza's conception of God as eternal in his own right; as a being who exists necessarily in virtue of his infinite nature. Spinoza's conception of God is therefore a positive one. He is a being who embodies all possible perfections, including existence. He is therefore unchanging, unconditioned and lacking in all limitations.[145] In Spinoza's view, to conceive Divine eternity correctly is just to understand that temporal extension and temporal contrasts cannot apply to Divine existence. I shall return to Spinoza's position shortly.

The second conception of God's eternal existence is directly criticized by Netwon. In a draft variant to the General Scholium (1713) he says that God 'endures from eternity and is present from infinity to infinity. His duration is not a *nunc stans* without duration, nor is his presence nowhere'.[146] In one of the Des Maizeauz drafts Newton offers a reason for his opinion: 'For a *nunc stans* is a moment wch always is & yet never was nor will be: which is a contradiction in terms'.[147] Lastly, in the 1690s manuscript he says: 'Let them therefore consider whether it is more agreeable to reason that God's eternity should be all at once [*totam simul*] or that his duration is more correctly designated by the names Jehova and "He that was and is and is to come"'.[148]

Newton contrasts his conception of eternal duration as a 'mode' or 'affection' of Divine existence, with the view that God exists in a "standing" or non-successive eternity. His criticism of the latter conception is shrewd. In effect he claims that it is a hybrid concept. To speak of God as existing in a 'frozen', unchanging present, is to think of his existence as being temporal in character. For on this view God exists in an ever present "now", which is in itself a temporal notion. Moreover, it is isolated from the structure of tenses in which it is naturally embedded. But to exist "now" makes no sense apart from implicit reference to past existence and to continuing existence in the future.[149] Successiveness is built into the very concept of time itself. Accordingly, those who propose that God exists in a *nunc stans*, are entitled to their view only so long as they covertly presuppose the notion of past and future existence. Otherwise existing in an ever-present "now" fails to make sense, since it lacks the conceptual boundaries indicated by the use of past and future tenses. Accordingly, supporters of the *nunc stans* conception of Divine eternity cannot discard these tenses and at the same time claim that God exists in an abiding present.

In Newton's view it is best to avoid the arid conceits of the schools.[150] They present conceptual difficulties, and abrogate the Biblical conception of God's existence by conceiving him as an abstract entity.[151] According to the Biblical dispensation, to say God is eternal is to say that he possesses an intelligent life that exists through unending duration.[152] God is named Jehova and 'He that was is and is to come'. To Newton this conception provides a proper motivation for Christian belief and worship, since God is conceived as an individual person who is able to act intentionally and purposively.

With this background in mind let us turn to Newton's positive conception of Divine existence. I begin with an important text from the 1690s manuscript:

> To exist in time and place, since this is the common nature of all things, does not argue imperfection. For the duration of a thing is not its flow, or any change, but permanence and immutability in flowing time. All things endure in so far as they remain the same at any time. The duration of each thing flows, but its enduring substances does not flow, and is not changed with respect to before and after, but always remains the same.[153]

Newton here distinguishes between permanent and successive duration. This marks a distinction between the real duration of things and sensible changes that occur to them. To characterize this, he draws on a Medieval distinction adapted from Aristotle. This is the distinction between *res permanentes* and *res successivae*.[154] It is also employed by Gassendi, Charleton and More.[155]

In general, the distinction contrasts a "permanent thing" whose mode of existing at once includes complete being, and a "successive" thing, the nature of which includes successiveness in its conception. Of the latter kind are games, walks and journeys; in general, things which have being sequentially. Of the former kind are relatively permanent things as a man, a stone or a horse. These have complete being and are said to be in fact (*in facto esse*).[156] It is realities of this factual sort to which Newton refers when he says that the duration of a thing is its permanence and stability in time. Thus all created substances endure unchangingly since it is their *esse* to do so.

But things also have histories. Something is true of them now which is not true of them at another time. That is, something is true of them other than the mere fact that they progress in time. Thus the states which mark a thing's history, and which result from its modifications, are successive in character. They are continually coming to be in time. When such changes include local and uniform motions they provide a basis of successiveness required for measures of relative time. The nature of permanent things, on the other hand, constitutes a perfection necessary for their duration. As successive change presupposes something that is relatively unchanging, stability and permanence over time are grounded in the essential nature of *res per-*

*manentes*. So conceived, *res successivae* possess natures of a less perfect kind, insofar as they are dependent on the existence of *res permanentes*. It is on these grounds that Newton distinguishes the absolute quantities of existence of things from sensible measures of their changing states. On this basis his distinction between absolute and relative time is also well-founded.

It is clear that Newton makes a close conceptual link between individual duration and the 'perseverance of the existence of things'.[157] This framework is extended to God's existence. 'God is one and the same God, always and everywhere';[158] for 'to be the same always and everywhere is supreme perfection'.[159] However, there are important differences between the nature of God's enduring existence and that of created things. Certainly it is the *esse* of *res permanentes* to endure. But that they endure depends ultimately upon Divine will, upon the will of a God whose power and perfection is greater than all else that exists.[160] According to Newton's theory of infinity and perfection, the natures of infinite space and time are distinct from God's eternal nature. Employing the same theory, he distinguishes God's nature and the nature of his dependent creation. While the 'quantity of existence' of finite things is limited in time, God's existence is sempiternal.[161] He depends on nothing *ab alio* for his perseverence in existence. Thus finite things come into existence and pass away, but God endures as the same individual at all times whatsoever. His is a nature whose existence has neither beginning nor end.

An obvious question arises. If God is said to exist temporally, even omnitemporally, does this not allow that his nature has the potentiality to change? How then can Newton say 'that the supreme God exists necessarily; and by the same necessity he exists always and everywhere'?[162] This question can be answered on the presupposition that God either exists supra-temporally or that he exists in a *nunc stans*. For both views presuppose that God's nature is not conditioned by time, and that in virtue of his indivisible nature he is necessarily the same. However, Newton does not share the atemporal conception of Divine existence embraced by these positions. On the contrary, in his view there is a clear link between existing sempiternally and existing necessarily.

Let us first consider God's permanent duration in time. To claim, as Newton does, that God is the same always, is to say that he remains the same individual at any given time. In other words, his essential nature remains the same: he still is what he was and will continue to be what he is. Thus on Newton's view, to specify God's permanent duration requires that he is *same* individual at all times whatsoever. Hence the stability of Divine presence – the conception that God remains one and the same person – demands for its very coherence the notion of temporal continuance. The notion of permanent existence therefore requires reference to time as much as the conception of change. God's absolute duration, however, requires reference to existence at all possible times. Moreover, at all times whatsoever, the same statements remain true of God.

For Newton, then, real change and permanent duration are in accordance with the nature of the individual that is said to change or to remain the same with respect to time. That something can change in time is not imperfection, if having that capacity is compatible with its essential nature. God, however, does not change with respect to time. Nor can mere temporal existence affect the essential perfections of his nature. Walter Charleton, whom Newton read in his early formative period, puts the point thus: '*Divine nature* [is] so Constant and Perfect, that in the eternal flux of time it can know nothing of Innovation or corruption . . . God is a *Pure, Simple, Homogeneous* substance, and so not subject to the invasion of any contrary'. Thus with respect to God, time *per se* 'or the succession of duration is not the cause, that induceth corruption'.[163] Finite things can and so change because they have a capacity for instantiating contrary properties. Moreover, they begin to exist at a time and cease to exist at another. This, however is not true of God's indivisible nature. Thus mere continuance in time is not a kind of change. Indeed, it itself is presupposed by change. For Charleton, as for Newton, time is neither cause, attribute nor accident. It is a feature of that objective nexus required for the actuality of all existing individuals.

Given these principles, Newton claims that God endures permanently with respect to time. Certainly Divine existence is successive in character, and thus extended and divisible. However, in regard to his enduring substance, God remains one and the same individual. Throughout its duration each human soul remains 'the same indivisible person'. Similarly, though there are 'successive parts in duration' and 'coexistent parts in space', neither 'the one nor the other is in the person of a man, or his thinking principle; and much less can they be found in the thinking substance of God'.[164] Thus nothing can be added or taken away from the perfections of Divine nature, even though with respect to space and time God's duration is divisible.[165] Divine nature is intrinsically simple, and has an innate capacity to exist all at once as an entire and permanent being. 'The most perfect idea of God is that he is one substance, simple, indivisible, live and making live, and necessarily existing everywhere and always . . .'.[166] Though God's existence involves reference to temporal successiveness, his essential nature endures the same. 'To be bounded in time and in place, or to be changeable does argue imperfection, but to be the same always and everywhere is supreme perfection'.[167]

But what case can Newton make for God's existing *qua* necessary being if he said to exist sempiternally? According to Spinoza's conception, to speak of God as enduring in time as to allow that his nature has the potentiality to change. For *now* something can be said to be true of God, that was not true of him yesterday: that is, that he has endured for a longer time. Since, however, Newton does not conceive God's essence and existence to be mutually implicative, omnitemporal existence does not in itself imply the mutability of Divine nature. For existence *per se* is not a first-level attribute. However, more is required of Newton in order to justify his conception. God must, of course, be conceived as an individual whose essential nature is unchanging. But Newton must also be able to say, on the basis of his position, in what appropriate sense God can be said to exist necessarily. Now for Newton to say that God is a necessary being, requires that two conditions be met. First that Divine existence be such that it lacks beginning and end. Second, that nothing can either prevent God from exercising his capacities or can in any way affect his essential nature. In other words, certain *causal* conditions must be satisfied by Newton if he is to have a coherent conception of Divine necessity.

In order to clarify his response to these requirements, let us consider what Newton is not saying. He is not maintaining that God's necessity can be explicated in terms of his omnitemporality. Newton states categorically that 'the supreme God exists necessarily; and by the same necessity he exists always and everywhere'. God's omnitemporality is therefore a result of the intrinsic necessity of his nature. Consequently, Newton is not committed to the following chain of reasoning. At every time God exists in the sense that he is present at that moment. If he exists at all moments of time, he cannot fail to be at any given time, so he is omnitemporallly present. Therefore to say that God exists omnitemporally, is to say that he exists necessarily at all times.[168]

The argument is, of course, patently analytic. However, it probably underscores Newton's reason for not attempting to derive necessity from omnitemporality. Simply to hold that God exists at all possible times is not to say that he is necessary being. This in itself does not entail that at no time is it possible that God can fail to exist. But this is what Newton claims when he says that God necessarily exists always. The modality embodied in the notion that at no time can God not possibly exist, cannot itself rest on the claim that God exists at all times whatsoever. That God cannot but exist at all possible times is a claim about his nature, not a claim about sheer existence omnitemporally. The latter claim is consistent with merely contingent existence. It is in virtue of his nature, then, that God is conceived as a necessary being, and that infinite time is a consequence of his existence.

Now according to Newton's conception, God is a being who always has and always will exist. So it is natural enough that he should think of God's existence as not being caused. As there was never a time at which God's existence was preceded by anything else, he can be said to be an uncaused being. This means that he is causally independent of conditions, state-of-affairs, and circumstances external or other than his nature. Thus Divine existence at any given time is solely explained by reference to God's existence at an earlier time. So that God exists omnitemporally is a fact about his unending and permanent duration.

However, if Newton's position is not ultimately to collapse into tautology, it must be the case that God cannot be changed with respect to his essential prefections. The fact that time is incapable of affecting God's nature does not prohibit the possibility that something can. It is well to remember that in Newton's view God is 'to be celebrated not so much for his essence as for his actions, the creating, preserving and governing of all things according to his good will and pleasure'.[169] Moreover, as God is an intelligent and living being who possesses a sempiternal life, he is capable of acting intentionally. However, if a being is said to possess life and intelligence, and is thus able to act purposively, its existence must be temporal in character. For to have a life is to exist with respect to time as being present is from the past and is moving into the future. So God, having no beginning nor end of existence, is fully capable of existing and acting omnitemporally. He 'is thereby able to act in all times & places for creating & governing the universe'.[170] Moreover, he is being who can be truly affected by prayer, rather than one who decrees from eternity to grant requests. Newton's God is therefore a Lord of Dominion who can act with providential benevolence. As such he is a God worthy of worship.

On the assumption that God possesses his certain perfections infinitely, dependent creation cannot cause him to cease existing. For there is nothing in the nature of space and time, or in that of created things, which can bring this about. As there is nothing that caused God to exist, so there is nothing that can cause him to go out of existence. Similarly there is nothing that can prevent God from exercising his essential capacities. The natures of created things, even the infinite nature of space and time, are inferior to God's intrinsic perfections. In terms of his theory of how the concepts of perfection and infinity relate Newton says; 'not everything eternal and infinite will be God, or will God be prevented from the eternal and infinite exercise of his omnipotence in the creating and governing of things, by the imperfect nature of created things'.[171] Thus with respect to the first-level attribute 'omnipotence', Newton claims that God cannot and will not change. In virtue of his nature is unalterably disposed to express his intrinsic power where appropriate. Moreover, so long as God's actions are not conceived as extending to the impossible (that is, undoing what is done) or involving an imperfection (that is, lying or deceiving), there is nothing in creation that is able to impede the exercise of Divine omnipotence. In this sense, God cannot change or be changed with respect to his infinite power. So being infinitely powerful is an essential part of what we understand by the infinity of Divine nature.

Nor is there any reason to doubt that Newton thinks of God's other attributes in a similar way. Take omniscience, for example. According to Newton's theory of perfection, God is infinitely wise. This means that he is wise in the most perfect and excellent manner, that is, in the highest degree. Thus to say that God is infinitely wise is to characterize his intrinsic nature. It is to say that God cannot but act wisely. Nor is there anything that can prevent God from acting in a way which is appropriate to his nature. God can neither change, nor be changed, with respect to the attribute of omniscience.

At this point Newton's position on necessary being can again be usefully contrasted with Spinoza's. In general terms, Spinoza holds that God exists in the manner of an eternal truth. In other words, he is conceived as a being who is not subject to those instantiations of properties that are consistent with the notion of temporal change. For Spinoza the necessity of God's eternal existence is entailed by his essential nature, so that temporal characteristics are inappropriate in explaining his manner of existing. God's existence is therefore grounded in his individual nature, and understood to be atemporal in character. So conceived Divine nature is such that it cannot change.

But what does Spinoza mean by saying that God 'cannot change? This has to be understood in terms of how he conceives the relation between Divine substance and its attributes. As discussed in Section 3, the doctrine of essential predication states that the attributes of God are really one and the same with Divine essence and existence, and one amongst themselves.[172] In Spinoza's view, this means that God is strictly identical with each of his defining attributes as they are identical with one another.[173] God is whatever He is, not through another being but through his very self. All the Divine attributes are therefore identical with Divine being. In other words Divine substance *is* its real essence. Thus Spinoza's conception of substance involves an ontological claim about how attributes and substance relate; they are identical and thus cannot be construed on the model of properties which are said 'to attach' to things. That is, the absolute infinity of Divine substance is not made up of an infinity of attributes. For the absolutely infinite being is not the sum of its attributes, as this implies composition. Rather, each attribute is conceived through itself and expresses the same infinite reality. So in reality, God's attributes, including his existence, are indistinguishable from his absolutely simple and unitary nature. Were this not so, Divine nature would be liable to change, variation and differentiation, all of which constitute imperfection.[174]

What is meant, however, by saying that God's defining attributes are identical? This is to be understood in terms of Spinoza's conception of Divine substance. It is that 'which is in itself and conceived through itself', such that there is an intrinsic relation between God's existing *in se* and the mind's being able to conceive him *per se*.[175] Thus the grounds for conceiving God are strictly dependent on his infinite and intrinsic reality, and are independent of the grounds involved in conceiving the nature of other existents. An individual whose nature and existence is conceived as depending on nothing beyond itself is eternal; it is the ground of its own existence and free of all limitation. Since the reality of God's nature transcends limitation, his defining attributes are infinite and conceived through themselves.[176] If they are infinite they cannot change; for if they could, they would be limited and hence not infinite. Now if an eternal God is his infinite attributes and is the intrinsic ground of his own immutable nature, there can be no reality independent of Divine existence by which to show that God's attributes are not identical. Thus

Spinoza's conception of the identity of God's infinite attributes is simply this. There are no conditions whatsoever under which the identity of God's essential attributes fails to obtain. These conditions are satisfied immediately in that the only possible world is the one constituted by God's actual existence.[177] So on the supposition that God is identical with each of his defining attributes he ceases, *per impossible*, to be the same individual if this identity condition fails to obtain. It is in this sense that Spinoza holds God's nature to be essentially *esse sine mutatione*; not in the sense (Newton's sense) in which Divine nature is said to be identically the same at all possible times. In the end Spinoza's claim that God exists timelessly in the manner of an eternal truth,[178] is merely a means of expressing his conception that God's essential attributes are individually unchanging and together identical. And it is no part of his conception that 'being timeless' is a special property that attaches to entities of a certain sort.

It is clear that Newton's view of Divine immutability is different from Spinoza's. This difference turns on their respective ontologies of Divine nature. For Spinoza God is immutable just because he is identical with each of his essential attributes under all conditions whatsoever. Newton, on the other hand, does not hold that God is identical with Wisdom, Power and Goodness themselves. According to his theory of perfection, characteristics such as being wise, powerful and good are attributes that Divine nature instantiates in the highest and most complete manner, namely, infinitely, where infinity is understood as a second-level characterization of each of God's defining attributes. Unlike Spinoza, who is able uniquely to specify Divine immutability *in alio*, or in terms of God's intrinsic nature alone, Newton's position obliges him to consider the condition of Divine existence in relation to the nature of other existents. So for Newton, God is an immutable being in that nothing whatsoever can causally change or in any way affect his defining attributes. He is fully capable of being identically one and the same person at all possible times, since there are no conditions whatsoever under which he can be caused to be otherwise than he is.[179] Unfortunately, neither Newton nor Spinoza explain a serious difficulty: what is meant by saying that God's properties and attributes are "in" his substantial nature? Simply to show that God's properties are causally inseparable from his nature, gives no account of how they are connected "in" God. But without this account, the relevant kind of causation cannot be understood. Equally, in Spinoza's case, to show *that* God's attributes are identical with his nature, is not to show how they are "in" his nature.

However, Newton and Spinoza (and also Descartes as discussed in Section 3) differ not only in their conception of necessary being, but on what *kind* of knowledge this involves. In Newton's view (as well as in Gassendi's) to say that God exists necessarily is to claim a *posteriori* knowledge; whereas for Spinoza it is an *a priori* truth. However, Newton's claim that knowledge of God cannot rest on a purely *a priori* basis is not be confused with the claim that the mind makes no contribution to the nature of abstract concepts. Rather, what Newton denies is that knowledge of God can rest only on the possibility of directly knowing real essences (see Section 3). Through its capacity for forming abstract concepts, the mind can and does structure experience; nevertheless, the content of every concept, including that of God, has its ultimate source in sensory experience. It is no wonder, then, that Newton developed a systematic natural theology in which a science of nature plays an indispensible role. Equally, Spinoza is not committed to rejecting sensory experience in demonstrating God's existence; in fact, he presents *a posteriori* proofs based on his claim that created existence (knowledge of which is derived from experience) is dependent on conditions beyond itself.[180] However, in Spinoza's view such arguments are at best methodological, and no substitute for the claim that the necessity of Divine existence is an *a priori* truth. So in the end the difference between Newton's and Spinoza's views of Divine knowledge rests on their respective conceptions of necessity. For the latter the necessity of God's existence is inherent in the intrinsic actuality of God's essential nature, a nature that is conceivable in and through itself. For Newton, God is a necesary being in that it is causally impossible that he be otherwise in any respect whatsoever. Thus it is no objection to Newton to say that he is attempting to establish a logical truth on the basis of contingent premises, anymore than it can be objected to *Spinoza* that his conception of Divine necessity is logical in the purely linguistic sense of that term. That is, Spinoza does not make the mistake of constructing necessary *being* from the necessity that attaches to certain statements.

We can now connect Newton's conceptions of Divine necessity and Divine actuality. Let us first summarize his general position on the notion of actual existence. To say that anything is an actual being, is not to say anything about its nature or essence. It is to claim that it is in a state-of-being-actual. This is a fact about it, namely, the fact that it is an individual with a determinate identity which occupies a position in objective time and space. To say that something actually exists is true only if it has a real nature; and it has a real nature only if it is true that it actually exists. Thus being actual is an inseparable part of the fact that something exists as an individual nature.

Newton applies these conceptions to God's nature and existence. Since Divine existence has neither beginning nor end, it is not restricted with respect to space and time. God's eternal nature is such that there is no time at which and no place in which causally he can possibly fail to exist. In this sense infinite time and place are absolute affections of his unending existence. So in Newton's view, to say that God exists is not to say that existence is a Divine perfection *per se*. Rather it is to refer to God's state-of-being-actual with respect to infinite time and space. Thus to be an actual being with respect to infinite time and space is an inseparable fact about Divine existence. And when Newton speaks of infinite time and space as 'unbounded consequences' or as 'emanative' effects of Divine existence what he means in this. They are an inseparable part of the fact that an eternal and omnipresent being exists. It is in this sense that infinite time and space exist because God exists.

But can God be a necessary being if he is said to exist with respect to infinite space? In the *Principia* Newton considers the notion that things occupy places in objective space. He argues that 'the place of the whole is the same as the sum of the places of the parts, and on that account it is internal, and in the whole body'.[181] Again, likening the external surface of a body truly moving to a shell he says: 'if the shell moves the kernal will also move, as being part of the whole, without any removal from near the shell'.[182] Thus, when a body is said to occupy a place completely, each of its interstitial parts individually occupy a part of the whole place. As a distinct individual, they and the body translate together from one objective place in absolute space to another.

These statements embody Newton's criteria for claiming that something is a place-occupying entity. Not only does the whole body require a place that it fully occupies, but each of its parts also require their places. This means that the finite or infinite divisibility of place-occupying bodies entails the finite or infinite divisibility of the space which contains them. In short, if anything is contained in space, necessarily its parts are contained in space.

But surely this conception prevents Newton from claiming that God exists with respect to infinite space? For taken strictly, this involves saying that these criteria also apply to God. To say that God exists in an infinitely extended space is therefore to conceive him as a space-occupying and extended being. But this contradicts Newton's claim that God is an indivisible being.

Consider Newton's conception of Divine omnipresence. He says that God 'is omnipresent not virtually only, but also substantially; for power cannot subsist without substance'. Moreover, God is omnipresent in the sense that there is no place in which he cannot possibly exist. 'Whatsoever necessarily exists, exists always and everywhere, with the same law of necessity in all places and times whatsoever'.[183] In a more religious frame of mind Newton says: 'The father is immovable, no place being capable of becoming emptier or fuller of him than it is by the eternal necessity of nature'.[184] And although he claims that God is actually omnipresent, Newton denies that space is an attribute of Divine nature: rather it is an affection of God's existence alone. Moreover, in both the General Scholium and the Des Maizeaux drafts Newton denies that God is literally immense – which is to say, that he denies that quantity can be ascribed to God's nature. So on the one hand, newton maintains that God is omnipresent and, on the other hand, rejects the view that he is everywhere in the manner in which an extended being occupies a place. Yet he insists that God exists with respect to infinite space. How are these statements to be reconciled?

In the first place Newton must deny that God exists in space in the manner of an extended being. This conception entails that a "part" of God would exist in each "part" of an infinitely extended space. But God is an incorporeal being: his nature is at once whole and undivided. To say that an indivisible God possesses *partes extra partes* which everywhere exists in the "parts" of an infinite space is a flat contradiction.

There are two lines of response open to Newton. Each depends on the sense in which an entity is taken to be indivisible. We might construe Newton as saying that nothing can be added or taken away from God's nature. Thought his duration takes time, though he is everywhere present in space, does not mean that God is extended and divisible in the manner of physical things. On this interpretation, God is indivisible in the sense that his essential nature is in fact unalterable. Nothing can be added to it, or separated from it. God is unchangingly and inseparably an *ens unum per se*. Consequently, since Divine nature cannot be separated from itself, the relation of entailment existing between the "parts" of physical things and their places does not hold. Moreover, space itself is also without "parts" in that it cannot be separated from itself. So that from the point of view of the nature of space the entailment relation between things and places also fails to apply to Divine presence.

This is essentially Henry More's position. In fact he goes further. God is essentially extended. But all the parts of Divine extension are 'inseparably and indiscerpibly one'. Moreover, extension is intrinsic to God's essential nature. In virtue of that nature he is an extended spirit who constitutes the place of all created things. More attempts to justify these claims by distinguishing between physical and spiritual extension. 'For the coexistence of parts, as they call them, of a spirit, are connected by a condition absolutely necessary and plainly essential'.[185] Unfortunately, this is little more than a lame attempt to suggest that extension, when attributed to God's nature, is a transcendental attribute. In the end it was Joseph Raphson, not Newton, who attempted to give an internally consistent account of More's position.[186]

It is already clear that Newton does not agree with More's approach. He can certainly accept that the "parts" of space are actually inseparable, that the perfection of infinite space cannot affect God's presence, and that god is an *ens unum per se*. However, Newton nowhere says that God is an extended spirit or that infinite space is the innermost amplitude of Divine essence. Nor does he suggest that space and time are transcendentally ascribed to Divine being. In fact, he simply claims in *De gravitatione* that 'any being has a manner proper to itself of being in spaces'.[187] In what manner then is God related to infinite space?

It turns on the sense in which God is taken to be an indivisible nature. There can be no doubt that in Newton's view God is an incorporeal being. If I am right in holding that he does not accept More's conception of God as an extended spirit, there is only one alternative. God is strictly indivisible: he cannot possess parts in any sense whatsoever. Moreover, as Newton claims that the individual soul everywhere exists in the body undividedly, so he holds that Divine presence is everywhere exists in the body undividedly, so he holds that Divine presence is everywhere with respect to the expanses of infinite space. How is this reconcilable with his claim that God is everywhere the *same* unique individual? We can plausibly reconstruct the following argument. To say that God is everywhere with respect to space, is to say that one and the same individual exists at *each* place in extended space. To make such a claim is a contradiction only with respect to the manner in which extended things exist spatially. For they cannot as complete beings exist at once in each "part" of the place they occupy. But there is no contradiction in holding that an essentially non-extended being is capable of so existing.[188] On the contrary, to deny that it can, is to think of God as an extended spirit. God is therefore omnipresent just in the sense that he remains numerically and unalterably the same individual at all places whatsoever. The conception appears paradoxical. However, Newton would claim that it only seems so, if we persist in imagining God's presence in space as we think of bodies individually occupying their determinate places.[189] Moreover, this conception is consistent with Newton's view that infinite space is a consequence of God's existence. As the concept of spatial immensity is inappropriate to Divine nature, so too is the notion of spatial inclusion. Space *per se* is in fact infinite, and hence positively lacks boundaries.[190] Thus infinite space is not a "receptacle", in which God is "contained" omnipresently, in the manner of place-occupying bodies. I conclude, therefore, that altogether there are no grounds pertaining to Newton's conception of how God relates to space which allow the denial of his necessary existence.

## 6. CONCLUSIONS

I have attempted to unearth the basic assumptions of Newton's natural philosophy. Moreover, I have tried to relate them to the philosophical concerns of the period. But they have also been examined in connection with the fundamental categories of Newton's thought – God, existence, space, time and necessity. Without the framework which these provide, his theories of matter, motion and force are without context and lack ultimate legitimation. This is the case whether we are concerned with the alchemical, physical, chemical or dynamical aspects of his thought.

If the claims in this paper are correct, Newton's general picture of the world has to be understood in the context of his conception of God's nature and existence. Crucial to this conception is Newton's view of God as an active and intelligent agent who exists omnitemporally and omnipresently. Accordingly, the doctrine that space and time are real existents, is rooted in Newton's conception of Divine existence. Since all created things exist along with God in the same absolute spatio-temporal framework, the absolute motion of things is with respect to that alone. Moreover, the intelligibility of this is itself grounded in the intelligibility of Divine existence. They stand or fall together. Hence to claim that absolute space and time are reifications of Newton's mathematical quantities, or that absolute motion requires the postulation of these entities, is to miss the fundamentally theological significance of Newton's reasoning. Moreover, it is to put the cart before the horse. This is not to deny the importance of Newton's attempt to show, by means of the whirling bucket and the rotating globes, that real effects of real motions truly exist. And from these effects to distinguish the measures of real and absolute space from sensible and relative space. Nor is it to deny the importance of Newton's views on the topological structure of space. However, these concerns function at a different level in his thought. They find their place within the larger and more metaphysically oriented concepts of his general natural philosophy. Unless this is appreciated, the importance of Newton's reading in the philosophy and theology of his period will be missed. Newton's aim goes beyond explaining the phenomena of nature, important though this is. Ultimately, he wished to render the Creator and the whole of creation intelligible to human understanding.

## ACKNOWLEDGEMENTS

I wish to thank the National Science Foundation, Washington, for financial support, without which this study could not have been completed. For their encouragement and criticism, I owe much to George Molland, James Bogen, Miles Burnyeat and Allen Gabbey. Most of all I am indebted to Peter Machamer. His constructive criticism sustained me through some difficult periods. Needless to say, none is responsible for any errors of fact or interpretation.

## NOTES

[1] Ralph Cudworth, *The True Intellectual System of the Universe* (notes and dissertations of Dr. J. L. Mosheim, trans. John Harrison: 3 vols., London, 1845), vol. 3, 231–232. (Hereafter '*True intellectual system*'.)

[2] *De gravitatione* is transcribed and translated by A. R. Hall and M. B. Hall, *Unpublished Scientific Papers of Isaac Newton* (Cambridge, 1962), 90–156. (Hereafter, '*De gravitatione*'.) See also Isaac Newton, *Philosophiae naturalis principia mathematica* (London, 1687), 5–11. Isaac Newton's *Philosophiae naturalis principia mathematica* (fasimile of the third edition (1726, ed. A. Koyré and I. B. Cohen), vol. 1 (Cambridge, 1972), 47–53. The General Scholium is reproduced in vol. 2, 759–765. Since this work gives the variants of the three editions published in Newton's life time it will be cited henceforth unless otherwise indicated.

[3] J. E. McGuire, 'Newton on Place, Time, and God: An Unpublished Source', *British Journal for the History of Science* **11** (1978), 114–129. (It is in the Cambridge University Library (hereafter CUL), Ms. Add. 3965, section 13, folios 545r–546r.) See also J. E. McGuire, 'Predicates of Pure Existence: Newton on God's Space and Time,' and the reply of John Carriero, 'Newton on Space and Time: Comments on J. E. McGuire': both papers are in *Philosophical Perspectives*

*on Newton*, Phillip Bricker and R. I. G. Hughes (eds.), MIT Press, Cambridge, Mass. 1990. Immediately following the publication of the *Principia* (1687) Newton began reconsidering its mathematical, physical and philosophical commitments. For the importance of the early 1690s in appraising Newton's intellectual achievement, see D. T. Whiteside, *The Mathematical Papers of Isaac Newton*, vol. 7 (Cambridge, 1976). For a brief survey of some methodological and philosophical issues emerging from the first edition of the *Principia*, see Chapter 6 below, and J. E. McGuire '"Newton's Principles of Philosophy": and intended preface for the 1704 *Opticks* and a related draft fragment', *Brit. Hist. of Sci* **5** (1970), 178–187.

[4] 'Newton on place, time and god' (note 3).

[5] The most sustained expression of the importance of More for the development of Newton's views on space is in A. Koyré's *From the Closed World to the Infinite Universe* (Johns Hopkins Press, Baltimore, 1957), chapters 5 and 6, 125–189. (Hereafter '*Closed world*'.)

[6] 'Tempus et Locus sunt omnium rerum affectiones communes sine quibus nihil omnino potest existere. In tempore sunt omnia quoad durationem existentiae & in loco quoad amplitudinem praesentiae. Et quod nunquam nusquam est, id in rerum natura non est' (CUL, Ms. Add. 3965, section 13, folio 545r).

[7] Lucretius, *De rerum natura* (edited with translation and commentary by Cyril Bailey: Oxford, 1947), vol. 1; see Liber I, lines 418–550, 196–203.

[8] *Ibid.*

[9] J. E. McGuire and P. M. Rattansi, 'Newton and the "Pipes of Pan"', *Notes and Records of the Royal Society* **21** (1966), 108–144. This paper analyses Newton's use of the ancient philosophers, including Lucretisu, in an attempt to legitimize his philosophy of nature in the early 1690s.

[10] For example, see Franciscus Piccolomineus, *Libri ad scientiam de natura* (Venice, 1596), Bk. I, cap. 7, 141; Hieronymus Zanchius, *De operibus Dei* (Hanau, 1597), 348–355; and Joann Magirus, *Physiologia peripatetica* (Leipzig, 1611), Lib. I, cap. VIII, 92 (Newton made substantial notes on a later edition of this work); F. Burgersdijk, *Idea philosphiae* (Oxford, 1637), 26–27; Robert Baron, *Metaphysica generalis* (London, 1658), 15, 32, and 67ff; and Benedicto de Spinoza, *Spinoza opera* (ed. Carl Gebhardt: Heidelberg, 1925), vol. 2, *Ethica*, pars Tertia, 139–148, and vol. 1, *Cogitata metaphysica*, Pars I, 233–249.

[11] *De gravitatione*, 103: the translation is mine. When Newton speaks of space and duration as 'affections or attributes' of beings, he is not predicating them, *qua* attributes, of the things themselves, but only of their 'quantity of existence'. In the scholium on space and time of the *Principia*, Newton also characterizes duration as the 'perseverence of the existence of things' (*duratio seu perseverantia existentiae rerum*); *Principia*, vol. 1, 48.

[12] For late Renaissance attempts to distinguish time, duration and motion see Sarah Hutton, 'Some Renaissance critques of Aristotle's theory of time', *Annals of Science* **34** (1977), 345–363. Hutton discusses a number of passages from the works of Renanissance writers in which they argue that duration can only be distinguished conceptually from both the permanence and motion of things. This is Newton's position. Things are in either a state of resting or a state of moving with respect to space and time. However, he also holds that the duration of things must be conceived with respect to their 'quantity of existence' in absolute time and space, where to speak of something's 'quantity of existence' is not to ascribe a real property. Moreover, absolute time and space are themselves said to endure independently of all changes pertaining to sensible things. In fact Newton goes so far as to say: 'Tempus absolutum . . . alioque nomine dicitur duratio' (*Principia*, vol. 1, 46). Concerning absolute space he says that it 'semper manet similare & immobile' (*Principia*, 46).

[13] *De gravitatione*, 104 and 99; the translation is mine.

[14] *Principia* vol. 1, 46.

[15] *Ibid.*, 46.

[16] *Ibid.*, vol. 2, 761.

[17] *Ibid.*, vol. 1, 52.

[18] *De gravitatione*, 103; 'Sic quantitas existentiae Dei secundum durationem aeterna fuit, et sccundum spatium cui adest, infinita'.

[19] *Principia*, vol. 1, 46–53. CUL, Ms. Add. 3965, 13 (c. 1692–93), folio 545r, summarizes and discusses these distinctions.

[20] *Ibid*., vol. 1, 48–49: 'But because the parts of space cannot be seen or distinguished from one another by our senses, in their stead we use sensible measures of them. For from the positions and distances of things from any body which we consider as immobile, we define all places: and then with respect to such places [*ad praedicta loca*], we estimate all motions, in so far as we consider bodies as transferred from some of those places into others. And so, instead of absolute places and motions we use relative ones; and that without inconvenience in human affairs'. Notice Newton's use of intentional language when describing how we use relative motions in conceptualizing our experience of things at rest, or moving with respect to one another. This probably shows the influence of Descartes whom Newton read thoroughly. The first part of *De gravitatione* subjects Descartes's concepts of motion, space and place to careful scrutiny (see pp. 91–99). Newton's treatment of relative concepts owes a great deal to Descartes's influence.

[21] *Principia*, vol. 1, 47–48 and 53: 'But how we are to obtain the true motions from their causes, effects and apparent differences [relative measures], and the converse, is explained in what follows. For to this end it was that I composed it'. For Newton 'true' or absolute motions demand the existence of absolute space and time.

[22] *Ibid*., 49.

[23] *De gravitatione*, 101–102. Newton argues that while we cannot *imagine* infinite extension, we can,however, *understand* its positive properties clearly enough. In this respect he differs from Descartes. In fact, Descartes is the target of Newton's criticism in *De gravitatione* (as well as earlier in his *Quaestiones quaedem philosophicae*; see notes 25 and 190 below). Descartes argues that we can and do have a positive idea of God's infinite perfections. Of extension, however, we have only a negative idea. We cannot determine the limits and boundaries of material extension; so far as we know, therefore, it is indefinite, but not in fact infinite. He applies this reasoning to the number of the stars and to the question of matter's divisibility (*Oeuvres de Descartes*, vol. 8, (ed. Charles Adam and Paul Tannery: Paris, 1905), *Principia philosophiae*, Pars Prima, section 26 and 27, 14–15). See Chapter 4 for a discussion of Descartes, Newton and Aristotle on infinity.

[24] *Principia*, vol. 1, 52.

[25] See Richard S. Westfall, 'The foundation of Newton's philosophy of nature', *Brit. J. Hist. Sci.* **1** (1962), 171–182. Westfall rightly emphasizes the Gassendist orientation of Newton's burgeoning thought. The evidence for Newton's having read Charleton's *Physiologia Epicuro-Gassendo-Charletoniana* (London, 1654) is in his earliest notebook labelled *Quaestiones quaedem Philosophicae* (CUL Ms Add. 3996, folios 88–135; *vide* folio 88). The *Quaestiones* were probably written between 1661 and 1665.

[26] (London, 1671), part I, cap. 2, Section I.

[27] (London, 1668), *Dialogue* I, 96. The Latin version of the *Dialogues* is in More's *Opera omnia* (London, 1679). The quotation reads: 'Quamobrem Extensio sive Amplitudo, intrinseca est Essentialis proprietas Entis quantenus Ens est' (section 25, 663). Since *De gravitatione* may well have been written as late as 1670, it is possible that Newton could have read it in time to have his views influenced by More.

[28] *Enchiridion*, part I, cap. 2, section I.

[29] *Ibid*., cap. 2.

[30] *Quaestiones*, CUL, Ms. Add. 2996, folio 89r: 'And y[t] Matter may be so small as to be indiscerpible the excellent Dr. More in his booke of y[e] soules immortality hath proved . . .'.

[31] *Enchiridion*, Part I, cap. 2, section 10. 'Duratio igitur pertinet ad Ens quatenus Ens'.

[32] *Ibid*., Cap. 27, section I.

[33] *De gravitatione*, 103: 'Nullum ens existit vel potest existere quod non aliquo modo ad spatium refertur'. Like More, opposes Descartes's conception of God as existing everywhere in virtue of his power alone. As we shall see, however, there is a closer relationship between the positions of Newton and Descartes than has been realized. For Descartes' view see his letter to More of 5 February 1649: '. . . we clearly conclude that no incorporeal substances are in any strict sense extended. I conceive them as powers or forces, which although they can act upon extended substances, are not themselves extended; just as fire is in white hot iron without itself being iron. Some people indeed do confuse the notion of substance with that of extended matter. This is because of the false prejudice which makes them believe that nothing can exist or be intelligible without being also imaginable . . .' (*Descartes' Philosophical Letters*, trans. and ed. Anthony Kenny: Oxford, 1970, 239).

[34] *De gravitation*, 99, 103 and 104.

[35] Henry More, *The Immortality of Soul* (London, 1659), Bk. I, Cap. 6, 31–35.

[36] *Ibid.*, 32–33.

[37] *Enchiridion*, part I, caps. 2, section 2.

[38] *De gravitatione*, 131. Newton amplifies this remark: 'But it is usually believed that these spaces are nothing: yet indeed they are true spaces. Although space may be empty of body, nevertheless it is not in itself a void; and something is there, because spaces are there, although nothing more then that' (translation is the Halls', p. 138). Newton nowhere argues for the intelligibility of incorporeal space and incorporeal dimensions in the manner of Henry More. Nor does he attempt to show that the concepts of dimensions and quantity can be applied to incorporeal entities. (See Charleton's *Physiologia*, chap. 6, 65 for a discussion of this point.) He simply takes the idea of a void, extended space to be un umproblematic and primitive notion, which requires reference to incorporeal reality. As we will see, space is a direct consequence of God's omnipresence. Philosophically, Newton's position is similar to Pascal's. In his letter to Le Pailleur, Pascal defends the reality of void space: '*Ni Substance ni accident.* Cela est vrai, si l' on entend par le mot de *substance* ce qui est ou corps ou esprit; car, en ce sens, l' espace ne sera ni substance, ni accident; mais il sera espace . . . parce que pour être, il n'est pas nécessaire d'être substance ou accident'. This is in the correspondence arising from the publication of Pascal's *experience nouvelles touchant le vide* (Paris, 1647); see *Oeuvres complètes* (preface by Henri Gouhier, ed. Louis Lafuma: Paris, 1975), 210. For a discussion of Newton's early and mature views on void space see Chapter 3.

[40] *Ibid.*, 99; translation mine.

[41] *Enchiridion*, part I, cap. 8, section 9. For a misleading discussion of this topic see A. Koyré, *Closed World*, chapter 6, 147–154. Koyré wrongly states that More identifies infinite space with God's essential being.

[42] *Enchiridion*, part I, cap. 10, section 14. When attributed to the being of God duration is absolute and infinite: 'Atque certe, ut nihil dissimulem, solidissimu videtur hoc Argumentum quo demonstretur Mundanam Materiam non posse esse absolute infinitum, sed duntaxat indefinitam, prout videtur alicubli pronunciare *Cartesius*, nomenque infinite soli Deo reservare. Quod ferio ac constanter afferendum est tam de *Duratione Dei* quam de ejus *Amplitudine*. Utraeque enim absolute infinitae sunt: Mundi vero indefinitae solummodo esse possunt' ('Accordingly, so as not to dissimulate in anyway, this [More's previous discussion in cap. 10 to show that physical things are indefinite] seems to be the most solid argument for demonstrating that the matter of the world cannot be absolutely infinite, but only indefinite, as Descartes has somewhere said, and to God alone reserve the name of infinite. Which must be asserted as well as of *Duration* as of the *Amplitude of God*').

Notice that Newton's position opposes More's. More asserts that duration and extended space are infinite only when ascribed to God. Otherwise they are indefinite. This 'Cartesian' line of reasoning is only present in More's mature thought. In his earlier writings, the world is simply infinite, and the Cartesian distinction between 'infinite' and 'indefinite' is held to be ill-founded.

We shall shrotly consider Newton's reasons for rejecting the Cartesian dichotomy between infinity and the indefinite. Descartes, of course, does not ascribe duration and extension to God's essence.

43 *De gravitatione*, 99–100. Thus it is clear that when Newton speaks of space as 'an affection of all being such as subsists', he is not speaking of being as a possible existent, but rather stating that space pertains to all being *generically* considered.

44 *Ibid.*, 99; translation mine.

45 *Ibid.*, 132; translation by the Halls.

46 *Physiologia*, Lib. 1, cap. 4, 57 and cap. 8, 92. Magirus's distinctions falls within the framework of his classification of common affections into extrinsic and intrinsic characterizations. An intrinsic character says something about the defining nature of a thing, whereas an extrinsic character does not do so necessarily.

47 *De gravitatione*, 99; translation mine.

48 I have used Craig B. Brush's translation of the passage as it captures the 'flavour' of Gassendi's style; see *The Selected Works of Pierre Gassendi* (London and New York, Johnson Reprint, 1972), 384. For the original Latin passage see Pierre Gassendi, *Opera omnia*, vol. I (1658: reprint Stuttgart-Bad Cannstatt, 1964), *Syntagma philosophicum* (hereafter cited '*Syntagma*'), Lib. 2, cap. I, 1982: 'Id nempe, quia nulla substantia, nullum accidens sit, cui non competat esse alicubi, seu quopiam in loco; & esse aliquando, seu aliquo tempore; atque ita quidem, ut tametsi talis substantia, taleve accidens pereat; non ideo minus constare Locus, aut fluere Tempus persevert'. While there is no direct evidence that Newton read Gassendi, he did read Charleton. In the latter's *Physiologia* there is a passage similar to the one quoted from Gassendi (*Physiologia* Lib. I, 66). Patrizi's *Nova de universis philosophia* (Venice, 1593) was the first systematic attempt in Western philosophy to show that Aristotle's categories do not apply to space and time. However, Patrizi still treats them as a kind of substance. Gassendi does not. Nor does Newton, though he does at times think of space and time as analogous to substance.

49 *Syntagma*, Lib. 2, cap. I, 179.

50 *Enchiridion*, part I, section 7, and cap 6, sections 1, 2 and 3.

51 Although he is highly critical of the principles on which he takes Aristotle to have based his categories, Gassendi nowhere completely rejects the fundamental division of reality into substance attribute. Rather he denies that it applies to the reality of space and time. See *Exercitaliones paradoxicae adversus Aristoteleos (Paradoxes contre les Aristoteliciens)* (trans. and annotated by Bernard Rochot: Paris, Vrin, 1959), Lib. I, *Exercitatio* 6, Articles 1 and 2, *Exercitatio* 3, Art. 1, 2 and especially article 3, 'Divisionem substantiae et accidentis non congruere'. The *Exercitationes* were first published in their entirely in 1649. The first book was published in 1624.

52 *Syntagma*, Lib. 2, cap. I, 182.

53 *Ibid.*, 182–183.

54 *Physiologia*, Chap. 6, 25; *De gravitatione*, 103.

55 *De gravitatione*, 99; translation mine.

56 A. Koyré and I. B. Cohen, 'Newton and the Leibniz – Clarke Correspondence', *Archives inter. d'hist. sci.*, nos. 58–59 (1962), 63–126. (Hereafter cited 'Leibnize – Clarke correspondence'.)

57 *Ibid.*, 96–97.

58 *Ibid.*, 98.

59 See also, *Ibid.* for a similar analysis.

60 *Syntagma*, Lib. I, cap. 7; and *Physiologia*, chap. 7, 75–76. The identification of the Aristotelian catgories of 'when' and 'where' with Time and Space occurred during the period of Middle Platonism. For an excellent account of this, see John Dillion, *The Middle Platonists* (Cornell Univ. Press, 1977), 133–135, 178–182.

61 G. G. Leibniz, *Opera Philosophica* (ed. J. E. Eardmann: Berolini, 1839–1840). 'Lettres

entre Leibniz et Clarke', Article 44, 768 deals with Leibniz's claim that space and time "part" of the essence of God according to Newton's ontology. Article 45, 768 states that 'On a bien oui dire que la propriéte soit dans le sujet; mais on n'a jamais oui dire que le suject soit dans sa propriété. De même, Dieu existe en chaque temps: comment donc le temps est-il dans Dieu; et comment peut-il être une propriété de Dieu ?'.

62 'Leibniz – Clarke correspondence', 97.

63 For a discussion of the notion of substance in seventeenth-century thought see E. M. Curley, *Spinoza's Metaphysics: An Essay in Interpretation* (Harvard Press, 1969), chap. 1.

64 J. L. Mackie, 'The riddle of existence', *The Aristotelian Society, supplmentary volume*, **50** (1976), 247–265. Besides this useful article see W. C. Kneale, 'Is existence a predicate?', *The Aristotelian Society, supplementary volume*, **15** (1936).

65 *Ibid.* Mackie also makes a similar point about *existere*. His paper came to my notice after my essay had been completed.

66 'Leibniz – Clarke correspondence', 99.

67 *Principia* vol. 2, 761: 'Non est aeternitas & infinitas, sed aeternus & infinitus: non est duratio & spatium, sed durat & adest. Durat semper & adest ubique, & existendo semper & ubique, durationem & spatium constituit. Cum unaquaeque spatii particular sit *semper*, & unumquodque durationis indivisible momentum *ubique*, certe rerum omnium fabricator ac dominus non erit *nunquam, nusquam*'; italics are Newton's. God's uniquity and sempiternality are clearly stressed.

68 'Leibniz – Clarke correspondence', 101.

69 *Ibid.*, 99.

70 *Ibid.*

71 For a classic discussion of the different consequences that follow from holding that God's attributes are identical with him as opposed to the doctrine that attributes are inherent in, and inseparable from, Divine nature see: *A Defense of True Liberty* (London, 1655) by Bishop John Bramhall, in which he criticizes Thomas Hobbes's necessitarianism. (See especially pp. 191–200.) Bramhall argues forcefully against Hobbes that 'whatsoever is in God, is God' (p. 199). Hobbes's position is similar to Newton's.

72 *Immortality*, Bk. I, chap. 6, 34–35.

73 *Divine dialogues*, Dialogue I, 96–100.

74 *Ibid.*, 94.

75 *Enchiridion*, part I, cap. 8, section 14.

76 *Ibid.*, cap. 8, section 9.

77 *Divine dialogues*, Dialogue I, 10.

78 *Enchiridion*, Part I, cap. 8, section 14.

79 *De gravitatione* 99; translation mine.

80 *Principia* vol. 2, 761: 'durat ab aeterne in aceternum, & adest ab infinito in infinitum'.

81 *Unpublished Scientific Papers of Isaac Newton*, 355: 'An Deus erit nusquam cum momentum temporis sit ubique?'.

82 *De gravitatione*, 104; translation mine. 'Denique spatium est aeternae durationis et immutabilis naturae, idque quod sit aeternis et immutabilis entis effectus emanativus. Siquando non fuerit spatium. Deus tunc nullibi adfuerit, et pronide spatium creabat postea ubi ipse non aderat, vel quod non minus rationi absonum est, creabat suam ubiquitatem'.

83 *True Intellectual System*, vol. 2, chap. 5, section I, 536.

84 *Enchiridion*, part I, cap. 8, section 8 and 13.

85 *De gravitatione*, 102.

86 *De gravitatione*, 102–103; I have altered the Hall's translation slightly. 'Sed video quid metuit Cartesius, nempe si spatium poneret infinitum, Deum forte constitueret propter infinitatis perfectionem. At nullo modo, nam infinitas non est perfectio nisi quatenus perfectionibus tribuitur. Infinitas intellectus, potentiae, faelicitatis &c summa infinitus ignorantiae, impotentiae, miseriae &c summa imperfectio; et infinitas extensionis talis est perfectio qualis est extendi'.

[87] CUL, Ms. Add. 3965, 13, folio 546r. The passage from which this is quoted is as follows: 'et quae nec bona sunt nec mala par aeternitatem et infinitatem nec bona evadent nec mala sed in suo tantum genere perfecta, id est diutissima et maxima. Res omnis per aeternitatem & infinitatem non melior redditur aut perfectionis naturae sed suo tantum genere durabilior et vel major numerosior, in honorem Dei conditoris'.

[88] *Principia*, vol. 2, 761; 'Et ex dominatione vera sequitur deum verum esse vivum, intelligentum & potentem; ex reliquis perfectionibus summum esse, vel summe perfectum . . . omnia regit; & omnia cognoscit, quae fiunt aut fieri possunt'. For a similar view see the *Opticks* (Dover edition, New York, 1952), Query 28, 370.

[89] CUL, Ms. Add. 3965, 13, Folio 546r.

[90] CUL, Ms. Add. 3965, 13, folio 545v: 'Spatium ex aeternitate et infinitate sua nec Deus erit nec sapiens nec potens nec vivum sed duratione et magnitudine tantum augebitur'.

[91] *De gravitatione*, 99.

[92] *Oeuvres de Descartes* (ed. Adam and Tannery), vol. 7, 66.

[93] *Ibid.*, vol. 8, Pars Prima, Article 14, 10.

[94] *Ibid.*, vol. 7, 323.

[95] *Ibid.*, vol. 7, 66–68.

[96] CUL, Ms. Add. 3996, *Quaestiones* folio 83r. Newton makes the same observation in another lengthy note on the *Fifth MEditation*: 'Ax: tis a contradiction to say, that thing doth not exist, whose existence implys noe contradiction, & being supposed to exist must necessarily exist . . . Tis only from the essence of it that a thing can perpetuate its existence without extrinsecall helpe. Wch essence being sufficient to continue it must bee sufficient to cause it, there being ye like reason of booth' (folio 83v; Newton's spelling is preserved).

[97] *Oeuvres de Descartes*, vol. 7, 109: 'Sed plane admitto alquid esse posse, in quo sit tanta & tam inexhausta potentia, ut nullius unquam ope equeritut existeret, neque etiam nune egeat ut conservetur, atque adeo sit quodammodo sui causa'.

[98] *Ibid.*, vol. 7, 66.

[99] *Ibid.*, vol. 7, 323: '. . . quod existit pluresque perfectiones habet, non habet existentiam ut perfectionem singularem unamque ex eo numero, sed ut illud, quo tam ipsum quam perfectiones existentes sunt, & sine quo nec opsum habere, nec perefectiones haberi dicuntur. Hinc neque existentia perfectionum instar existere in re dicitur, neque, si res careat existentia, tam imperfecta (sive privata perfectione) dicitur quam nulla'.

[100] Pierre Gassendi, *Disquisitio metaphysica* (1664: trans. and annotations by Bernard Rochot: Paris, Vrin, 1962), 497: 'Existentiam, non esse proprietatem, nesque esse ullius determinati generis rerum'. (Hereafter cited '*Disquisito*'.)

[101] *Opera omnia*, vol. I, *Syntagma*, Lib. 2 , cap. I, 179–184.

[102] These are in his *Quaestiones quaedem philosophicae*, folio 83r–v.

[103] For a discussion of this see the introduction by Adam and Tannery, *Oeuvres de Descartes*, vol. 8, v–xviii.

[104] Many of these "small" editions of Descartes works are incomplete. The volume containing the objections and replies is missing from many of the sets which I have examined. However, it is most likely that Newton used the 1664 English reprint of Descartes's work. I was unable to locate a copy. On the question of Newton's pagination not matching the editions which I have examined, it could well be that he is citing them from memory and thus approximately. With this in mind it is also possible that he used Trinity College library's copy of Descartes's work, *Opera philosophia* (ed. 3, Amsterdam, 1656).

[105] *Oeuvres de Descartes*, vol. 7, 'Quintae Responsiones', 382–383: '. . . non video cujus generis rerum velis esse existentiam, nec quare non aeque proprietas atque omnipotentia dici possit, sumendo scilicet nomen porprietatis pro quolibet attributo, sive pro omni eo quod de re potest praedicari, . . . Quim etiam existentia necessaria est revera in Deo proprietas strictissimo modo sumpta, quiua illi soli competit, & in eo solo essentiae partem facit'.

[106] *Ibid.*, vol. 7, 383.
[107] *Ibid.*, 'Nec verum est *essentiam & existentiam in Deo*, quemadmodum in triangule, *unam absque alia posse cogitari*, quia Deus est suum est, non autem triangulus. Nec tamen inficior quin existentia possibilis sit perfectio in idea trianguli, ut existentia necessaria est perfectio in idea Dei' (italics in text).
[108] For Descartes, the objects of mathematics are true and real entities (*vera entia*). The difference between them and the objects of physics is that the latter are not just true and real entities, but they actually and specifically exist *per se* and extramentally. See *Descartes' Conversion with Burman* (trans. with introduction and commentary by John Cottingham: Oxford, 1976), 90–92. Cottingham gives a useful analysis of Descartes's notion of an intrinsic link existing between the essential features of things. As Cottingham indicates, Descartes's notion of an intrinsic link does not give him a systematic way of distinguishing unitary from composite entities.
[109] *Meditationes, Oeuvres de Descartes*, vol. 7, 64: '. . . talis figura nullibi gentium extra cogitationem extiterit meam existat, nec unquam exiterit, est tamen profecto determinata quaedam ejus natura, sive essentia, sive forma, immutabilis & aeterna, quae a me non efficta est, nec a mente mea dependet'.
[110] *Ibid.*, vol. 7, 64.
[111] *Ibid.*, vol. 7, 66 and 67; see also *Principia*, vol. 8, Pars Prima, articles 18, 19 and 20.
[112] *Meditationes*, vol. 7, 66–67; see also *Principia*, vol. 8, Pars Prima, article 14.
[113] *Ibid., Meditationes*, 67: 'Atqui ex eo quod non possim cogitare Deum nisi existentem, sequitur existentiam a Deo esse inseparabilem, ac proinde illum revera existere; non quod mea cogitatio hoc efficiat, sive aliquam necessitatem ulli rei imponat, sed contra quia ipsius rei, nempe existentiae Dei, necessitas me determinat ad hoc cogitandum'.
[114] See Alan Gabbey, 'Anne Conway et Henri More. Lettres sur Descartes', *Archives de Philosophie* **40**, cahier 3 (1977), 379–404. The letters from which the quotation is taken was written in 1650.
[115] *An Antidote Against Atheisme* (London, 1653), chap. 8, 20–21.
[116] *Disquisitio metaphysica*, 490–196. For a fine study of Gassendi's skepticism, epistemology and nominalism see O. R. Bloch, *La philosophie de Gassendi* (Martinus Nijhoff, The Hague, 1971), chapters 3, 4 and 5.
[117] *Syntagama, Opera omnia*, vol. I, sect. I, Lib. 4, cap 2, 3, and 5.
[118] *Principia*, vol. 8, Pars Prima, article 52: 'Verumtamen non potest substantia primum animadverti ex hoc solo quod sit res existens, quia hoc solum per se non afficit, sed facile ipsam agnosecimus ex quolibet ejus attributo, per communen illam notionem quod nihili nulla sint attributa, nullaeve proprietates aut qualitates'.
[119] *Ibid.*
[120] *Ibid.*, Pars Prima, article 15.
[121] *Ibid.*, article 18; see also *Meditationes – Quinta*, vol. 7 , 64–65.
[122] *Principia*, vol. 8, Pars Prima, article 17. In *Meditations* three and six, Descartes makes significant use of the notions of eminent and formal causation (vol. 7, 40–44; and for *Meditation* six see vol. 7, p. 79). To speak of something as a formal cause is to accord it real existence and the power to affect both the mind and other existing things. Thus material objects are capable of causing the content of our ideas in a way quite independent of our volition. To do so, they must be as "real" *qua* cause, that is, formally, as is the 'representative' or 'objective' reality in the idea *qua* effect. In other words the cause must be as perfect with respect to its kind, as the idea it causes is with respect to its kind. This is to day, that the cause must have a nature appropriate to the effects it causes, and where the formal reality embodies a real essence and the objective reality is that real essence as it is occurrently considered by the mind. An eminent cause is at a level higher than a similar cause functioning at the formal level. For example, God is the ultimate cause of material things. However, the perfection of material things – the fact that they are necessarily extended – is with respect to God's mind an idea eminently contained. That is, God as the cause of material things contains that reality in the most perfect

manner. For he contains only an eternal idea of the essence of created extension, since the imperfection of created extension cannot in any way be attributed to Divine nature. Newton was certainly acquainted with this doctrine. In expounding Descartes's view on necessary existence he employs these notions (CUL, Ms. Add. 3996, folio, 83v). In his notes on causality he paraphrases it (Ms. Add. 3996, folio 49r). Finally in *De gravitatione*, 109 he uses the doctrine to show that thought and extension are reconcilable in Divine nature as he contains them eminently.

123 Descartes extended argument for these conclusions is found in the *Sixth Meditation (Oeuvres de Descartes*, vol. 7, 71–83).

124 *Disquisitio*, In *Meditationes V. Dubitatio* 1, 469–479. See particularly Art. 2, 479: 'Naturas rerum aeternas, ac realeis nullas esse; et quas, ut ideas, habemus in mente, ab ipsa mente ex inspectis rebus singularibus fieri' ('There are no natural principles of things which are eternal and real; and those which we have in the mind are ideas created by the mind itself upon examining particular things').

125 *Ibid.*, 469–170; see also *Dubitatio* I, article 5, 488–489.

126 I have slightly amended Brush's translation (see *The Selected Works of Pierre Gassendi*, note 48, 262). The Latin reads (*Disquisitio, Dubitatio* 2, Article 3, 507): 'Nonne hoc ipsum probandum est, pertinere ad ejus naturam, ut exsistat? Dicesne forte id remanere ex facta a te descriptione probatum? At describere qualem rem concipias, est ne demonstrare exsistere ipsam? An id probas enumerando omneis Dei perfectiones, aeternitatem, immensitatem, summan potentiam, sapientiam, exsistentiam, etc. At quid agis aluid, quam ponere inter perfectiones existentiam, ut esse exsistentiam, videlicet idem per se ipsum probes?'.

127 *Disquisitio, Dubitatio* 2, Article 2, 501.

128 *The Selected Works of Pierre Gassendi*, 262; Brush's translation. *Disquisitio, Dubitatio* 2, Article 3, 507: 'An probas, quia caeteris rebus non alia exsistentia, quam possibilis, atque contingens convenit; ipsi autem Deo actualis, atque necessaria? At nonne haec ipsa est quaestio, an exsistentia Deo conveniat, quaecumque tandem illa sit, cum si quis universe neget, etiam actualem, est necessarium neget?'.

129 See Bloch, *La philosophie de Gassendi*, ch. 4, for an interesting analysis of Gassendi's anticipation of Kant on the question of existence and its negation. For a fine discussion of the Medieval critics of Anselm's ontological argument, see Julius R. Weinberg, *Ockham, Descartes and Hume: Self-Knowledge, Substance and Causality* (Madison, Wisconsin 1977), Part I, chapters 1 and 2, 3–21. Gregory of Rimini explicitly states Gassendi's point that substantive descriptions do not necessarily include proper referents.

130 *De gravitatione*, 99, 106, 107, 109, 110 and 112. Newton's strategy is to attack the traditional scholastic notion of material substance construed as an unintelligible substratum. In his view, generic extension can be adequately grasped by both the mind and the senses. Thus the metaphorical picture of properties inhering in an unknowable substrate can be discarded. On the positive side, Newton argues that God directly endows extended space with a causal basis adequate for producing ideas of material phenomena in the mind.

131 *Principia*, vol. 2, 763. In the General Scholium, Newton's epistemology does take on a Lockean patina. But that is all. It is manifest that his rejection of the notions of inhering and subsisting occurs earlier in *De gravitatione*. It probably represents Newton's acceptance of the move away from explanation through essences current in this period. As such his views are probably a result of his reading of Charleton, Boyle and Glanvill, all of whom are represented in the notes of the 'Quaestiones' (CUL, Add. Ms. 3996). Moreover, he may have also read Gassendi. Furthermore, his rejection of innate dieas and explanation via essences antidates any influence from Locke. In the *Quaestiones* we read: 'The nature of things is more securely & naturally deduced from their operacons one upon another & upon e[e] [our senses. And when by y[e] former experiments we have found y[e] nature of bodys, by y[e] latter wee may more clearly find y[e] nature of o[?e] senses' (folio 28v).

132 *Principia*, vol. 2, 763. The implicit epistemology is similar to the passage from the

*Quaestiones* in note 131. It is clear, then, that Newton's conception of the sensory basis of knowledge was generally formed as early as the *Quaestiones* (1661–65). Moreover, Newton's position is Gassendist in orientation and anti-Cartesian. To be sure, More does say in *The Immortality of the Soul*, Bk. I, Chap. 2, Axiom VIII, that the 'naked essence' of anything is inconceivable. However, he does not relate this conception, as does Gassendi, to the causal basis of what we claim to know.

133 This is a question that still demands research. Many Dutch theologians, whether or not of Remonstrant persuasion, held that God exists sempiternally and also substantially in the place of his creation. For example, see Conrad Vorstius, *Tractatus theologicus de Deo* (Steinfurt, 1606); Hieronymous Zanchius, *De operibus Dei* (Hanau, 1597); and David Derondon, *Philosophia contracta* (Geneva, 1664). For later in the century, see Philip Van Limborch, *Theologia Christiana* (Amsterdam, 1686). For a general study, see Rosalie L. Colie, *Light and Enlightenment: A Study of the Cambridge Platonists and the Dutch Arminians* (Cambridge, 1957), especially ch. 5 and 6.

134 See Colie *ibid.*, for an account of the relationship between the Dutch Arminans and the Cambridge Platonists. For Newton's indirect concern with Spinozistic doctrine see 'Leibniz-Clarke correspondence', 93.

135 *Spinoza opera*, vol. 1, *Cogitata metaphysica*, Pars 2 Cap. I, 'Quid sit Aeternitas?', 251–252. The 'Cogitata' was first published in 1663.

136 *Ibid.*, Pars I, cap. 4, 244. See also Pars 2, cap. I, 250: ' . . . duratio est affectio existentiae, non vero essentiae rerum; Deo autem, cujus existentia est de ipsius essentia, nullam durationem tribuere possumus'.

137 *Ibid.*, Pars 2, cap. 1, 250.

138 *Ibid.*, 252: 'omnes res creatas, dum praesenti duratione, & existentia fruuntur, futura omnino carere, nempe quia continuo ipsis tribui debet; at de earum essentia nihil simile potest dici. Verum Deo, quia existentia est de ipsius essentia, futuram existentiam tribuere non possumus; eadem enim, quam tum haberet, etiamnum ipsi actu tribuenda est; vel, ut magis proprie loquar, Deo infinita actu existentia competit eodem modo, ac ipsi actu competit infinitus intellectus'.

139 *Ibid.*, 251.

140 *Cogitata metaphysica*, vol. I, Pars I, cap. 4, 244.

141 *Ibid.*, 244.

142 *Ibid.*, Pars 2, cap. I, 250.

143 *Ibid.*, 250.

144 *Ibid.*, 250–251.

145 *Ibid.*, 251. For Spinoza's mature expression of this view, see *Spinoza Opera*, vol. 2, *Ethica*, pars I, prop. 8: 'Omnis substantia est necessario infinita'. The writing of the *Ethics* was completed about 1665. The first publication was in the *Opera posthumus* (1677). That something is unlimited in a technical term. With respect to any attribute under which God is conceived, he is infinite; for he in fact possesses completely the perfection expressed through each of these. If it lacks with respect to a kind of perfection, a thing is thus limited (for example, a triangle lacks the perfection of infinite extension, since it is bounded in that respect).

146 *Unpublished Scientific Papers of Isaac Newton*, 357: 'Duratio ejus non est *nunc stans* sine duratione neque praesentia ejus est nusquam'.

147 'Leibniz-Clarke correspondence', 97.

148 CUL, Ms. Add. 3965, 13, folio 546r: 'Viderint igitur utrum rationi magis consentaneum sit aeternitatem Dei esse totam simul, an durationem per nomina Jehovae & "Οωνκαιοηνκαιερχομέυου" rectius designari'. Newton first declined the Greek phrase in the nominative case, but then changed it to the genitive inflection to agree with *Jehovae.*

149 For a fine analysis of this ponit see G. E. L. Owen, 'Plato and Parmenides on the timeless present', *Monist* **50** (1966), 332–333. Also see W. Kneale's important 'Time and eternity in theology', *Proceedings of the Aristotelian Society* (1960/61), 87–108.

150 CUL, Ms. Add. 3965, 13 folio 546r: 'In mysteria pronum est humanum genus neque aliquid aeque sanctum & perfectum ducit ac quod intelligi non potest. Quod tamen in connceptu Dei periculosum est et ad ejus existentiae explodendam conducit. Theologorum interest ut conceptus ille reddatur quam fieri possit facilis et rationi consentaneus ne cavillationibus obnoxius sit & inde in dubium vocetur'.

151 *Ibid.*, folio 546r.

152 *Principia*, vol. 2, 763.

153 CUL, Ms. Add. 3965, 13 flio 545r: 'In tempore et loco esse, cum rerum omnium natura communis, imperfectionem non arguit. Nam Duratio rei non est ejus fluxio vel mutatio quaelibet sed permanentia & immutabilitas in tempore fluente. Res omnes eatenus durant quatenus manent eaedem in omni tempore. Duratio rei cujusque fluit, at substantia durans non fluit non mutatur secundum prius et posterius sed eadem semper manet'.

154 For discussion of the uses of the distinction in medieval thought see Herman Shapiro, *Motion, Time and Place According to William Ockham* (The Fransiscan Institute, St. Bonaventure, N. Y., 1957), ch. 3, 36–83; William A. Wallace, *Causality and Scientific Explanation* (University of Michigan Press: Ann Arbor, 1972), vol. 1, 56–57; and John Murdoch and Edith Sylia, 'Walter Burley', *Dictionary of Scientific Biography*, vol. 2 (1970), 610.

155 For Charleton see *Physiologia* Bk. I, chap. 7, 79–80; for Gassendi see *Opera Omnia*, vol. I, *Syntagma*, Lib. I, cap. 7 and for More *Enchiridion*, Pars, I, cap. 6. These discussions are close to Newton's, both in detail and in language. A source for seventeenth century discussions of the distinction between permanent and successive duration is F. Suarez, *Disputationes metaphysicae* (1605), Disput. 50, sect. 7

156 The *locus classicus* for the basis of this distinction is Aristotle. See *Physics*, Bk. III, 6, 206a, 20–30 (Leob Classical Library, London, 1929).

157 *Principia*, vol. 1, 48.

158 *Ibid.*, vol. 2, *General Scholium*, 762.

159 CUL, Ms. Add. 3965, 13, folio 541r: 'idem esse semper et ubique, perfectio est summa'.

160 *Principia*, vol. 2, 762–763.

161 *Ibid.*, 762.

162 *Ibid.*, 762: 'Deum summum necessario existere in confesso est; Et cadem necessitate *semper* est & *Ubique*'.

163 *Physiologia*, Bk. I, chap. 8, 82.

164 *Principia*, vol. 2, 761–762: 'Partes dantur successivae in duratione, co-existentes in spatio, neutrae in persona hominis seu principio ejus cogitante; & multo minus in substantia cogitante Dei'.

165 The same point is made by Descartes in his conversation with Burman. In response to Descartes's claim that thoughts take up time, Burman asks whether our thoughts will then be extended and divisible? Descartes replies: 'Nihil minus. Erit quidem extensa et divisibilis quod durationem, quia ejus duratio potest dividi in partes; aed non tamen, est extensa et divisibilis quoad suam naturam, quoniam manet inextensa' (*Entretien avec Burman*, trans. and annotated by Ch. Adam: Paris, Vrin, 1975, 10). Descartes does not extend the point to God's nature (there is however no reason to prevent his doing so). But if a case can be made for its appropriateness with respect to the nature of thought (that is, the fact that taking place in time is not essential to its nature), there is on Descartes's showing no difficulty in principle with the case of God's permanent nature.

166 CUL, Ms. Add. 39645, 13, folio, 542r (1690s): 'Perfectissima Dei Idea est sit substantia una simplex indivisibilis, viva et vivifica, ubique semper necessario existens . . . '.

167 *Ibid.*, folio 541r: 'In tempore quidem et loco finiri aut mutabilem esse imperfectionem arguit, sed idem esse semper et ubique perfectio est summa'.

168 For discussion of necessity and omnitemporality see Jaakko Hintikka, *Time and Necesity* (Oxford, 1973). Hintikka's discussion is important. However, he does scant justice to the many

passages where Aristotle employs necessity and possibility independently of the notion of something's always being so. Often Aristotle's primitive notion is that something could not possible be other than it is.

[169] The passage is from the yahuda manuscripts at Jerusalem. I quote from Frank E. Manuel's *The Religion of Isaac Newton* (Oxford, 1974), 22. Unfortunately Manuel does not date the passage. From its content, it could well have been written in the 1690s.

[170] 'Leibniz-Clarke correspondence', 101. Newton's point is that since God is in time he can act in the world according to his pre-ordaned plan and in accordance with the essential nature.

[171] CUL, Ms. Add. 3965, 13 folio 546r (1690s): 'Unde nec aeternum et infinitum omne Deus erit, neque Deus aeterno et infinito omnipotentiae suae in rebus creandis et regendis exercitio ex imperfecta rerum creatarum natura prohibebitur'.

[172] The phrase 'essential predication' is not to be taken linguistically. It means that what is predicated *of* something is a feature essential to that thing that if it were removed the thing would no longer exist as such. For Descartes's acceptance of this view of substance (including Divine substance) see *Principia*, part I, 63.

[173] *Spinoza opera ethica*, vol. 2, Pars I, prop. IV, 47, prop. XIX, and Prop. XX, 64–65.

[174] *Ibid.*, Pars I, Prop. XI, 52–54: see especially the scholium, p. 54. For an account of Spinoza's doctrine of attributes see H. H. Joachim, *A Study of the Ethics of Spinoza* (New York, Russell & Russell, 1964), 65–73.

[175] Spinoza, *ibid.*, Definiton II, 44.

[176] *Ibid.*, Prop. X, 51–52. See also Definition VI, especially the *Explanation.*

[177] *Ibid.*, Propositions IV, V, VI, VII and VIII including Scholia I and II, pp. 47–51.

[178] *Ibid.*, Prop. XX, ccorollary I, 65.

[179] Newton's position runs into a serious difficulty. On any account of God's nature we want to say that he has absolutely every perfection and that he is thus uniquely the same being. The standard way of arguing this, is to show (however successfully) that having every perfection implies supremacy; and that supremacy implies uniqueness, otherwise there would be more than one supreme being. It seems that on his account of the infinity of God's perfections (that is, that God possesses them to the highest possible degree while thinking beings instantiate them deficiently), Newton is unable to rule our the possibility that individuals could exist which equally exemplify the manner in which God possesses his perfections. The conception that God transcends causal change in no way helps, since it does not necessarily show that only *one* unique individual satisfies the claim.

[180] *Spinoza opera ethica*, vol. 2, Pars I, prop. XI.

[181] *Principia*, vol. I, 47.

[182] *Ibid.*, 50.

[183] CUL, Ms. Add. 3965, 13, folio 541v: 'Quicquid necessario existit illud semper et ubique existit, cum eadem sit necessitatis lex in locis et temporibus universis'.

[184] H. McLachlan, *Sir Isaac Newton's Theological Manuscripts* (Liverpool, 1950); quoted from McLachlan's reproduction of the 'Twelve articles' of faith, p. 56.

[185] *Enchiridiom, Part I, cap. 28, sec. 13:* 'Coexsistentiae enim partium, quas vocant, Spiritus, conditione quam maxime necessaria planeque essentiali connectuntur, cum sit *Ens simplicissimum*, sive *Ens unum per se, & non per aliud*'. See also sections 10 and 11 of chapter 28. In section 10 More denies that an *ens unum per se*, although extended, contains any notion of parts in its idea. That its parts 'all jointly and unitedly exist together' in a spirit, is a law 'contained in the very Idea, or nature of every *spirit*'.

[186] *Closed World*, chap. 8.

[187] *De gravitatione*, 103–104: 'et insuper ens quodlibet habere modum sibi proprium quo spatijs adest' (translation mine).

[188] See the True intellectual system, vol. 2, 258–259 for essentially the same approach to the problem of God's relation to space.

[189] *De gravitatione*, 101. Newton argues that while we cannot imagine infinite space we can understand the concept. What he says here applies equally to the conception of God's presence in an infinite space.

[190] It is not clear whether an *actually* infinite space can be said to have "parts" or indeed true places. Since it lacks both boundaries and center, there is no way of distinguishing its intrinsically undifferentiated homogeneity. Nor is it clear that an extrinsic reference frame can be imposed on it, since infinite space lacks a *fundamentum* capable of supporting the conceptualization implied. Thus Newton's claim that infinite space possesses true places is not at all clear. His adherence to the view that extended space is in fact infinite, occurs as early as an entry ('Of quantity') in his *Quaestiones* (CUL, Ms. Add. 3996, folio 76 r). As usual Newton is attacking Descartes's belief that the mind is capable only of ascertaining extension as indefinite: 'If extension is indefinite onely in greatness & not infinite, $y^n$ a point is but indefinitely little & yet we cannot comprehend any thing lesse. To say $y^t$ extension is but indefinite (I mean all $y^e$ extension wch exists & not soe much onely as we can fancy) because we cannot perceive its limits, is as much as to say God is but indefinitely perfect because wee cannot apprehend his whole perfection' (bracekets are Newton's). Although with respect to the finite mind extension is undetermined, it does not follow, Newton claims, that extension is not in fact infinite; any more than it follows that God is indefinite because the mind is incapable of determining his complete nature.

CHAPTER 2

# ATOMS AND THE 'ANALOGY OF NATURE': NEWTON'S THIRD RULE OF PHILOSOPHIZING

As in the case of ancient atomism, the revival of atomistic doctrines in the seventeenth century gave rise to questions regarding the existence and properties of primordial entities. With their insistent emphasis on sensory experience as a foundation of physical knowledge, the English experimental empiricists were faced with various epistemological problems. The majority, such as Robert Hooke and Henry Power, believed that atoms could in principle be observed: hence their problem was not the nature of knowledge of the microscopic, but rather how our knowledge could be extended into that realm.[1] Thinkers such as Henry More and Ralph Cudworth tended to the view that, even if we could observe atoms, it would not add to our knowledge of first principles.[2] Newton, and to some extent Locke, were confronted with a more difficult problem. Simply stated it is this: are there justificatory principles by which to sanction inferences from what is observable to what in principle is unobservable? If the basis of natural knowledge is strictly confined to what is actually or possibly observable, how can any claims be established about the nature and existence of atoms? The following study will refer to these questions as the problem of transduction;[3] moreover, it will be concerned primarily to analyse Newton's use of the phrase 'the analogy of Nature'. In its best known context that phrase appears in the Rules of Philosophizing prefixed to the third Book of the *Principia* in 1713. There for the first time in print Newton states:

> The qualities of bodies which cannot be intended and remitted, and which apply to all bodies on which it is possible to set up experiments, are qualities of all bodies universally:
>
> For since the qualities of bodies are only known to us by experiments, we are to hold for universal all such as universally agree with experiments; and such as are not liable to diminution, can never be quite taken away. We are certainly not to relinquish the evidence of experiments for the sake of dreams and vain fictions of our own devising; nor are we to recede from the analogy of Nature, which uses to be simple, and always consonant to itself.[4]

There are a number of problems which arise from this passage. One I have dealt with elsewhere.[5] The phrase 'such as are not liable to diminution, can never be quite taken away' is an outcrop from an earlier draft of the third rule in which the 'analogy of Nature' is not mentioned. This is important. In this early draft (see Draft 2 in the appendix below) Newton, I have argued, is propounding a purely conceptual thesis about the essential qualities of phenomenal things which is connected with the intension and remission criterion: being extended, being impenetrable, being hard, being endowed with inertia, and being either at rest or in motion are integral elements in our conception of matter.[6] Since, as we shall see, he ultimately identifies matter with both bodies and atoms, in consequence of this thesis the problem of transduction should have been irrelevant. But the 'Cartesianism' of this line of reasoning only partially satisfied the strong empiricist and experimental bent of Newton's mind. As is apparent in Drafts I and 3 (see appendix), generalizations from experimentation and experience as well as the 'analogy of Nature' are clearly stressed, though the doctrine of essential qualities, namely those which cannot be intended and remitted, is maintained. Moreover, the drafts clearly indicate that Newton moved from the view that qualities found in all bodies universally did not have to be invariant, to the view that those which fall under the intension and remission criterion were essential and thus immutable. This relates to a minor problem considered in the last section of the paper. Strictly considered, the rule only warrants the extension of those qualities which cannot be intended and remitted; yet it plays a vital role in the universalization of gravitation in the third Book of the *Principia* which does not satisfy that criterion.

One of the central aims of this study will be to show that the third rule is not primarily an induction rule, but chiefly a transduction rule. In this regard an understanding of what Newton meant by 'the analogy of Nature' is fundamental. I shall contend that the ancient metaphysics of the macrocosm and the microcosm along with the connected doctrine of the great chain of being, transformed to be sure to suit a new purpose, formed the basic framework in terms of which some English thinkers, including Newton, discussed transduction. This doctrine will be characterized as the metaphysics of analogy, meaning thereby that it implicitly or explicitly constituted a view of the way things are. It is in terms of this world-view that these thinkers justified analogical reasoning, and consequently I would not wish to imply that every seventeenth-century thinker who reasoned analogically justified it in terms of this doctrine, or indeed thought such reasoning needed justification: Gassendi, Hooke and Boyle are undoubted examples of the latter two positions. Nevertheless, the metaphysics of analogy, at least in England, had not become at the end of the seventeenth century simply a tool employed by the poetic imagination,[7] nor had it become simply a substratum supporting certain doctrines as they were developed in eighteenth-century natural theology.[8]

On the Continent the approach to the problem of transduction was different and, apart from Descartes and the Cartesians, it was usually discussed in terms of the traditional metaphysical distinction between simples and compounds. In the hands of Leibniz, this distinction between two sorts of entity was transformed into a theory of reality which considered atomism and its consequences as a crass and naive metaphysics. Compared with the English philosophical approach, that on the Continent was decidedly more subtle and sophisticated. The former, however, gave rise to a more powerful and dominant scientific tradition during the eighteenth century. Whether they accepted it or not, many thinkers discussed Newton's third Rule of Philosophizing and considered it favourably or unfavourably as a methodological basis for inferential knowledge of the invisible world.[9]

In the history of philosophy there are many examples to show that there is a general problem about connecting observables to unobservables, for example Kantian metaphysics, in which the noumenal world is not a possible object of experience but yet is held to exist. There are also many particular exemplifications of the problem: the transition from sense-experience to material objects; from material objects to the entities of science; from the behaviour of people to their inner states; from the traces of the present to reconstructions of the past. The problem in each case is characterizing the tie between two different sets of statements. They also exemplify inferences from statements of one category to those in a different category, the latter going beyond the evidence with which we began. Now transduction as conceived in the seventeenth century was not simply a type of trans-categoric inference. Induction exemplifies the latter, since all inductive inferences go beyond the evidence from which we begin. The distinctive feature of transductive inferences, I wish to maintain, is that the conclusion not only goes beyond the evidence we already have, but goes beyond any evidence we might possibly acquire. Put otherwise, it is not that transductive inferences go beyond the initial evidence, but that the evidence is appropriated to a problem to which it strictly does not apply.

In what follows an attempt will be made to analyse the problem of transduction as it was concevied by various seventeenth-century thinkers, and to consider the various techniques of justification which they developed. Related to this, as a consequence of a more general awareness of the metaphysical distinction between simples and compounds, is the problem of deciding whether a theoretical conception of matter in natural philosophy is to be identified with simples in nature or with visible bodies.[10] If English thinkers were fully aware of this problem, I have encountered few signs in the literature. It was, however, in the late seventeenth century generally recognized on the Continent.[11] Had Newton, for example, seen the problem in clear perspective – thought there is evidence to suggest that occasionally he did – the attribution in the third rule of the essential characteristics of observable bodies to what is unobservable should have seemed less straightfoward. As we shall see, however, in Newton's mind the rule played a more important role than simply exemplifying a metaphysical distinction: it was crucial as a background to the general argument for universal gravitation.

## I

In this section I shall discuss the metaphysical distinction between simples and compounds. By the early seventeenth century, the traditional arguments for rejecting the Aristotelian conception of matter had led most thinkers to accept a distinction between elemental natures and what is derivable from them.[12] Elements were generally conceived as immutable and unchanging in nature. As a result of the revival of ancient atomistic doctrines, they were also conceived as being continuously uniform, only *differentiae* such as size and shape distinguishing one particle from another. From the various combinations of these homogeneous particulate substrata, compound aggregates were conceived to be generated. Though perceptible bodies seemed to be uniform wholes, they were in reality compounds derived from the immutable constituents of nature. In themselves, being mere aggregates of elements, compounds had no separate reality. Elements were therefore basic in the sense that they could exist independently of the latter. Thus a basic ontological difference between the two sorts of entity was clearly maintained, constituents of visible bodies being conceived as the only substantial and fundamental units.[13] Later in the century Locke, though he maintained this distinction, held in the fourth Book of his *Essay concerning Human Understanding* that composite bodies exist independently of the perceiver as well as their component parts. Moreover, he considered them as a starting point for our knowledge of nature. Newton seems to have adhered to the same position.

One of the most interesting and learned of the Continental discussions of the distinction between elements and compounds is found in Gassendi's *Syntagma.*[14] Based on a comprehensive exposition of the principles of Epicurus, Gassendi's approach to atomism is essentially historical. In passage after passage, the learned of all ages speak on behalf of the atomistic view of reality, perception and knowledge. Typical is the following, in which the ontological distinction in question is made:

> And again . . . because all things are compounded of them, they are not only described as *archai*, Principles or Primordials, and *stoicheia*, Elements, but also as *prota somata*, Primary Bodies, and *prota megethe*, Primary Magnitudes, as well as the Seeding or Seminary, *panspermia*, and the Genital Seeds of things, and *tōn ontōn hyle*, the Underwood or Raw Material of things that are.[15]

Gassendi argues that the order, regularity and permanence of experience presuppose the immutability of the basic constitutes of nature. This basic Epicurean argument involves a number of claims. The constituents are the only true and substantial unities: all else is merely aggregation and resolvable into these. The components being intrinsically unresolvable and permanent, it is a paralogism to suppose that what is true of composites is true of them.[16] We can only say what they are like by analogy. Using the example of motes in a sunbeam, Gassendi argues in Epicurean fashion as follows: we have to imagine that the physical minimum stands to the atom as the perceptible minimum stands to the perceptible whole, since the physical minimum differs only in size from the perceptible minimum. So just as there is a minimum in visible things, from this we are to infer that there is a minimum in the things which are not visible but conceivable.[17] Thus the conception of atoms is, as Epicurus stated, 'by analogy with the things of the visible world'.[18] Surprisingly enough, thinkers such as Locke and Newton did not draw to any significant extent on the arguments of Epicurus for the existence of indivisible minima in nature. Nor did they seem to profit from Gassendi's discussion of these arguments and their significance, although there is little doubt that both thinkers probably read Gassendi and certainly his epigone Walter Charleton.

Gassendi also adduces further arguments independent of the last to distinguish atoms from composites. He attributes to them a highly complex quality: *pondus*. This is specified as a sort of innate power whereby they are continuously impelled to move.[19] Nothing is at rest in Gassendi's world. In his analysis of their explanatory power, he also conceives atoms as having a *sensus naturalis*, an unconscious sense of awareness.[20] Being truly the 'Genital Seeds of things', they are not only incessantly active, but also sentient in a minimal degree; nature at the elemental level is thus vitalistic and dynamic, embodying powers derived from divine agency and providentially controlled. In this way Gassendi attempts to give an account of the generation of perceptible bodies.

In an intended defence of his atomic doctrine, designed for the first six propositions of the third Book of the *Principia*, Newton, probably in part influenced by Gassendi, took a historical approach in elucidating this ontological distinction.[21] In passages where he discusses the doctrines of ancients such as Epicurus and Lucretius, he states:

> Now they would have it that atoms were solid bodies, endued with various shapes and indivisible not mathematically but only physically: for each atom also consists [they said] of innumerable smaller parts, parts which are not combined by having conjoined, but are from eternity closely and without pores, solidly compacted that they cannot be wrenched asunder and separated from each other by any forces of nature. Things which are brought together by the conjoining of atoms can [they said] be again divided and resolved into those same atoms, because the atoms do not conjoin into solidity but are in mutual contact only at some mathematical point, the remaining spaces being left in each case between them vacant: but the atoms themselves, because of their solidity and the plenary contact of their parts and their extreme density, cannot be diminished or by any means be decreased or even assume any increase, but abide eternally as the immutable seeds of things: and this is the reason why the species of objects are conserved in perpetuity.[22]

Thus, like Gassendi, he considers atoms as the generative principles of nature, 'immutable seeds', from which arise the variety of phenomena. Being fundamental to the explanatory power of the atomic hypothesis, they are ontologically different from the phenomena generated. The 'classical scholia' intended for the first six propositions which develop the argument for the universalization of gravitation, afford clear evidence that Newton was not only deeply learned in the history of atomistic doctrines, but that he also comprehended the ontological distinction between these two sorts of entity thought to exist in nature.

The most detailed discussion of the new ontology in England, though apparently owing little to Gassendi, is to be found in Boyle's essay *On the Origin of Forms and Qualities*.[23] He states unambiguously that 'the matter of all natural bodies is the same; namely, a substance extended and impenetrable'[24] and that the 'generality of philosophers'[25] agree with this view. Boyle also affirms that this common matter is separated by motion into corpuscles 'because this matter being in its own nature but one, the diversity we see in bodies must necessarily arise from somewhat else, than the matter they consist of'.[26] He declares later that these corpuscles are absolutely independent, so that if 'all the rest of the universe were annihilated, except any of these intire and undivided corpuscles, it is hard to say what could be attributed to it, besides matter, motion (or rest), bulk and shape'.[27] Thus corpuscles of this sort convening together 'compose any distinct body, as a stone, or a metal'.[28] Boyle unmistakably draws the distinction between immutable elements and compound bodies which arise from them. There was of course for Boyle, and later for Locke, the problem of what caused bodies to cohere into 'primitive concretions'. Both thinkers held that this was unknowable. Boyle, unmistakably draws the distinction between immutable elements and compound bodies which arise from them. There was of course for Boyle, and later for Locke, the problem of what caused bodies to cohere into 'primitive concretions'. Both thinkers held that this was unknowable. Boyle, in an attempt to avoid the question, claimed that concretions brought together by cohesion from the *minima naturalia* were below 'the discernment of sense, and though not absolutely indivisible by nature into the original particles or *prima naturalia* that compose it . . . they very rarely happen to be actually dissolved or broken but remain intire in a great variety of sensible bodies'.[29] The ontological problem of the mode of cohesion was only in principle soluble by Newton's theory of gravitation, which held that natural phenomena individually and in a variety of mutual relations were bound together by this single force.

Boyle and Newton, especially the latter, were largely instrumental in transforming the metaphysics of simples and compounds into an effective natural philosophy with a plausible basis in experience. Throughout his writings on matter, published or otherwise, Newton makes a consistent distinction between *corpus* and *particula*; the former being a compound body, the latter an immutable primordial. In a series of manuscripts dealing with proposed changes in the prefatory section to the third Book of the *Principia*, and written about 1715, Newton defined body as 'everything that can be moved and touched, in which there is resistance to tangible things, and its resistance, if it is great enough, can be perceived'.[30] Clearly these criteria do not directly apply to the particles which compose compound bodies. There are abundant passages which attest to Newton's explicit adherence to this distinction in the context of natural philosophical speculation. The following, from Query 31, is typical:

> . . . these primitive Particles being solids, are incomparably harder than any porous Bodies compounded of them; even so very hard, as never to wear or break in pieces, no ordinary Power being able to divide what God himself made one in the first Creation. While the Particles continue entire, they may compose Bodies of one and the same Nature and Texture in all Ages: But should they wear away, or break in pieces, the Nature of Things depending on them would be changed. Water and Earth, composed of old worn Particles and Fragments of Particles, would not be of the same Nature and Texture now, with Water and Earth composed of entire Particles in the Beginning. And therefore, that Nature may be lasting, the Changes of corporeal Things are to be placed only in the various Separations and new Associations and Motions of *permanent* Particles.[31]

Thus the primordials of nature are permanent and can only be altered by a fiat of God's will. Moreover, had they been created differently, the emergent world of sensible experience would also have been otherwise. Thus the passage clearly indicates the extent to which Newton conceived the intrinsic intelligibility of nature to be based on the immutability of the ultimate particles. All variety and change presuppose this. As passages from the draft Conclusion to the *Principia* and from *De Natura Acidorum* alone testify, this view of matter was held by Newton as early as 1687.[32]

More specifically, Newton conceived particles to differ from perceptible bodies in that they had associated with them stronger attractive forces:

> Now the smallest Particles of Matter may cohere by the strongest attraction, and compose bigger Particles of weaker Virtue . . . and so on for divers successions, until the Progressions end in the biggest Particles on which the Operations in Chymistry, and the Colours of natural Bodies depend, and which by cohering compose Bodies of a sensible Magnitude.[33]

Unlike Gassendi, Newton does not conceive his atoms to be vitalistic. But they do have forces closely associated with them; the smaller they are, the more intense the force. At the level of perceptible bodies, attractive forces are the weakest. This is not simply a matter of greater distance. For in Query 21, Newton argues that, 'The exceeding smallness of its [Aether] Particles may contribute to the greatness of the Force by which those Particles may recede from one another . . .'.[34] The basis of his argument is analogy. For example, in small magnets, attraction is greater in proportion to mass than in larger magnets. Therefore, in the limit, the forces associated with the smallest primordial particles will be the greatest.

The ontological distinction between primordials and perceptible bodies as conceived by Newton is brought out most clearly in his discussion of the inner constitution of bodies in the first eight propositions of Part III, Book II, of the *Opticks*.[35] His main concern is to show that their constitution is causally responsible for the colours of natural bodies; and that the orders of colour can be arranged according to the size of the constituent particles, by analogy with spectral rings on bubbles and glass plates. In Proposition VII,[36] it is suggested that small corpuscles are transparent because their particles are arranged in an irregular structure allowing little or no light to be reflected. Newton ends his discussion by surmising that microscopes may be constructed which will enable the 'greatest of corpuscles' to be seen, adding, however, 'which if we shall at length attain to, I fear it will be the utmost improvement of this Sense. For it seems impossible to see the more secret and noble Works of Nature within the Corpuscles by reason of their transparency'.[37]

This is an interesting passage; in it Newton attaches a clear sense to the notion that the smallest particles of matter are in principle physically impossible to see. His argument hinges on the conception that matter is almost vacuous, that bodies are 'much more rare and porous than is commonly Believed'.[38] It is not only that the smallest particles because of size are beyond sensory experience, but even were the senses aided by instrumentation they could not be seen because the transparent corpuscles, of which they are constituents, cannot interact with light. Though it penetrates into the vacuities of worlds within worlds inherent in the increasingly smaller corpuscles, because of these vacuities and the smallness of the particles, light is not reflected back. As Newton tells us in Query 6, light 'enters the Bodies, and is often reflected and refracted within them, until it be stifled and lost?'[39] Thus the world of indivisible primordials is without colour and hence beyond observation. The chemical and optical techniques which enabled Newton to determine the relative magnitudes of invisible corpuscles do not apply to the ultimate particles. Therefore, because of these reasons embodied in his natural philosophy, they have a different physical status from other levels of matter.

In view of his awareness of this physical distinction, it may be wondered why Newton did not recognize that any procedure whereby characteristics of macroscopic bodies are transferred to elements is manifestly invalid. It must be remembered, however, that he was not interested in the philosophical aspects of his ontology for their own sake. Newton was primarily concerned to investigate the natural world, and he tended to view the third rule as describing a methodological procedure compatible with his experimental empiricism. From this point of view, he did not focus on justifying knowledge of invisible matter, but rather on the plausibility of the rule as a guide to practice. It will be necessary to return to his metaphysical assumptions when considering the 'analogy of nature'[40] as part of the conceptual background to the rule. First, however, I want to consider Newton's approach to transduction in terms of his conceiving the rule as a justification of scientific practice. This involves considering Newton's concept of physical law in relation to his doctrine of induction.

In a letter to Cotes, Newton refers to his laws as having been deduced from phenomena and afterwards 'made general by induction'.[41] In two different but related ways, he conceived his 'physical' axioms to be restricted in scope. Being trained in classical logic,[42] Newton adhered to a fundamental principle that all strictly inductive inferences proceed from observed members of a class to at least one other member of the same class. This principle demands that generalizations based on such inferences have direct inductive support. Thus the properties of a class of perceptible phenomena can be generalized to include unobserved members which are in principle observable. It is this principle which lies behind Nwton's persistent claim that his laws are generalized 'by induction'. Thus the laws are clearly limited to experimental and observable phenomena, either actual or physically possible. Secondly, a consideration of what is meant by deducting[43] axioms from phenomena will further illuminate the character of the laws as statements about motion. As Mandelbaum has pointed out, this notion emphasizes the way in which a statement has been established to be the case.[44] In all places where he speaks of 'deducing' or 'inferring' statements from nature, Newton implies that such statements were formulated by a careful, and continually repeated, scrutiny of experimentally-produced phenomena. Thus axioms are a category of statement which have been, as it were, 'extracted from' nature, as opposed to being confirmed by deductive testing. Newton's use of the strong verb 'deducere' in the *Principia* indicates his psychological attitude towards his laws.[45] They were closely tied to their evidence and for Newton seemed to reflect the actual structure of reality. Such statements are the foundations of the *Principia*, and in turn support deductive consequences which are in principle testable.

It is clear from the above analysis that Newton held the laws of motion to be primarily substantive, factual statements. This was his implicit position in the polemic occasioned by his 1672 paper on light, his view in the nineties, and clearly evident in his letter to Cotes of 1712/13.[46] Of course laws could be used as definitional principles, as in the first Book of the *Principia*; but in their primitive status they are constitutive statements about the world. Because of his attitude towards how his laws had been established, and his strict conception of induction, Newton considered it necessary not only to justify extending the laws to imperceptible particles, but even to planetary motion. But there is also an important epistemological consideration which lies behind his position. As his definition of body shows, touch, not sight, is the most authoritative of the senses, and for Newton the natural criterion for physical reality. Bodies mutually interact through resistance and resist contact with the human body. The presupposition of Newton's definition is that some material things have to be tangible and visible for the mind to have any conception of a material thing at all. As we shall see, he held that these literally tangible and visible bodies provide a basis for analogical arguments by which material status can be accorded to both celestial bodies and imperceptible particles. Thus atoms owe their status as physical objects existing in the natural world to their disposition to affect other objects. And the natural world of material objects, whether or not it offers perceptible resistance to other objects including the human body, is for Newton a system of things acting and reacting on one another. There is to be supposed, therefore, an analogy of tangible effect between the visible and invisible realms of nature, even though strictly speaking invisible entities are intangible since they are not direct objects of tactile sensation.

As late as 1715 in proposed revisions of the Rules of Philosophizing as they appear in the second edition, Newton felt constrained to show that heavenly bodies were 'phenomena'. As we have seen, in these proposed emendations he defined a body as something which was tangible as well as visually perceptible. Can this be said of celestial bodies? Apart from the fact that the argument for universal gravitation in the first section of the third Book turns on an appeal to analogies between terrestrial and celestial phenomena, Newton explicitly mentions that the planetary bodies cannot be straightforwardly assimilated to the terrestrial case: 'For the argument from the appearances concludes with more force for the universal gravitation of all bodies, than for their impenetrability; of which among those in the celestial regions, we have no experiments, nor any manner of observation.'[47] In the same manuscript where the definition of body is given Newton states, after a long, rather Aristotelian list of terrestrial objects:

> I add the heavenly bodies. These emit and reflect light, and are numbered among phenomena, and are weighted down by their parts pressing heavily upon them, and assume a round shape, and in their motions observe the laws of bodies.[48]

It seems extraordinary that at this late date he should so laboriously outline what he takes to be criteria of materiality, and so emphatically state that planetary bodies satisfy it. Moreover, there is what can be described as a residual Aristotelianism in his remarks: he still feels constrained to divide the visible world into terrestrial and celestial matter. It would be a mistake, however, to suppose that Newton is making a sharp categorical distinction. His concern is primarily to indicate how our knowledge of these two sorts of body differs with respect to sensory experience, the difference in the mode of achieving it, and why they are both exemplifications of material substance. These remarks were written thirty years after the first edition of the *Principia*! Thus the problem of justifying the application of the laws of motion to the heavens, as Newton conceived it, was not straightforward. Even less so was that of attributing qualities such as impenetrability to imperceptible particles, and of subsuming them under the laws of motion; though it is not clear that Newton recognized any difference between extending known laws to the unobserved bodies and particles, and extending qualities to the same. In the first draft of the rule, however (see appendix), the term 'proprietates' was added after 'leges corporum omnium'. Again, we must not forget that Newton made the problem especially difficult methodologically, by insisting on the way his laws had been established and by his strict conception of induction. Thus it should be no surprise that it occupied his mind as early as the nineties when he was drafting versions of Rule III.[49]

The importance of having a statement such as the third rule prior to the sustained argument for gravitation of the third Book almost certainly occurred to Newton even before the publication of the first edition of the *Principia*. There is evidence to suggest that he did not satisfy himself on dynamical grounds that the inertial law could be extended to the heavens until shortly before the early stages in the writing of that treatise. On fly-leaf in his copy of Wing's *Astronomia Britannica* (1669), Newton used a combined inertial and vortical component in analysing the moon's motion.[50] These comments were probably written in the early 1670s. Thus, he had probably not yet conceived of the application of the inertial principle in its dynamical sense to the heavens, though by 1665 there is a clear formulation of the principle in his *Waste Book*.[51]

Again, there is no evidence to show that he had reduced the status of centrifugal force, for general dynamical purposes, from being a 'real' *conatus* of matter, to being an 'apparent' force, until the 1680s.[52] Prior to this, he appears to have made three physical assumptions in analysing constrained orbital motion: a tangential tendency from the centre, a centrifugal *conatus* away from the centre, and a centripetal tendency to the centre. In his mathematical treatment of the magnitude and measurement of these forces, all three were never given the same dynamical weight at any given time. For instance, in one of his early treatments of circular motion, the mathematical analysis considers the constraining force as an instantaneous deflection from a tendency to move alone a tangential path.[53] Still, there is no indication that Newton had reduced centrifugal force to impotency. Thus the tangential force was not yet conceived in is true inertial sense. For Newton, this meant that a body unaffected by external forces would move naturally in absolute rectilinear motion with reference to absolute space. Thus the existence of a centrifugal tendency cannot be reconciled with the rectilinear principle of inertia conceived in this sense, since the latter implies centripetal force in accounting dynamically for constrained orbital motion.

I am not suggesting, however, that about 1679 Newton began to apply a different principle of inertia to the heavens. It was the same concept. But he could only assure himself that it applied in its absolute, and dynamical sense, after he had seen that from a true rectilinear path, elliptical motion was necessarily directed to the centre. Fundamentally, this involved seeing that a planet's trajectory did not result from a physical balance or imbalance between a centrifugal tendency and gravitational attraction; and that given elliptical orbits, Kepler's third law was a necessary and sufficient condition for centripetal force varying as the inverse square of the distance. A dynamical question of this magnitude needed clarification, before a generalized and systematic mathematical treatment of planetary motion could be made. Only in the 1680s, after having given a general account of Keplerian motion, was

Newton in a position to show that a dynamical planetary model which combined a continuous inertial force, with a discontinuous centripetal force, and which embodied the second and third laws, truly applied to the system of the world.[54] Though Newton's physical intuition led him to accept the truth of the model, he still felt the need to justify the extension of the laws to the heavens. For in the third Book he was dealing with the physical world in terms of his basic epistemological and ontological commitments. And the mathematical assumptions of the first Book were not sufficient. This was to be primarily the burden of the third rule. That these considerations were paramount in Newton's mind is clear from the fact that in the earliest draft of the rule he wrote simply: 'The laws of all bodies in which experiments can be made are the laws of bodies universally.'[55]

## II

The last section has considered a variety of different doctrines, metaphysical, physical and logical, which indicate Newton's awareness of the ontological distinction between natural elements and derivative compounds. In its metaphysical form, the distinction is in some respects an exemplification of what Passmore has called a 'Two-Worlds' theory.[56] Doctrines of this sort make an absolute ontological distinction between the mind and the body, the eternal and the temporal, the supernatural and the natural. Once entities are separated into sharply divided realms of existence, the problem is traditionally for the dualist, bridging the gap between the two worlds. All attempts to do this in effect destroy the ontological contrast which the theory supposedly establishes. For example, such intermediary entities as plastic natures, hylarchic principles and animal spirits, half spiritual and half material, either break down the gap they are supposed to bridge, or remain on one or the other side of it; in either case they destroy the supposition.

The duality between perceptible and imperceptible matter need not be conceived as polar opposition. As we shall see, neither Locke nor Newton had any such conception. Descartes saw the problem, but attempted a solution in terms distinctly his own. Leibniz, by considering the dichotomy in its strict sense, concluded that matter could be identified with either elements or bodily compounds but not with both. Each adopted a solution which is instructively different from that of Newton and Locke. These solutions clearly bring out the more metaphysical approach of the Continental natural philosophers against the more empirical approach of the English.

I shall analyse two Continental thinkers whom Newton read in some detail – Descartes and Leibniz – as well as an interesting passage from Galileo which bears on this distinction. A fuller account, which goes beyond the scope of this study, would involve a consideration of the thought of Malebranche, whom Newton probably read;[57] the sceptical critics of Descartes, Foucher, Huet and du Hamel; the extreme scepticism of Bayle; and lastly, reacting partly to the sceptical thought of these thinkers, a new theory of reality expounded by Berkeley.[58] Malebranche and the occasionalists subjected the Cartesian contrast between the real and the perceived to sceptical examination, and thus reduced the distinction between primary and secondary qualities to the world of appearance, raising doubts about the existence of a material reality independent of mind.[59] Descartes's critics, by denying a difference in ontological status between primary and secondary qualities, again tended to abolish the reality of an external material world.

Bayle brought 'la crise pyrrhonienne' to a head by reducing everything to the realm of appearance. It is clear that these thinkers, by either denying the distinction between the perceptible and the real, or by attempting to reinterpret the distinction as did Malebranche and Berkeley, explicitly avoided the epistemological difficulties embodied in the Lockean dualism of ideas and independently existing particles as their cause. Thus they held that any attempt to bridge the gap between perceptible things and a realm of existence beyond the senses was inherently untenable.

Descartes attempted to obliterate the gap by founding his account of nature on a unitary conception of matter. Making *res extensa* the necessary and sufficient condition of materiality, Descartes accepted the consequence that 'extension in length, breadth, and depth that constitutes the space is absolutely the same as that which constitutes the body.'[60] Thus the external world constitutes nothing more than a monistic, homogeneous spatial substance from which the differentiation of plurality follows derivatively. This conception is clearly different from the atomistic view of Boyle and Newton, which conceives compounds as a synthetic aggregation of primary and immutable constituents. Moreover, the individuation of Cartesian extension into particles is apparent rather than real: 'A space, or intrinsic place, does not differ in actuality from the body that occupies it; the difference lies simply in our ordinary ways of thinking.'[61] It is obvious that Descartes's unitary conception of matter is irreducibly geometrical. *Res extensa* entails that matter is indefinitely divisible; thus the distinction between the elemental and the compound has no true significance in the Cartesian scheme. Though Descartes's ontology overcomes the 'two-world problem' with respect to matter, it raises well-known difficulties about individuation, namely, the arbitrariness of having to appeal to God in order to explain the movability of extended things which in themselves do not entail motion.[62] Conceptually, of course, Descartes's position is incoherent, as was shown by Henry More with respect to motion, force and infinity.[63]

Despite this metaphysical background, Descartes's main purpose in his *Principles* was to show how his doctrine of nature could explain the variety of phenomena. Basic to the sorts of explanation that he gives is an appeal to the intestine motions of insensible particles. In sections 201 and 203 of the fourth Book of his *Principles*, Descartes raises the problem of transduction. Noting that he may not have the approval of those who take the senses 'as the measure of what can be known', he avers that we cannot doubt the existence of a multitude of bodies so small as to be undetectable by sensation.[64] Descartes, however, feels constrained to offer a justification of this claim:

> . . . my assigning definite shapes, sizes, and motions to insensible particles of bodies, just as if I had seen them, and this in spite of admitting that they are insensible, may make some people ask how I can tell what they are like. My answer is this. Starting from the simplest and most familiar principles which our minds know by their innate constituion, I have considered in general the chief possible differences in size, shape, and position between bodies whose mere minuteness makes them insensible, and the sensible effects of their various interactions. When I have observed similar effects among sensible objects, I have assumed that they arose from similar insensible bodies; especially as this seemed the only possible way of explaining them.[65]

Ontologically, since matter, whether it be perceptible or otherwise, is of one and the same kind, there is no problem for Descartes. But scientific explanations must take account of the world of experience: to explain this, however, we must have recourse to insensible particles. Since Descartes does not tie knowledge to the observable, he is able to say that our knowledge of imperceptible matter must fall under the criteria of clear and distinct ideas. Therefore it too is essentially extended, admitting only differences in size, shape and position. All knowledge that we can have of matter must be consonant with these ideas alone.[66] Secondly qualities, such as colour, in no way correspond with external reality. The force of Descartes's contention is strikingly evident in the following passage where he asserts that there could not truly be an infinity of worlds since they 'could not but consist of one and the same kind of matter. . . . For we clearly understand that matter, whose nature consists merely in being extended substance, already occupies every imaginable space where other worlds would have to exit; and there is not to be found in us any idea of any other sort of matter'.[67] Thus if there were worlds differently constituted from the actual world, the *differentiae* could only be modes of extension, size, figure and position. Hence all matter – remote, imperceptible, or otherwise – must be extended. The question whether a general conception of matter is to be identified with visible bodies or with particles does not arise for Descartes, since the only *differentia* in nature is merely the divisibility of extension.

Leibniz's treatment of the problem is more complex. Here I can only sketch the basic characteristics of his position. Of all seventeenth-century thinkers, he perhaps most clearly saw the ontological difference embodied in the conception of phenomena construed as compounds of basic constituents. In the first place, Leibniz agrees with contemporary opinion that every body is extended and impenetrable. Moreover, he emphasizes that this is what we mean by matter; thus matter is explicitly identified with bodies.[68] But in

Leibniz's view, bodies are merely *phaenomena bene fundata*; regular and coherent appearances in nature as against being mere phantasms of imagination. Thus extension and impenetrability become sensory appearances conceived much in the same way as secondary qualities by the mechanical philosophers.[69] Having accepted atomism in his early thought, it is well known that Leibniz rejected it by the early 1680s. More basic than his belief that the atomic doctrine violated the law of continuity and the identity of indiscernibles, was his criticism that it embodies an incorrect relationship between part and whole.[70] For example, according to the atomic doctrine, the extension of compound bodies is a mode of the extension of the constituent parts. In a letter to De Volder, he clearly states his position:

> You ask whether in my judgement 'the active principle is extension, or a mode of extension, or a substance distinct from extension'. I answer that it appears to me this principle is the substantial and constitutive basis of extended things themselves or of matter, that is, of that whose nature sustains not only extension and antitypy but also action and resistance. Extension itself is for me an attribute which results from many substances existing continuously at the same time. The original force can consequently be neither extension itself nor a mode of the same. Nor does it act upon extension but in the extended.[71]

Thus Leibniz makes a clear distinction between the basic constituents of reality, and matter identified with, and conceived of, as purely phenomenal objects. For him, substantial reality must be a truly irreducible unity, at once real and indivisible. Extension does not satisfy this criterion since it has parts and is divisible. Neither do atoms satisfy it; to avoid being divisible they would have to be infinitely hard. But for Leibniz this is an incomplete and relative term. Moreover, both views violate the doctrine of continuity which is basic to Leibnitz's thought. This conception is, in fact, the key to understanding Leibniz's doctrine of the unity of reality which constitutes phenomenal bodies.

> I do not think that substance is constituted by extension alone, since the concept of extension is incomplete. Nor do I think that extension can be conceived in itself, but I consider it an analyzable and relative concept, for it can be resolved into plurality, continuity, and coexistence or the existence of parts at one and the same time. . . . But it would appear from this that something must always be assumed which is continuous or diffused, such as white in milk . . . and the resistance in matter. For, by itself, continuity (for extension is nothing but simultaneous continuity) no more constitutes substance than does multitude or number, where something is necessary to be numbered, repeated, and continued. So I believe that our thinking is completed and ended in the concept of force rather than that of extension.[72]

Thus attributes like extension can no more complete the essence of an extended thing than number 'suffices for things that are enumerated'. Extension presupposes repetition, the continuity of a plurality of things coexisting and diffused throughout nature. But what is the character of this reality which is the ground for phenomenal matter? Leibniz substitutes an intensive continuum of force for the extensive plenum of Descartes. The relationship of the part to the whole is no longer conceived to be a quantitative aggregation, the latter being regarded as a static consequence of the former. Rather it is a dynamic one, with every unit or monad expressing in itself its relation to the entire universe. Because of its continuous degrees of intension, a force contains the whole potentially; and each monad represents nature from its particular point of view. Thus matter as bodily phenomena is merely one such representation. By viewing matter as phenomenal and developing a non-quantitative ontology, Leibniz avoided problems arising from the distinction between element and compound. His basic insight was to see that the distinction conceived in atomistic terms was untenable. There can be no real whole without a true unity of the constituents. Only substance construed as an intensive centre of force, a monad having no parts and thus indivisible can satisfy this: '. . . there can be nothing real in nature except simple substances and the aggregates resulting from them'.[73] Otherwise the whole, the visible world, would be an arbitrary unity. Though Leibniz's analysis of the problem is entirely in terms of the metaphysics of simples and aggregates, its immediate consequence is his basic justification for abandoning the doctrine of atoms and the void along with the empiricism which the English associated with it. With the rejection of these doctrines Leibniz was in fact repudiating transduction.

In the first day of his *Discorsi e Dimostrazioni Matematiche Interno a Due Nuove Scienze*,[74] Galileo has an interesting discussion of this distinction. In reply to Sagredo's question, 'are we then to believe that substances become fluid in virtue of being resolved into their infinitely small divisible components? Salvati sums up his reply by saying:

> Seeing that water has less firmness than the finest of powder, in fact has no consistence whatsoever, we may, it seems to me, very reasonably conclude that the smallest particles into which it can be resolved are quite different from finite and divisible particles; indeed the only difference I am able to discover is that the former are indivisible [né saprei ritrovarci altra differenza, che l'esser indivisibili]. The exquisite transparency of water also favours this view, for the most transparent crystal when broken and ground and reduced to powder loses its transparency; the finer the grinding the greater the loss; but in the case of water where the attrition is of the highest degree we have extreme transparency. Gold and silver when pulverized with acids [acque forti] more finely than is possible with any file still remain powders, and do not become fluids until the finest particles of fire [gl'indivisibili de Fuoco] or the rays of the sun dissolve them,[75] as I think, into their ultimate, indivisible, and infinitely small components [credo ne i lor primi altissimi componenti, infiniti, indivisibili].[76]

Though many of Galileo's examples, namely, the pounding of substances into powders, applying acids to them, and the behaviour and transparency of water occur in many later writers, there is no evidence that major scientific thinkers such as Hooke, Boyle and Newton were influenced by Galileo in their discussions of the component parts of phenomena. These operations were common knowledge in the seventeenth century. While it is not clear how this analysis of indivisible components relates to Galileo's doctrine of 'mathematical atoms'[77] presented earlier in the first dialogue, he undoubtedly makes a clear distinction between compound bodies and their ultimate physical particles. Though the difference between the two is that the latter are indivisible, this does not preclude the possibility that there may be other differences discovered. He also holds that what can be affirmed of perceptible things does not necessarily hold for their indivisible parts. It is interesting to observe that Newton was in a position to accept these sorts of argument, since his particles were indivisible and 'infinitely' hard, qualities not attributable to compound bodies. Galileo also remarks of his particles that despite their smallness and their being 'imperceptible to our sight and touch' they are 'nevertheless finite in size, possess shape, and capability of being counted [tuttavia son eglino ancor quanti, figurati e numerabili]'.[78] Accordingly, at least at this stage in his thought, Galileo seems to maintain that in principle we can have knowledge of primordials. For him the problem of transduction does not arise, though he makes a distinction between phenomena and their particles. Nor in this context does the problem arise as to whether primordials or observable bodies are to be considered matter.

## III

Before discussing more fully the epistemology and ontology of the third rule – which will be primarily based on considering the phrase 'the analogy of Nature' – it will be helpful to analyse Newton's use of the concepts 'axiom', 'rule' and 'hypothesis' in the 1690s. Rule III of the 1713 edition of the *Principia* was called both an axiom and a hypothesis before Newton, some time prior to 1700, settled on the category of rule. Why did this shift take place? In a period of some eight years, Newton moved from calling the statement a hypothesis to calling it an axiom, a hypothesis again, and finally a rule. If an unambiguous answer can be given for this series of transformations, a clearer picture of the epistemological status of the rule may emerge.

The earliest extant version of the rule, probably written at the latest early in 1690, is found among a group of *errata* on an end sheet of Newton's personal annotated copy of the first edition of the *Principia*, now in the Library of Trinity College, Cambridge. Labelled 'Hypoth III', it reads: 'The laws (and properties) of all bodies on which experiments can be made are the laws (and properties) of bodies universally.'[79] The words 'et proprietates' occurring twice in this form, were added afterwards. Originally, Newton was thinking only about the 'universal' application of his laws to matter and the problem of universalizing essential and nonessential qualities seems not yet to have occurred to him, though it is connected with the general application of laws or axioms. On the margin of p. 402, in the same copy of the *Principia*, beside Hypothesis III which states that all bodies can transform into one another, Newton wrote the new 'Hypoth III'. He began with 'Proprietates', cancelled this, and then wrote 'Qualitates corporum quae intendi et remitti . . .'.[80] Hence it is clear that the version of the statement annotated in the text was written later than the draft on the end sheet. The new statement is then completed as it is on the end sheet, except that the terms 'leges' and 'proprietates' do not occur. It remains a puzzle, however, why Newton at this stage dropped an explicit mention of the transduction of laws. For our immediate purpose it is important to note that the statement is called a hypothesis and that it now includes the intension and remission criterion for distinguishing essential from non-essential qualities.

Another version, again written in the early nineties, is found on a loose manuscript sheet. After some false starts, it states that: 'The qualities of perceptible bodies which cannot be intended and remitted are the properties of all bodies.'[81] This statement is labelled an axiom. At first blush it seems puzzling that Newton should so classify it, when by the late nineties it would be called a rule in one of his own copies of the *Principia.* Could he have thought that it had the same status as his laws of motion? There are two reasons why he may, for a time, have held this view. After all, he was attempting to remedy the logical structure of the prefatory section to the third Book. If statements labelled 'hypotheses' in the first edition could now be called 'phaenomena', is it so surprising that the new ideas of this statement, themselves in the initial process of clarification, should be called an axiom? Secondly, he may have considered the statement to be basic to the third Book, in the way in which the laws of motion were basic to the whole treatise: thus it was labelled an axiom. Nevertheless, if Newton did hold this view, he appears to have quickely changed his mind. In neither his annotated nor inter-leaved copy of the first edition of the *Principia* is any version of the statement called an axiom. These emendations were all made in the nineties.

In Newton's inter-leaved copy (now in the Portsmouth Collection of the University Library, Cambridge) on p. 402, Hypotheses I and II of the first edition are changed to 'Rules', and 'Hypoth III', the transformation statement, is crossed out. The heading on the printed page has been altered from 'HYPOTHESES' to read 'REGULAE PHILOSOPHANDI'.[82] On the interleaf, Newton has written out the new 'Hypoth III'. In the statement he had first written 'sunt proprietates', then changed to 'pro qualitatibus', and the opening sentence of the remarks to the statement was changed accordingly to 'Nam qualitates' from 'Nam proprietates'.[83] Some time later, Newton changed the designation 'Hypoth III' to read 'Reg III'. Thus, as his ideas developed, Newton moved from calling the statement a hypothesis about 1690, to calling it an axiom, for a time a hypothesis again, and then finally a rule. It cannot be supposed, however, that at any stage Newton indifferently called a statement an 'axiom'. After all, it is a fundamental category of his philosophical schema. It will therefore be instructive if we can understand more fully why he considered putting it in this category.

In the first edition, all the statements in the introductory section to Book III were labelled 'Hypotheses'. This book, in contrast with the first two, concerns itself with the 'system of the world', and therefore there is little doubt that Newton used the term in the sense in which sixteenth- and seventeenth-century writers spoke of the Copernican or Ptolemaic hypotheses of the world.[84] These hypotheses, some of which were later to be renamed phaenomena, were considered by Newton to be the basic postulates of his system of nature. This interpretation has been strengthened by an observation by Cohen that Newton may have been following the usage of Christiaan Huygens in the second part of his *Horologium Oscillatorium* of 1673.[85] Huygens begins this section with a list of three Hypotheses, the first of which is the law of inertia. Here again, 'Hypothesis' is used in the sense of the axioms or postulates on which a conceptual system is based. It is easy to understand why Hypotheses I and II of the first edition should be changed to 'Rules' in the second.[86] They are clearly the usual sort of methodological principle. Being compatible with phenomena, though not in Newton's sense 'deduced' from them, they could with equal facility be called either 'Hypotheses' or 'Rules'. Newton probably decided, however, that naming them rules would more effectively indicate their methodological character. This is also true of 'Rule IV' added to the third edition.[87] It, too, is a principle of procedure which underlies Newton's contention that his experimental empiricism is grounded on phenomena.

The situation is not so straightforward with the axiom and its successor, the third rule. This statement is more than a procedural principle designed to justify the application of the mathematical machinery of Books I and II to the physical world. It is more than an epitome of an epistemological position: it also embodies an ontological theory about perceptible and imperceptible matter. Still, if Newton could call, as late as 1689,[88] his *Leges Motus* either 'Axiomata' or 'Hypotheses', should we be surprised that early in its conception Newton should call this statement either an axiom or a hypothesis? After all, he had probably not yet fully appreciated the extent to which the statement, though purportedly based on phenomena, embodied stronger epistemological claims about the world than the laws of motion. Though it made claims about the nature of imperceptible matter, it was as essential to the structure of the third Book as were the laws of motion for the complex general argument for universal gravitation. Therefore at an early stage the statement could be considered as either an axiom or a hypothesis in the sense of a well-confirmed physical postulate which no one could rightly dispute.

More light may be thrown on the problem if we next consider why the statement was finally called a rule. In revisions of the rules in the second edition of the *Principia* published in 1713, Newton formulated a definition of what he meant by a rule:

A Rule I call every Proposition that is (established from) gathered from Phenomena through the argument of Induction and agrees with them.[89]

Though this was written later than the 1690s when the concept of rule was first employed, there is no reason to suppose that the definition does not reflect Newton's ideas as they crystallized in the nineties. Notice that Newton does not say that rules are 'deduced' from phenomena; though of course he allows that they can be generalized like other statements by induction. This is important. As early as the controversy arising from his paper on light in 1672, Newton spoke of the properties of light as being derived from experiments.[90] Much later, as we have seen, in a letter to Cotes of 28 March 1713, he said in a celebrated passage that the 'First Principles or Axioms of motion were deduced from the phaenomena & made general by induction: w$^{ch}$ is the highest evidence that a proposition can have in this philosophy'.[91] Thus it is clear that in the nineties, this was also his view of how laws and properties were established from the phenomena. It is therefore plain that both axioms and rules are related to phenomena; not, however, in quite the same way.

An axiom is 'deduced' from the phenomena. To express this, Newton generally uses the verb 'deducere' which literally means 'to draw out from' or 'to extract from', and, as was noted above, in English he uses either 'infer from' or 'deduce from'. In contrast to this, in the definition of a rule he wrote 'stabilitur' [to establish from], deleted it, and wrote 'colligitur', from the verb 'colligere' meaning 'to gather into a whole'. This is analogous to the sense in which we speak of 'gathering' something from the facts. There are ample examples of this in Newton's published writing. For instance, in the Observations of Book II, Part I of the *Opticks*, we find such phrases as 'and from these Measures I seem to gather this Rule'.[92] In the 1706 *Optice*, the Latin is rendered 'videor mihi colligere hanc Regulam'.[93] In Observation 14 of the same Book, we find 'and hence I seem to collect that the thickness of the air . . .',[94] which in the Latin reads 'hinc id porro videor mihi colligere'.[95] In both cases the verbs are used in the context of making cautious statements. In both Latin and English they are grammatically weaker than 'deducere' and 'deduce'.

Accordingly, the verbs 'deducere' and 'colligere' characterize the different ways in which axioms and rules are related to phenomena. In saying that rules 'agree with' phenomena, Newton contends that the evidence for them is considerable. In contrast an axiom, being strictly derived from phenomena, is more tightly related to the latter. After 1690, only eleven statements in Newton's natural philosophy are accorded this status – the laws of motion and eight axioms heading Book I, Part I of the *Opticks*. Still, in the early nineties, the difference in status between an axiom and a rule seemed to be more one of emphasis than of category. Newton was still in the process of consolidating his experimental empiricism.

The verbs 'deducere' and 'colligere' reflect a difference in the way in which 'axioms' and 'rules' are related to experience. In might, however, be thought that too much is being made of a mere verbal distinction, despite the fact that the latter verb is used in contexts in which Newton wishes to make a more tentative epistemic claim. In the remarks to the third rule and in all the draft precursors, Newton uses 'colligere' in its basic sense and not 'deducere'.

Thus it is important to understand why he does not speak of deducing the essential qualities of matter from experience. This in turn will help to clarify further the difference between an axiom and a rule, and ultimately to clarify why Newton called the statement a rule and not an axiom or a hypothesis. It is true that in speaking of his theory of colours to Oldenburg he says: 'You know the proper Method for inquiring after the properties of things is to deduce them from Experiments.'[96] Newton is here referring to a special physical property of white light, namely, that it is composed of rays of differing refrangibility. This phenomenon had to be brought about by experimentation so that in Newton's sense the properties could be deduced from it.

On the other hand it is quite natural to speak, as Newton does, of knowledge regarding the qualities of matter arising from reflection grounded in experience. That matter is impenetrable, for example, is something gathered from experience. The position is put clearly in a passage from an intended preface to the 1704 edition of the *Opticks*:

Another Principle is that matter is impenetrable by [other] matter. This is usually looked upon as a maxim known to us by y$^{e}$ light of nature altho we know nothing of bodies but by sense. We find by dayly experience that bodies resist one another as often as they come together & cannot by any force be made to penetrate one anothers dimensions. And this holding true without exception in all bodies here below in w$^{ch}$ we can come to make observations we conclude it to be the [property of] all bodies whatever. And such observations occurring every day to every man this property of bodies is acknowledged by all men without dispute. . . .[97]

Thus knowledge of the qualities of matter is not deduced from experimentally produced phenomena, such as the laws of motion and the properties of colours; rather, it is something that can be established by all thinking men. It was considerations of this sort which led Newton, in part, to place the statement in a new category which he now called a rule.

From Newton's writings there appear to be two counter-examples to this interpretation. Does he not mention in the very formulation of the third rule the use of experiments in determining the nature of bodies? And does he not say in the *Scholium Generale* referring to 'particular propositions' being 'inferred from phenomena':

Thus it was that the impenetrability, the mobility, and the impulsive force of bodies, and the laws of motion and of gravitation were discovered.[98]

In answer to the first point it is clear, for example, that so far as inertia itself is concerned, Newton conceived bodies as being naturally endowed with it. Speaking of *vis insita* which he held to manifest itself in an inertial state of rest or motion, Newton said in *De Motu Corporum*: 'Corporis vis insita innata et essentialis es.'[99] Thus matter is endowed with a property which is inherent, innate and essential; and by which 'Every thing doth naturally persevere in y$^{t}$ state in w$^{ch}$ it is unless it be interrupted by some external cause . . .'.[100] But in the formulation of the third rule, Newton is also concerned to include, as well as other laws, the law of inertia. Thus in contrast to the qualities of matter, the operations of its laws can only be established quantitatively by experimental determinations of the various forms of interaction. As to the second point, it must be seen in the context of the polemic with Leibniz. In his correspondence with Newton, Cotes had advanced reasons which led him to believe that the third law was only a hypothesis. Newton replied: 'that as in Geometry the world Hypothesis is not taken in so large a sense as to include . . . the First Principles or Axioms w$^{ch}$ I call the laws of motion'.[101] And in the same letter he instructed Cotes to insert into the General Scholium a statement similar to this in content. If Cotes could misinterpret the status of a law, what could be expected of the Continental critics? Aroused to a renewed consideration of what he took to be the irresponsible criticism of his rivals, Newton injudiciously stated that his theory of matter was inferred from phenomena. When he was formulating the third rule in the early nineties, he would never have contemplated seriously such a view.

We are now in a position to appreciate why Newton's calling the third rule both an axiom and a hypothesis early in its formulation, indicates that his ideas were not yet crystallized into their final form, and consequently that he had not characterized fully the basic philosophical categories of the third Book. Certainly in the nineties he began to take a harder line on hypotheses, as is evident by his discarding, as it was formulated, the old Hypothesis III on transformation. It was in this period that a hypothesis became an 'opinion which is neither a phenomenon, nor is demonstrated, nor is deduced from phenomena through the Argument from Induction'.[102] Though this was written about 1715 in conjunction with the definition of a rule, it nevertheless reflects Newton's opinion in the earlier period. This is more plausible when it is recognized that Newton's discussion of these terms, and especially his distinction between axioms and hypotheses, is probably related to a traditional conception, current in writers such as Descartes and Locke, that scientific knowledge (*epistemé*) is demonstrable. Thus an axiom or law is as near as natural philosophy comes to 'mathematical' certainly, or 'demonstration' from phenomena. Hypotheses, on the other hand, are opinions which seem, for Newton, to have little moral certainty. Certainly he never fully considered their role as a method of achieving a sort of physical knowledge which could be justified. Since Rule III was held to be consistent with experience, it could not fall into the category of a hypothesis as strictly understood. Not only was it an expression of common knowledge regarding the qualities of perceptible bodies, but it was an attempt to justify the affirming of essential qualities and the laws of motion to all matter whatsoever. Hence it made strong epistemological claims about the world. Newton almost certainly became aware of this as he considered the philosophy of the third Book in the 1690s. He undoubtedly concluded, therefore, that any statement which attempts to justify the application of axioms to a domain of phenomena which includes both celestial and microscopic matter, cannot in itself have the status of an axiom. Thus it was placed in a new category which Newton called, somewhat unsatisfactorily, a rule.

## IV

Having discussed the relative status of axioms, rules and hypotheses as they evolved in Newton's thought during the crucial period of the nineties, we are now in a position to analyse more fully the epistemology of the third rule. In another manuscript also from the early nineties, he clearly formulated the problem of transduction. In this intended emendation, he planned to reproduce the old 'Hypoth III' on transformation of the first edition with the new 'Hypoth III' as it was to appear, as a rule, in 1713. Newton's comments on the new hypothesis are of interest:

HYPOTH III. The qualities of bodies which cannot be intended and remitted and which apply to all bodies in which it is possible to set up experiments are qualities of all bodies universally. The same is to be understood of qualities of all bodies of the same kind. This evidently is the Foundation of all Philosophy. For otherwise one could not derive the qualities of insensible bodies from the qualities of sensible [ones]

HYPOTH IV. Things which agree as to all their known qualities with other kinds of things are not to be considered as new kinds of things

HYPOTH IV. That every body can be transformed into a body of any other kind, and take on successively all intermediate degrees of qualities.[103]

The most striking aspect of the explanatory comments is the phrase 'otherwise one could not derive the qualities of insensible bodies from the qualities of sensible [ones]'. As Newton conceives transduction, any claim to know the characteristics of the invisible realm of nature must be based on inferences from recurring sensory experience. Since, like Locke, he restricts the basis of natural knowledge to what is directly observable, the problem becomes one of justifying whatever means are used in attempting to bridge the gap between the perceptible and the imperceptible. With his unquestioned assumption that nature is uniform, Boyle's approach is similar to that of Newton:

. . . both the mechanical affections of matter are to be found, and the laws of motion take place, not only in the great masses, and the middle sized lumps, but in the smallest fragments of matter . . . and therefore to say, that though in natural bodies, whose bulk is manifest and their structure visible, the mechanical principles may be usefully admitted, [but] that [they] are not to be extended to such portions of matter, whose parts and texture are invisible, may perhaps look to some, as if a man should allow, that the laws of mechanism may take place in a town clock, but not in a pocket watch.[104]

Boyle clearly supposes that there is a similarity between the visible and the invisible realm of nature, such that all matter will be endowed with certain constitutive qualities, and that it will be governed by the same type of law. How did Boyle and Newton suppose that this procedure could be justified? In Boyle's work the question never arises in a general form. He tends to treat questions about the internal structure of matter in terms of conjectural analogies appropriate to the problem at hand. On the whole, he supposes that procedures which apply at the visible level can be transferred to the invisible: matter has the 'catholic' principle of motion and its 'general affections' in common.[105] On the other hand, Newton tends to conceive the problem in a general form. Rule III is meant to apply in all cases where transductive inferences are employed. And in his remarks on that rule, he attempts to offer a justification for transduction. As we shall see, an important aspect of his procedure is an appeal to what he calls the 'analogy of Nature'. The obscure reference to 'qualities of all bodies of the same kind' and its slight amplification in the first of the two 'Hypoth IV' above are connected with this notion. In the remainder of this study I shall elucidate these connections by attempting to throw light on the notion of the 'analogy of Nature' itself as it is used by Newton in various contexts. But this first involves considering the use of the expression by some of Newton's contemporaries, the works of whom he had read, and of outlining the metaphysics which lies behind it, though some of the thinkers I shall discuss, such as Power and Cudworth, do not use this expression, nor did they consider the problem of transduction. This metaphysics of analogy, however, is more widespread in later seventeenth-century natural philosophical thought than is generally recognized by commentators.

Imbued with the ideas of Sir Thomas Browne on the macrocosm and microcosm in connection with the chain of being, Henry Power in the preface to his *Experimental Philosophy* of 1664 related these doctrines to the potentiality of microscopes:

> . . . and there it hath often seem'd to me beyond an ordinary probability and something more than fancy . . . to think, that the least Bodies we are able to see with our naked eyes are but middle proportionals (as it were) te'xt the greatest and the smallest Bodies in nature, which two extremes lye equally beyond the reach of humane sensation; For as on the one side they are but narrow soules, and not worthy the name of Philosophers, that think any Body can be too great or too . . . vast in its dimensions; so likewise, are they inapprehensive, and of the same litter with the former, that on the other hand think the particles of Matter may be too little, and that nature is stinted at an Atom, and must have *non ultra* of her sub-divisions.[106]

Not only does this passage affirm Power's adherence to the continuous subdivisibility of matter but it indicates, because of a continuous chain linking bodies at all levels, that what applies to the macrocosm is reciprocally applicable to the microcosm. By means of this metaphysics, Power attempts to give a justification for the many analogical arguments that are to be found in his work.

One of the most sustained and elegant defences of the chain of being is to be found in Ralph Cudworth's *True Intellectual System of the Universe* of 1678. In Chapter V of Book I, where Cudworth mounts an elaborate attack on the atomistic athetists who attempt to derive '*the phaenomena of Cogitation (Life and Understanding, Soul and Mind)*' from '*Senseless Matter*',[107] he evokes the chain of being to justify his doctrine of the ontological priority of the spiritual over matter. He finds it an abomination that they 'conclude that there is no such *scale* or *Ladder* in *Nature*, no such *Climbing Stairs* of *Entity* and *Perfection*, one above the other, but the whole *Universe* is *One Flat and Level* . . .'.[108] In place of this conception, Cudworth strongly asserts:

> Wherefore there being plainly a *Scale* or *Ladder of Entity*; the *Order* of Things was unquestionably, in way of *Descent*, from *Higher* Perfection, Downward to Lower, it being as impossible, for a *Greater Perfection* to be produced from a *Lesser*, as for *Something* to be Caused by *Nothing*. Neither are the *steps* or *Degrees* of this Ladder (either upward or downward) Infinite; but as the *Foot, Bottom* or *Lowest Round* thereof, is *Stupid* and *Senseless Matter*, devoid of all *Life* and *Understanding*; so is the *Head Top* and *Summity* of it, a *Perfect Omnipotent Being*, Comprehending itself, and all Possibilities of thngs. A *Perfect* Understanding Being, is the *Beginning* and *Head* of the *Scale* of Entity; from whence things *Gradually* Descend downward; lower and lower, till they end in *Senseless Matter*. προγενέσατος νοῦςπὰντων *Mind is the Oldest of all things*, Senior to the Elements, and the whole Corporeal World.[109]

The priority of the spiritual over the world of brute nature is a constantly recurring theme with Cudworth. With equal fervour, the doctrine is embraced by Newton. In an early treatise *De Gravitatione*, a doctrine of levels of perfection in successive orders of reality is propounded: God, Eternity, Time and Space, immaterial forces, and dead matter.[110] Each of these entities is seen as an explication of God, with matter exemplifying the lowest possible manifestation of reality. Since all aspects of reality are dependent upon God not only for their existence, but even for their preservation, a universal inter-relatedness among all realms, moral, spiritual and physical is implied, even though in the natural world, intrinsic connections are not observable.[111] This metaphysics, only implicit in Newton, was elaborated, as we shall see, by his would-be disciple George Cheyne in 1715. Though Cudworth's treatise was published in 1678, some ten or so years after Newton's *De Gravitatione* was written, the former work was almost certainly ready for the press before 1670.[112] Being in Cambridge at this period, and probably in contact with men such as More and Cudworth, Newton may well have discussed with his contemporaries this view of the relation of nature to God. It is interesting to observe that like Cudworth, Newton does not connect the great chain of being with the doctrine of plenitude. Though there are 'Infinite Possibilities of things', the chain itself is not of infinite length.

The earliest direct reference to Newton's phrase which I have encountered is in Joseph Glanvill's *Sadducimus Triumphatus* of 1681.[113] There he speaks of the '*Analogy of Nature*, which useth not to make *precipitous leaps* from one Thing to another, but usually proceeds by *orderly steps* and *Gradations*'.[114]

Glanvill uses this principle in the context of arguing for the existence of immaterial spirits. As a classic statement of the doctrine of the great chain of being, the passage is worth quoting in full:

And that all the upper Stories of the Universe are furnished with Inhabitants, 'tis infinitely reasonable to conclude from the Analogy of Nature; since we see there is nothing so contemptible and vile in the World we reside in, but hath its living Creatures that dwell upon it; the Earth, the Water, the inferior Air, the Bodies of Animals, the Flesh, the Skin, the Entrails, the Leaves, the Roots, the Stalks of Vegetables; yea, and all kind of Minerals in the subterraneous Regions. I say, all these have their proper Inhabitants; yea, I suppose this Rule may hold in all distinct Kinds of Bodies in the World, *That they have their peculiar Animals*. The Certainty of which, I believe, the Improvement of microscopical Observations will discover. From which I infer, That since this little Spot is so thickly peopled in every *Atom* of it, 'tis Weakness to think, that all the vast Spaces above, and Hollows under Ground, are desert and uninhabited. And if both the superior and lower Continents of the Universe have their Inhabitants also, 'tis exceedingly improbable, arguing from the same Analogy, that they are all of the meer insensible Nature, but that they are, at least, some of the rational and intellectual Orders.[115]

Forcibly expressed is the conception of the imperceptible gradation of one order of being into another, so that the *scala natura* exemplifies a chain which links the insensible levels of nature with the intellectual and spiritual. In characteristic fashion, Glanvill declares a belief in common with Power that the instruments of science will discover the natural order beyond the senses. Nature is harmonious. In relation to the doctrine of the chain and in the light of our knowledge of nature's visible order we have, Glanvill implies, implicit knowledge of the invisible realms.

What is surprising, and seldom noticed, are the similarities in the views of Locke and Glanvill. A careful reading of the whole of Locke's *Essay* reveals that he accepted the corpuscular hypothesis and that he believed in the existence of immaterial hierarchies.[116] Within this conceptual framework he attempted to elaborate the significance of interpretations of nature current in his time. Yolton has recently observed that Locke was not primarily concerned in his treatise with justificatory questions: 'His enquiry into the original [sic], certainty and extent of knowledge is descriptive, not just justificatory.'[117] I do not wish to raise questions about this particular interpretation of Locke; I think there is, however, sufficient support for it in the *Essay*. Nor do I wish to raise the traditional polemics regarding the consistency of Locke's epistemology with respect to the existence of objects in the world. I do, however, wish to claim, more than Yolton will allow, that in his acceptance of the great chain of being, Locke was concerned to offer a type of justification for the idea and existence of insensible particles. In other words, he was aware of transduction as a justificatory problem.

What Locke has to tell us is to be found mostly in 4.16.12. He is primarily concerned to emphasize the use of analogy from what is known for understanding what cannot be observed. Locke specifically mentions and connects two kinds of reasoning from analogy: moving from observables to insensible particles: and 'The Existence, Nature, and Operations of finite immaterial Beings without us; as Spirits, Angels, Devils etc.'[118] Locke firmly believed that nature has no 'Chasms or Gaps'; above and below the chain of being progresses in 'easy steps, and a continued series of Things'.[119] Thus the great chain of being of the theologians was related in Locke's mind with the insensible particles of the Virtuosi. Moreover, he was deeply committed to the idea of a purposive universe: 'Nature never makes excellent things, for mean or no uses'[120] and it is within the context of such an interpretation of nature that Locke's adherence to the chain of being must be placed. His account of it begins with the analogy from experience to what is unobservable:

Thus, observing that the bare rubbing of two bodies violently one upon the other, produces heat, and very often fire itself, we have reason to think that what we call Heat and Fire, consists in a violent agitation of the imperceptible minute parts of the burning matter: observing likewise that the different refractions of pellucid Bodies produce in our Eyes the different appearances of several Colours; and also that the different ranging and laying the superficial parts of several Bodies as of Velvet, watered Silk, etc., does the like, we think it probable that the Colour and shining of Bodies is in them nothing but the different Arrangement and refraction of their minute and insensible parts. Thus, finding in all parts of the Creation, that fall under human Observation, that there is a gradual connexion of one with another, without any great or discernible gaps between, in all that great variety of Things we see in the World, which are so closely linked together, that, in the several ranks of beings, it is not easy to discover the bounds betwixt them, we have reason to be persuaded that, in such gentle steps Things ascend upwards in degrees of perfection. Tis a hard matter to say when Sensible and Rational begin, and where Insensible and Irrational end; and who is there quicksighted enough to determine precisely which is the lowest Species of living things, and which the first of those which have no life? . . . Observing, I say, such gradual and gentle descents downwards in those parts of the Creation that are beneath man, the Rule of Analogy may make it probable, that it is so also in things above us and our observation; and that there are several ranks of intelligent Beings, excelling us in several degrees of perfection, ascending upwards towards the infinite Perfection of the Creator, by gentle steps and differences, that are every one at no great distance from the next to it. This sort of probability, which is the best conduct of rational Experiments, and the rise of Hypothesis, has also its Use and influence: and wary Reasoning from Analogy leads us often into the discovery of truths and useful productions, which would otherwise lie concealed.[121]

This fascinating passage must be seen in its entirely. It is not, in the first place, an isolated aberration on the part of Locke. In 3.6.12, which is concerned with the distinction between nominal and real essence and the question of natural species, there is a lengthy discussion of the chain of being within the context of the doctrine of plenitude, and the above passage must be read in conjunction with it.[122] From the intrinsic evidence of the passage above, it is clear that Locke's notion of analogy-based hypotheses, and his distinction between knowledge and judgement, have to be seen in relation to his complete acceptance of the great chain of being: for 'when we consider the infinite Power and Wisdom of the Maker, we have reason to think, that it is suitable to the magnificent Harmony of the Universe, and the great design and infinite goodness of the Architect, that the species of Creatures should also, by gentle degrees, ascend upward from us towards his infinite Perfection, as we see they gradually descend from us downwards'.[123] Though this passage does not mention insensible particles it provides background to show that it is the perfection of the Deity, the evidence of his harmonious creation connected with the chain of being, that provides the context within which Locke conceives the problem of insensible particles. Analogical reasoning has force because of this doctrine embedded as it is in a total world-view so unmistakably articulated in these passages. Consequently, we cannot overlook Locke's adherence to the chain of being when discussing his doctrine of judgement in relation to analogical reasoning and the use of hypotheses. As in the thought of Henry Power, the microcosm is a reflection of the macrocosm. By virtue of the analogy of nature, we can be reasonably assured that in certain respects – their essential qualities – sensible and insensible objects are alike. It is therefore not surprising that Locke should insist that speculations about unobservables be related or derived from sensory experience, even though he did not think that we could have clear ideas of insensibles. I am not, of course, claiming that this particular justificatory account of the existence of insensible particles has any bearing on Locke's general epistemology of ideas, or the inconsistencies attributed to his position regarding knowledge of things other than ideas. Nor am I suggesting that he came by the idea of insensible particles through the great chain of being. The evidence would suggest that he took over the corpuscular theory from Boyle. These larger issues, however, lie beyond the scope of the present study.

Though never officially a member of the clan, the would-be Newtonian George Cheyne developed an elaborate metaphysics of analogy in his *Philosophical Principles of Religion* in 1715. Basing himself on Newton's philosophy of nature, he propounded a complex web of analogies between the natural, moral and spiritual worlds. One of his basic concordances compares God and the sun as attractive forces:

> . . . as the *Attraction* of the *Sun* on the *Planets*, makes them first move, and then describe regular *Orbits*, so this Divine *Attraction* in spiritual Beings, animates the Will, and actuates the Affections, and these do all the rest that is to be done in this present state: and as the discordant *Attraction* of some wandering *Comets* would certainly distract and disorder the *Harmony* of the Motions and Revolutions of the *Planets*, if they approach'd too near them; so gross irregular carnal Affections, earthy and sensual *Attractions* admitted too high, disturb and destroy the beautiful Progress of *Spiritual Beings*, towards the *Centre* and End of their Being.[124]

The framework within which Cheyne comprehended his doctrine of analogy is strikingly similar to the views of Newton as expressed in the Queries of the *Opticks*, the 1675 letter to Oldenburg, and various draft material relating to the General Scholium.[125] And we have no evidence to suppose that Newton would find anything inimical in Cheyne's elaboration of a position which he himself held implicitly. The concordance of Cheyne's views with those of Newton is best seen in the following passage in which he concludes after commenting on the 'exact Symmetry' to be observed in the works of nature that:

. . . this *Analogy of Things*, necessarily infers the Existence of the Author of these Things, and the Wisdom of the Contriver of this *Analogy*. These Things and this *Analogy*, cou'd come from nothing else but from their original *Ideas* and *Architypal Patterns*, in the *Divine Mind* or *Imagination*, and their Harmony and Proportion, can possibly arise from nothing but their being Representations of his *Ideas*, who is omniscient and does everything in *Number, Weight,* and *Measure.*[126]

Thus not only is nature harmoniously interrelated, but such symmetry is best explained by the 'analogy of things' which corresponds to the moral and spiritual orders. Moreover, the whole of this schema is but the ectype of the eternal and archetypal patterns in the mind of God. In a chapter which discusses Newton's doctrine of attraction and gravitation Cheyne, within the framework just expounded, declares in phrases reminiscent of the third rule:

For it's certain that Nature is uniform and consistent with itself, and wou'd not deprive one part of Matter of so Cardinal a Property [gravity] with which she had endow'd all the rest. The whole Foundation of *Natural Philosophy*, is *Simplicity*, and *analogy*, or a Simple, yet Beautiful *Harmony*, running through the Works of Nature in an uninterrupted Chain of Causes and Effects, with proper Limitations of Circumstances: And if these principles be superceded, or this Chain be broken, we can expect nothing but Absurdities and inconsistencies in Philosophy.[127]

The second part of his treatise clearly indicates that Cheyne associated his doctrine of a chain of causes with an acceptance of a 'Scale of Beings'. Thus in the Preface to the second part, he states:

That there is a perpetual *Analogy* (Physical not Mathematical) running on in a Chain, thro' the whole *System* of *Creatures*, up to their *Creator* . . . the Visible are images of the Invisible, the Sensible of the Insensible, the *Ectypical* of the *Architypical*, the *Creatures* of the *Creator*, at an absolutely infinite distance.[128]

Since all things are explications of God, the various orders of creation must be analogous to each other and to their Divine Prototype. Thus the chain allows Cheyne to maintain the orthodox dualism of creator and created, since there are an infinite number of degrees of reality between God and the lowest order of creation at 'an absolutely infinite distance'. There is therefore the implication that matter itself moves through imperceptible gradation from sensible objects to insensible particles: nature would be a monstrous distortion if the macrocosm were not reflected in the microcosm. With clear reference to Newtonian principles and Lockean empiricism, Cheyne declares in his discussion of the third rule that unless the metaphysics of analogy holds then 'all *Universal Properties* in Philosophy must be given up'.[129] Only in this way can the principle of gravitation and the essential qualities of matter be universalized. The very notion of natural uniformity is itself based on a rich conception of the metaphysics of analogy. And Cheyne has no doubt that progress through the successive realms of being must begin with the 'analogy of Nature' experimentally observed.[130]

V

Against this background I shall attempt to show that fundamental to Newton's treatment of the problem of transduction is the concept of the 'analogy of Nature'. In the third rule he says: 'We are certainly not to relinquish the evidence of experiments for the sake of dreams and vain fictions of our own devising; nor are we to recede from the analogy of Nature, which uses to be simple, and always consonant to itself.'[131] In conjunction with the manuscript briefly discussed in the last section, there are two points that require comment. First and most obvious is the fact that Newton is appealing to a principle of cosmic order. I shall return to this in the following section. Less obvious, but of considerable importance, is the fact that he is also appealing to the doctrine of the chain of being, conceived in terms of the analogy of the microcosm with the macrocosm. In order to establish the latter claim, I shall analyse various aspects of Newton's thought, and thereby hope to show that the third rule assumes its true significance against the background of the metaphysics of analogy.

Newton's earliest reference to the analogy of nature occurs in his letter of 1675 to Oldenburg. He observes that the vibrations of light rays in the production of colour behave

much after the manner, that in the sense of Hearing Nature makes use of aërial vibrations of severall bignesses to generate Sounds of divers tones, for the Analogy of Nature is to be observed. And further, as the harmony and discord of Sounds proceed from the proportions of the aereall vibrations; so may the harmony of some colours, as of a Golden & blew, & the discord of others, as of red & blew proceed from the proportions of the aethereall.[132]

Here the appeal is to a principle of simplicity: the mode of production of colours is analogous to that of sounds. Thus whatever understanding can be established of the one may be applicable to the other. Later in the *Opticks* this use of the analogy of nature is more clearly articulated:

For as sound in a Bell or musical String, or other sounding Body, is nothing but a trembling Motion, and in the Air nothing but that Motion propagated from the Object, and in the Sensorium 'tis a Sense of that Motion under the Form of Sound; so Colours in the Object are nothing but a Disposition to reflect this or that sort of Rays more copiously than the rest; in the Rays they are nothing but their Dispositions to propagate this or that Motion into the Sensorium, and in the Sensorium they are Sensations of those Motions under the Forms of Colours.[133]

Thus again analogies between the source of a colour or sound, their propagation as motions, and their action on our sensory apparatus arre drawn: and the two processes are thereby made mutually comprehensible. Again in the *Opticks*, the analogy of nature is stressed in terms of 'attractive powers':

For it's well known, that Bodies act one upon another by the Attractions of Gravity, Magneticism, and Electricity; and these Instances shew the Tenor and Course of Nature, and make it not improbable but that there may be more attractive Powers than these. For Nature is very consonant and conformable to herself.[134]

This is an interesting passage. Not only is the uniformity of natural operations stressed with reference to the analogous attractive properties of gravity, magneticism and electricity, but Newton thinks it highly probable that there exists a hierarchy or chain of such phenomena discoverable but not yet known. The 'Tenor' and consonance of nature is undoubtedly of fundamental importance in his philosophy of nature. Though forces may have many properties which differ, Newton seems to conceive the possibility that the form a continuous chain with attractive powers of varying degrees. Thus each of these forces will have an analogy 'with the forces of greater bodies'. Apart from the suggestion of a possible hierarchy of forces as links in a chain, Newton's use of the analogy of nature in the passages considered generalizes the notion of the simplicity of causes: 'For Nature is pleas'd with simplicity, and affects not the pomp of superfluous causes.'[135] The main burden of the analysis will now shift to the problem of relating the analogy of nature to the doctrine of a chain of being. This will first involve a consideration of Newton's views regarding continuity in nature.

Though he did not accept the doctrine of plenitude – that there is the greatest possible variety of kinds in nature – Newton developed early in his thought a conception that natural kinds are able gradually to transform into one another. As I have shown elsewhere, in the 1670s he propounded various theories whereby an aethereal substance functioned as a primitive *arché* out of which all natural things originated and to which they will return by various processes of transmutation.[136] Transmutation is in fact the key to understanding how Newton regarded any natural kind as imperceptibly changing by a gradual process into another natural kind. Typical of many passages written prior to the first edition of the *Principia* in 1687 is the following concerned with composite bodies:

Their particles coalesce in new ways and by means of the attractive forces of contiguous ones they come together more densely: For which reason that rare substance water can be transformed by continual Fermentation into the more dense substances of animals, vegetables, salts, stones and various earths. And finally by the very long duration of the operation be coagulated into mineral and metallic substances. For the matter of all things is one and the same, which is transmuted into countless forms by the operations of nature, and more subtle and rare bodies are by fermentation and the processes of growth commonly made thicker and more condensed.[137]

The gradual transformation of composite bodies results from the changing configurations of their constituent particles. Thus compound bodies are merely the emergence of natural kinds, receptive to chemical agents such as fermentation, which are causally produced by the alteration of their inner constitution itself comprising particles of the same uniform ontological type.

The whole process, moreover, is conceived in dynamic terms. Thus Newton's conception of the chain of being is that of a continuous process at once distinguishable from the more static doctrine of plenitude which asserts for theological reasons that there are no gaps in the species or kinds in nature. This is exemplified in Leibniz's thought, with his insistence on a plenitude of variety as the greatest indication of God's goodness and perfection. Newton's characterization of the chain allows the animal to merge imperceptibly into the vegetable and the latter into the mineral until the process ends in the density of the diamond. There is evidence, too, that he conceived the chain as ascending through vapours, exhalations, and chemical spirits to animal spirits and the higher realms of spirituality.[138]

With Newton's early optical researches in mind, Lohne has emphasized that Newton's thought regarding nature was guided by seven unquestioned principles.[139] Three which he singles out are clearly related to the sort of thinking under discussion. The first, arising out of his aethereal speculations of the mid sixteen-seventies, is embodied in Newton's statement that '. . . nature is a perpetuall circulatory worker'.[140] The second, as formulated by Lohne is closely connected with Newton's theory of matter: 'There are no "jumps" in the operations of nature. All gradations of velocities exist and also all intermediate sizes of corpuscles.'[141] This principle in no way indicates a commitment to the doctrine of plenitude, since in Newton's view, both varying velocities and the constitution of corpuscles, as indeed all bodies, require the existence of void space. The third principle singled out by Lohne is the transformation hypothesis of the beginning of the third Book in the first (1687) edition of the *Principia*. That every body can 'take on successively all intermediate degrees of qualities' is unmistakably a declaration of continuity in nature. In the 1690s, however, Newton began to restrict transmutation to the secondary qualities of matter:[142] it is they alone that can manifest continuous degrees of intensity and, by transmuting into one another, can change any given object into another sort of thing. This is illustrated, with respect to quantity of matter, in the only direct reference to the hypothesis in the 1687 edition of the *Principia*:

> . . . if the aether, or any other body . . . were to gravitate less in proportion to its quantity of matter, then because there is no difference between that and other bodies but in form of matter, by successive changes of form, it might be transmuted into a body of the same condition with those which gravitate most in proportion to their quantity of matter (by Hypoth III). . . .[143]

Though Newton's argument is designed to show that the proportion of gravitation to quantity of matter is not affected by mere successive changes of any given body's shape, qualities which can change in this way constitute a truly continuous chain having a dynamic potential of gradually transforming themselves. Despite his more explicit restriction of transmutation to the secondary qualities of matter, Newton did not abandon the doctrine itself. Thus the principles which Lohne singles out as guiding Newton's early thought are largely those employed in the later development of his ideas.

The conclusion that I wish now to defend and to elaborate is that the analogy of nature with reference to the third rule was conceived by Newton against the background of his theory of the gradual transmutation of the one natural kind into another. Put simply, the gap between composite bodies and constituent particles is filled by a chain of bodies imperceptibly changing one into another until the elemental simplicity of voidless primordial particles is reached. Thus knowledge of essential qualities of sensible bodies is more securely transferred to the imperceptible in terms of the postulation of a chain which exists between the two orders. But most important of all, in terms of this conception, significance is given to the view that primordial entities differ only in size, not in kind, from sensible bodies. Moreover, they have in common only the essential qualities of the latter which we know from experience and which are determined by the intension and remission criterion. Though we cannot observe the primordial particles we can give content, by thinking through and by means of the underlying analogy, to a claim to know the nature of matter in general.

Against this background the rather cryptic sentence in the draft of the third rule reproduced in the last section has some significance. Immediately after a statement of the rule (called at this point a hypothesis) Newton adds: 'The same is to be understood of qualities of all bodies of the same kind.'[144] It is precisely the essential qualities which unite matter, perceptible or otherwise, and these qualities which unite matter, perceptible or otherwise, and these qualities are invariably associated with sensible bodies. Thus the theological chain of being is transformed by Newton to provide a general framework with which to comprehend material phenomena, and in so doing he combines it with the primordials of the atomic hypothesis. These are the '*Foot Bottom*, or *Lowest Round*' of nature existing in the invisible realm of the microcosm. Not only does the chain of being link the composite with the elemental, but it is the underlying support for Newton's statement that 'By no other means may one derive the qualities of imperceptible bodies from the qualities of perceptible [ones]'. Matter is unified by the analogy of nature; thereby, to Newton's satisfaction, the ontological gap between compounds and elements is bridged, and the problem as to whether bodies or atoms are to be identified with matter does not arise. Thus, though he recognized this distinction on conceptual grounds, the general tendency of his thought rendered it unimportant in considering natural philosophical problems.

The unification of matter is also exemplified, as has already been suggested, in Newton's doctrine of essential qualities. In *De Gravitatione*, they are clearly conceived as the immediate effects of God's will, and there is evidence to show that this remained Newton's view to the end of his life.[145] Epistemologically, his criterion for distinguishing the essential from the non-essential in matter turned on his conception of the intension and remission of qualities: non-essential qualities such as colour are 'intended and remitted', that is, they can manifest continuous degrees of intensity, all else being equal. Essential qualities not falling under this description cannot be increased or diminished, all else being equal; since they define the essence of matter, falling under the intension and remission criterion would entail that any given physical object could be more a body at one time, less at another, as something is now hotter, now colder. Theologically, since they depended on God's will, the immutability of the essential qualities was probably conceived as a physical correlation of the perfection of Divine Nature: that is to say, these qualities are an important expression in nature of Newton's general conception that reality 'images' God. This is the interpretation which Gowin Knight, writing in 1748, placed on the third rule: 'All immediate Causes, being the effects of God's will, must necessarily be constant, immutable, and irresistible to any finite force. From hence appears the Truth of Sir Isaac Newton's third Rule . . .'.[146] Thus behind the rule which seems merely to be a methodological statement lies a complete ontology based on the notion of the great chain of being and the Christian doctrine of the perfection of God. It is interesting to observe that Knight's interpretation of the intension and remission criterion is analogous to Descartes's doctrine that the preservation of motion in nature is a consequence of the immutability of Divine Nature. For both Descartes and Newton, the order of nature is guaranteed by this theological doctrine. Since for Newton the essential qualities of matter are a reflection of God's immutability, we can be assured that particles beyond observation are analogous in this respect to the composite bodies of everyday experience.

## VI

In order to clarify a number of issues in relation to this interpretation of the 'analogy of Nature' I shall begin by discussing Mandelbaum's account of the third rule. His main point is that Newton justified inferences concerning the essential qualities of matter in terms of a special interpretation of generality, namely, that he was convinced of 'their *universal* applicability'.[147] By this, Mandelbaum means that the third rule shows inferences from the observable to the unobservable to be 'valid *so long as the latter can be assumed to be of the same kind as those found within experience and this assumption can legitimately be made whenever we are dealing with characteristics which are found to hold without exception within our experience*'.[148] This is, of course, simply a special use of the principle of the uniformity of nature with respect to certain essential features of matter. For Mandelbaum, this special use is clarified by a dictum to the effect that 'characteristics which are invariably associated with experienced objects are also characteristic of all objects whatsoever'.[149] The third rule is in this sense a 'rule of induction'.

Mandelbaum's analysis of this particular use of the rule is of value. Newton himself recognized it. In a draft of a letter intended for Conti, written in 1715/16 during the Clarke-Leibniz controversy, he says of Leibniz:

He denys none of Mr. Newton's experiments. He denys not the Third Rule of Philosophy. And yet from experiments & that Rule universal gravity necessarily follows. But he denys the conclusion . . . That third Rule is the Rule of Induction. And without it no Proposition can become general in Natural Philosophy. Without it we cannot affirm that all bodies are impenetrable. And the argument by Induction for universal gravity is as strong as the argument for universal impenetrability.[150]

It is not surprising that, during a heated controversy, Newton should give the impression that the rule is merely a statement of general induction. At this level he is clearly concerned to maintain that as an inference it is reasonable and rational, which does not involve him in giving general criteria for deciding the nature of reasonable inference. Mandelbaum does not explicitly note, however, that the rule is meant to cover the generalization of both essential and non-essential qualities. In the passage above, Newton implies that there is as much evidence for universal gravitation as there is for impenetrability. But in the second edition of the *Principia* in 1713, three years earlier than the draft letter to Conti, Newton went further in asserting that 'the argument from appearances concludes with more force for universal gravitation of all bodies, than for their impenetrability';[151] this despite the fact that in the third edition he added that gravity is not 'essential to bodies'. This new addition to the rule was doubtlessly more immediately occasioned by the Cotes preface to the second edition, in which he spoke of gravity as a primary rather than an essential quality of matter.[152] In any event, strictly interpreted, the formulation of the rule as it appears in the second and third editions of the *Principia* gives warrant only for the generalization of what Newton holds, in terms of the intension and remission criterion, to be essential qualities. But he also uses it to generalize the application of laws and non-essential qualities such as gravity. This was clearly his intention in the earliest formulation, namely that 'The laws (and properties) of all bodies on which experiments can be made are the laws (and properties) of bodies universally', at which stage, as we noted above, universal qualities did not have to be invariant. From the evidence presented, it seems that Newton was not completely aware of this difficulty, despite the fact that he almost certainly recognized as being philosophical in nature his criterion for distinguishing and deciding which qualities are essential. Moreover, disciples such as Cotes and Clarke were clear that gravitation was not an essential quality of matter, but rather invariably associated with it.[153] Nor was Newton fully able to think of the term 'law' unambiguously, despite his clear distinctions between axioms, rules and hypotheses. On one level a law was uniform mode of order imposed by God on nature: with respect to experimental philosophy it was a general fact about nature. That the notion was somewhat confused in Newton's mind, however, is clear from the fact that he uses the concept contextually, in a traditional sense, to cover generalizations which are without exception sufficiently extensive, especially when based on induction: whereas the fourth rule of the 1727 edition of the *Principia* states that a law can be 'liable to exceptions'. Still, it is precisely because of his belief that the third rule sanctions the inferential step connecting gravitation with all bodies whatsoever, that Newton based his central arguments for the doctrine of gravitation at the beginning of the third Book.

In the second place, Madelbaum's analysis is not sufficient since he does not account for Newton's appeal to the 'analogy of Nature'. As I have attempted to show, it was an appeal to continuity in nature in terms of a doctrine of transmutation, itself conceived within the framework of the chain of being, and the analogy of the macrocosm and the microcosm. As it is presented, Mandelabum's analysis makes the rule both a characterization of uniformity in the observable aspects of phenomena, and the analogy of nature a primitive, unexplained justificatory principle. Since Newton was committed to the thesis that knowledge, though not the criterion of truth, originated solely from sensory experience, he saw the need for a non-empirical justificatory principle which would sanction inferential steps to the invisible realm of nature. He was satisfied that he found it in the 'analogy of Nature' and its associated metaphysics. Thus, like Locke, Newton saw justificatory questions as important and legitimate: but also like Locke, they were construed within a world-view which accepted the existence of insensible particles and the metaphysics of analogy.

Again, only in this way or, as with Gassendi, by the use of analogical arguments developed by Epicurus, were possibilities open to meet an important criticism of atomism. The argument was first used in England during the seventeenth century, in an important scientific context, by William Harvey in his *De Generatione*, and later employed by Stillingfleet in his widely read *Origines Sacrae*.[154] After quoting the whole of the argument as presented by Harvey, Stillingfleet in commenting gives the essence of it:

> If then none of these things which *bodies* are *resolved* into, and are supposed to be *compounded* of, either have been or can be proved to exist *separate* from or *antecedent* to those bodies which they compound, what then becomes of all our company of atoms which are supposed by their *concourse* in an *infinite Space* to be the *origin* of the *world*? I know not where to find them, unless *dancing* with the *School-men's* Chimeras in a vacuum, or in a Space as empty as the *infinite* are, viz. some *Epicurean's brains*.[155]

The polemic tone apart, this is a serous criticism of the atomic doctrine, especially when based, as with Newton, on experience and a rigorous formulation of induction. Moreover an Aristotelian could argue that atoms being homogeneous – not a composite concretion of formed observable matter – were occult in nature: especially since thinkers like Boyle and Locke could not explain how their atoms cohered to form various combinations, and went so far as to claim that the mode of cohesion was unknowable. It is clear that an empiricist can only hope to counter Stillingfleet's argument by analogy based on experience or by an appeal to principles that go beyond experience as those of the metaphysics of analogy. Gassendi took the former, Locke and Newton the latter. As to the second charge, a new philosopher could only point out, prior to Newton's demonstration that gravitation is the most general binding force of nature which is experimentally quantifiable, that their programme of invisible atoms in mechanical motion, as a means of studying phenomena, could not be justified by an Aristotelian theory of experience.

Apart from his recognition of the metaphysics distinction between simples and compounds. Newton, as we have seen, also adduced a number of natural philosophical arguments to show the difference between particles and composite bodies. Did he ever, therefore, consider the possibility that the laws and properties of the microcosm might differ in character from those of the macrocosm? There is a unmistakable evidence that such a possibility did occur to Newton. In a number of unpublished drafts written in 1705 for Query 23 of the 1706 Latin edition of the *Opticks*, he adumbrates a framework for a new physics of nature to be based on the effects of micro-forces.[156] In these drafts, Newton speaks of the possible formulation of laws governing such forces, which will be different in mathematical form and in content from the *axiomata* of the *Principia*. The basis for inferential knowledge of this invisible realm is primarily chemical: fermentation, combustion, putrescence, generation and heat, all of which generate 'vital motions' caused by 'very potent active' principles. Thus the possibility of a *Principia* of the microcosm was recognized by Newton, based on framework of laws different in character from the published *Principia*.[157] In his typically cautions fashion, Newton mooted the possibility to himself alone, though a few vague hints appear in the published Queries. In the final analysis, the 'analogy of Nature' was firmly entrenched in his mind, and in the absence of techniques more powerful than those at his disposal, these ideas were to remain as a mere possibility. The powerful hold that the metaphysics of analogy had on Newton's mind is evident in a hypothesis intended for a projected fourth Book to the *Opticks* and written in the early nineties. After drawing an analogy between the operations of long- and short-range forces, Newton comments:

> And if Nature be most simple and fully consonant to her self she observes the same method in regulating the motions of smaller bodies w^ch^ she doth in regulating those of the greater. This principle of nature being very remote from the conceptions of Philosophers I forebore to describe it in that book [*Principia*] least I should be accounted an extravagant Freak & so prejudice my Readers against all those things w^ch^ were y^e^ main designe of the Book.[158]

Thus it is clear that Newton had his 'principle of nature' in mind while he was composing his formidably mathematical *Principia* in the middle eighties. Though at a certain level the principle postulates the uniformity of nature, it is far more important to appreciate the significance of the macrocosm and microcosm context in which it is discussed in this passage. This type of analogy abounds in Newton's writings: without due consideration of its role in his thought, any analysis of his achievement will prove insufficient.

This study has attempted to contrast, in part, the approach of the Continental and English thinkers with respect to the epistemology of the non-observable. The analysis has based itself on the problem of transduction, which appeared in its most general form in the thought of Isaac Newton. On the Continent, the problem was avoided either by the use of Cartesian conceptualism, or by the development of an entirely new conception of reality based on appearance. With doubts concerning the knowability of a reality existing independent of mental appearance, thinkers such as Malebranche, Leibniz, Bayle, and later Berkeley – despite Locke's concern to give it a philosophical foundation – were attempting to explode philosophically the basis of the new science in England as it was generally accepted during the later seventeenth century. The English tradition with its acceptance of an ontology embodying a two-worlds theory of ideas and matter, tended to justify transduction by an explicit or implicit appeal to the metaphysics of analogy. Thus in broad outline many essentials of the Elizabethan world-picture[159] were enlisted in the service of Restoration natural philosophy at the same time as Continental scepticism was challenging the new science of nature. Nevertheless the third rule of philosophizing, as such, continued to be discussed or employed throughout the eighteenth century: in England, however, the 'analogy of Nature' was absorbed into the mainstreams of theological thought.

## APPENDIX

[The relative dating of the draft variants to Rule II is found in Chapter 6. In each case only the beginning sections of these drafts, related to the problem discussed in this chapter, have been transcribed.]

*Draft 1* N. Q. 16.200 Trinity College Library

Hypoth III leges (et proprietates) corporum omnium in quibus experimenta instituere licet sunt leges (et proprietates) corporum universorum.

Nam proprietates corporum non nisi per experimenta innotescunt ideoque generales statuendae sunt quotquot cum experimentis generaliter quadrant. Certe contra experimentorum tenorem somnia temere confingenda non sunt, nec a naturae analogia recedendum, cum ea simplex esse soleat et sibi semper consona.

*Draft 2* U. L. C. Ad. 4005.15 Folio 81$^r$
Axioma 4 Qualitates corporum sensibilium quae intendi et remitti neque sunt proprietates corporum omnium (quatenus experiri licet corporibus omnibus competunt, pro corporum omnium proprietatibus haberi debent).

Qualitates quae intenduntur (et remittuntur ut motus et quies) calor et frigus, humiditas et siccetas, lux et tenebrae, color et nigredo (opacitas et perspicuitas, odor bonus et malus) aciditas, amaritudo et dulcedo, volatilitas et fixitas, (vivacitas salus et agritudo) hic considerandae non veniunt. Quae vero intendi et remitti nequent (magnetudo) uti impenetravilitas, (soliditas) et mobilitas . . . & inertia (illa) qua motibus et motus mutationibus resistentur (soliditas, extensio) pro corporum omnium proprietatibus haberi solent.

*Draft 3* U. L. C. Adv. b. 39I p402A
Hypoth III [deleted and changed later to *Reg III*] Qualitates corporum quae intendi et remitti nequeunt quaeque corporibus omnibus competunt in quibus experimenta instituere licet, sunt proprietates [deleted and changed to 'pro qualitatibus'] corporum universorum habendae sunt.

Nam proprietates [deleted and changed to 'qualitates'] corporum non nisi per experimenta innotescunt ideoque generales statuendae sunt quotquot cum experimentis generaliter (neque possunt auferri quae minui non possunt) [this phrase is interpolated; differing slightly from the 1713 version: neither is found in Draft I] Certe contra experimentorum tenorem somnia temere confingenda non sunt, nec a naturae analogia recedendum cum ea simplex esse saleat & sibi semper consona.

*Philosophiae Naturalis Principia Mathematica* (Cantabrigiae, 1713), Book III, 357–358 *Regula III*

Qualitates corporum quae intendi & remitti nequent, quaeque corporibus omnibus competunt in quibus experimenta instituere licet, pro qualitatibus corporum universorum habendae sunt.

Nam qualitates corporum non nisi per experimenta innotescunt, ideoque generales statuendae sunt quotquot cum experimentis generaliter quadrant; & quae minui non possunt, non possunt auferri. Certe contra experimentorum tenorem somnia temere confingenda non sunt, nec a Naturae analogia recedendum est, cum ea simplex esse soleat & sibi semper consona.

## NOTES

[1] For a detailed discussion of this approach with a full analysis of a range of texts see Laurens Laudan, 'The Clock Metaphor and Probabilism: The Impact of Descartes on English Methodological Thought, 1650–65', *Annals of Science* **22** (1966), 81–97. I am grateful to Charles B. Schmitt, Laurens Laudan, Gerd Buchdahl and John Yolton for comments on an early draft of this paper. They are in no way responsible for its interpretation.

[2] See Henry More, *Divine Dialogues* (London, 1668) and Ralph Cudworth, *The True Intellectual System of the Universe* (London, 1678).

[3] In his study *Philosophy, Science and Sense Perception* (The Johns Hopkins Press, Baltimore, 1964), Maurice Mandelbaum has used the term 'transdiction' in designating this problem. To conform more with the usage of the terms 'deduction' and 'induction', I have decided to use the term 'transduction', i.e. 'leading across'. In no way do I wish to imply thereby that the latter is a special type of 'inference'; rather, the term is used to designate a problem recognized in seventeenth-century natural philosophy.

[4] Sir Isaac Newton, *The Mathematical Principles of Natural Philosophy*, translated by Benjamin Motte (London, 1729), vol. II, 203. I have altered the Latin of the statement of the rule in the interest of accuracy. The corresponding Latin is in the appendix to this paper. The original is found in Isaac Newton, *Philosophiae Naturalis Principia Mathematica* (Cambridge, 1713), 357.

[5] See Chapter 6 below for a full discussion of the use of the phrase 'cannot be intended and remitted' in relation to the problem of primary and secondary qualities.

[6] Another such integral element is homogeneity. Though never explicitly stated by Newton, this is a fundamental presupposition lying behind his identification of a quantity of matter with its *vis inertiae*, a proportionality questioned by Roger Cotes in a letter of February 1711/2. See Joseph Edleston, *Correspondence of Sir Isaac Newton and Professor Cotes* (London, 1850), 65–66.

[7] This is the received view as developed in the writings of Basil Willey, *The Seventeenth-Century Background* (London, 1934); M. Macklen, *The Anatomy of the World: Relations between Natural and Moral Law from Donne to Pope* (Minnesota, 1958); and G. P. Conger, *Theories of Macrocosms and Microcosms in the History of Philosophy* (New York, 1922).

[8] For this topic see Earl R. Wasserman, 'Nature Moralized: The Divine Analogy in the Eighteenth Century', *A Journal of English Literary History* **20** (1953), 39–76.

[9] See, for instance: Henry Pemberton, *A View of Sir Isaac newton's Philosophy* (London, 1728), 24–25; Colin Maclaurin, *An Account of Sir Isaac Newton's Philosophical Discoveries* (London, 1748), 22; Gowin Knight, *An Attempt to demonstrate that all the Phaenomena of Nature may be explained by two simple active Principles Attraction and Repulsion* (London, 1748), 7; George Cheyne, *Philosophical Principles of Religion: Natural and Revealed* (London, 1715), 36; Willem Jacob Storem-van's Gravesande, *Phyices Elementa Mathematica . . . sive Introduction ad Philosophiam Newtoniam* (Editio Secunda, Lugduni Batavorum, 1725), Praefatio, and Liber I, Caput I, 1–7; Benjamin Martin, *Pilosophia Britannica: or a New and Comprehensive System of the Newtonian Philosophy* (Reading, 1747), 1–42. In his first lecture, Martin discusses Newton's four rules in some detail. Thus when 'we take survey of the visible world', we conclude that all bodies 'consist of *one and the same sort of Matter or Substance*'. He concludes that '*Matter*, thus variously modified and configurated, constitutes an infinite variety of bodies, all of which are found to have the following *common Properties . . .*'. Thomas Rutherforth, *A System of Natural Philosophy* (Cambridge, 1748), Introduction. The first three rules are discussed as the only means of making 'knowledge real and universal'. Rule Three is seen primarily as an inductive rule. Richgard Helsham, *A Course of Lectures in Natural Philosophy*, published by Bryan Robinson, 5th edition (London, 1777). The four rules are quoted in full in the Preface. Robinson notes that 'This method and these rules, have been carefully observed by our author in these *Lectures* . . .' (p. viii). George Adams, *Lectures on Natural and Experimental Philosophy* (London, 1794), vol. II. In a lengthy lecture entitled 'In the method of Reasoning in Natural Philosophy' Adams concludes his discussion of the history of method by analysing Newton's first three rules. Of the third rule he says that it allows us 'by analogy to extend our conclusion to all other bodies, and thus make it universal: a way of reasoning, that is agreeable to the harmony of things, and to the old maxim, ascribed to *Hermes* . . . (p. 134). For this maxim see note 40 below, on Maclurin. Adams is drawing on Maclaurin's treatise. Tiberius Cavallo,

*The Elements of Natural or Experimental Philosophy* (London, 1803), vol. I, 7. Quotes Newton's four rules and connects them with the generality of the three laws. William Nicholson, *An Introduction to Natural Philosophy* (London, 1805), vol. I. In his introduction he discusses intuition, demonstration, and probabilistic knowledge, saying of natural philosophy that it is based on analogy: 'To give stability to this science, it is necessary to admit no probabilities, as first principles of analogy, but those which possess the strongest and most incontrovertible resemblance to truth. For this purpose, the following rules are adopted': the first three rules are then quoted without comment. Joseph Pristley, *Disquisitions Relating to Matter and Spirit*, 2nd edition (Birmingham, 1782), sections I and II; Etienne Bonnot de Condillac, (*Euvres Complètes de Condillac*, tome troisième (Paris, 1803), chapitre XII, 327–358; Petrus van Musschenbroek, *Elementa physicae* (Lugduni Batavorum, 1734), 6, and *Introductio ad philosophiam naturalem* (Lugduni Batavorum, 1764), vol. I, 14; Roger Joseph Boscovicch, *A Theory of Natural Philosophy* (Latin-English Edition, London, 1922), sections 40–42. Boscovich is critical of the third rule on the grounds that induction does not support the transference of many physical properties to the parts of phenomena. The *Theory* was first published in 1758. Brian Higgins, *A Philosophical Essay Concerning Light* (London, 1776), 18–19; James Hutton, *Dissertations on Different Subjects in Natural Philosophy* (Edinburgh, 1792), part III, 279–300; William Whewell, *On the Philosophy of Discovery* (London, 1860), chapter XVIII, 181–200, section 14. Whewell is mainly concerned in this chapter with the notion of *vera causa* as discussed in Newton's first rule. He points out, however, that the third rule's conception of universal laws in inconsistent with the stricture of the fourth rule that a 'law may be inaccurate'. See also *Philosophy of the Inductive Sciences* (London, 1840), vol. I, 416, where Whewell observes that the third rule is circular. James Challis, 'On Newton's "Foundations of all Philosophy"', *Phil. Mag.* **26** (1863), 280–292. In this particular article Challis is attempting to base a theory of science on Newton's third rule.

These discussions are the most interesting in the literature I have examined. Most of these writers are concerned with the problem of generalizing properties of matter, rather than with Newton's criterion for deciding essential qualities. Many other writers such as Halley, Stirling, Ferguson, Desaguilers, Wright and Worster mention or refer to the rule; these with the above references are more than sufficient to establish the widespread influence of this aspect of Newton's methodology throughout the eighteenth and early nineteenth centuries. I am grateful to Laurens Laudan for the references to Condillac and Musschenbrock, and to Peter Heimann for calling attention to Higgins.

[10] This question is discussed with great vigour and in detail by James Hutton in his *Dissertations on Different Subjects in Natural Philosophy* (Edinburgh, 1792), part III, 'Physical Dissertations on the Powers of Matter and Appearance of Bodies', and in his *An Investigation of the Principles of Knowledge, and of the Process of Reason, From Sense to Science and Philosophy* (Edinburgh, 1794), vol. II, section XII. Hutton makes it clear by referring to unnamed contemporaries that the problem of identifying matter with bodies, or with some sort of unperceived theoretical entity was a live issue in the late eighteenth century. His own position is unambiguous. Identifying matter with intensive powers he states: 'Matter, as a thing distinctly different from body, has not been observed by us, nor ever can be made the subject of our observation. Our immediate knowledge, or all external information, comes necessarily from sensible and perceptible bodies, which being formed of matter, must be distinctly different from it', *Dissertations*, part III, 306. A similar distinction is made by the Newtonian Thomas Exley, *Principles of Natural Philosophy: or, a New Theory of Physics, Founded on Gravitation, and Applied in Explaining the General Properties of Matter* (London, 1829). Exley makes a distinction between 'The theoretical constitution of matter', section II, 6–45, and 'The theoretical construction of bodies'. In Postulate I (p. 3) he states: 'Let it be granted that an atom of matter consists of an indefinitely small sphere of repulsion, which is the central part of an indefinitely extended concentric sphere of attraction'; drawing on this postulate, Proposition I of section II states: 'If matter be consti-

tuted, as in *post.* I., it will be capable of rest and motion, as will also the bodies consisting of *such matter*'. [Italics mine.] Exley's distinction is really that between ordinary matter and 'theoretical' matter; since in Definition I of section I he defines matter as 'that substance which we perceive by the senses'. For a discussion of these theories see P. M. Heimann and J. E. McGuire, 'Newtonian Forces and Lockean Powers: Concepts of Matter in Eighteenth-Century Thought', *Historical Studies in the Physical Sciences* **3** (1971), 233–306.

[11] The basis for this statement will be provided in this study.

[12] There is much literature on this, though more research is desirable. The standard sources are: Kurd Lasswitz, *Geschichte der Atomistik vom Mittelater bis Newton*, vol. I (Hamburg and Leipzig, 1890); J. R. Partington, *A History of Chemistry*, vol. II (London, 1961). See also M. Boas [Hall], 'The Establishment of the Mechanical Philosophy', *Osiris* **10** (1952), 412–541; R. Hooykaas, 'The Experimental Origin of Chemical Atomic and Molecular Theory before Boyle', *Chymia*, **2** (1949), 65–80; Thomas S. Kuhn, 'Robert Boyle and Structural Chemistry in the Seventeenth Century', *Isis* **43** (1952), 12–36; Tullio Gregory, 'Studi sull'Atomismo de Seicento. David van Goorle e Daniel Sennert', *Giornale Critico della Filosofia Italiana*, Fasc I (1966), 44–63; W. Subow, 'Zur Geschichte des Kampfes zwischen dem Atomismus und dem Aristotelismus im 17. Jahrhundert (Minima naturalia und Mixtio)', in N. A. Figurowski *et al.*, *Sowjetische Beiträge zur Geschichte der Naturwissenschaft* (Berlin, 1960), 161–191; E. J. Dijksterhuis, *The Mechanization of the World Picture*, trans C. Dikshoorn (Oxford, 1962), part III, section V; Robert H. Kargon, *Atomism in England from Hariot to Newton* (Oxford, 1966).

[13] For a discussion of this distinction in relation to Galileo, Gassendi, Descartes and leibniz see Ivor Leclerc, 'Atomism, Substance, and the Concept of Body in Seventeenth-Century Thought', *Filosofia Della Scienza*, No. 27 (Turin, 1968) (printed also in *Filosofia*, 1968).

[14] Petrus Gassendi, *Opera Omnia*, Band I, *Syntagma Philosophicum* (Stuttgart-Bad Cannstatt, 1964).

[15] *Ibid.*, *Physicae*, sectio I, liber III, caput V, 279–282: 'Et rurùs, quod omnia ex iis componantur, dici non modo ἀρχὰς *Principia, Primordia*, & στοιχεῖα *Elementa*; verùm etiam πρῶτα σώματα *Corpora prima* & πρῶτα μεγὲθη, *Primas Magnitudines*; itémque *Seminium, Seminarium*, πανσπερμιὰν & *Genitalia Semina rerum* ac τῶν ὄντων ὕλη *Rerum Syluam*, siue *Materiam*'.

[16] *Ibid.*, caput VI, 266–267; caput VIII, 279–282.

[17] *Ibid.*, caput VI, 268.

[18] David J. Furley, *Two Studies in the Greek Atomists* (Princeton, 1967), 24. The sentence is from Furley's translation of Epicurus's *Letter to Herodotus*. Furley's treatment of the use of analogy in the atomic theory of Epicurus is important, and throws light on Gassendi's thought.

[19] Gassendi, *op. cit.*, *Syntagma*, caput VII, 273 and 279.

[20] *Ibid.*, caput VII, 280–282.

[21] J. E. McGuire and P. M. Rattansi, 'Newton and the "Pipes of Pan"', *Notes & Records of the Royal Society* **21** (1966), 108–143.

[22] MS. U. L. C. Add. 3965. 6. Fiolo 270$^r$: 'Atomos vero corpora solida esse voluerunt et figuris varÿs praedita, non mathematice sed physice tantum indivisibilia. Nam et atomum unumquemque ex pertibus innumeris minoribus constare sed quae non ex conventu conciliatae sunt, at arcte et sine poris et intestitÿs ab aeterno stipatae et adeo condensatae et in solidum compactae ut in solidum compactae ut nullis Naturae viribus ab invicem avelli et separari possinit. Quae ex atomorum congressu conflantur iterum dividi posse et in atomos eodem resolvi propterea quod atomi non in solidum congrediantur sed se mutuo in puncto tantum mathematico contingant, spatÿs reliquis inter se vacuis relictus: sed atomos ipsos ob soliditatem et plenum partium contactum ac densitatem summam nec dividi posse nec alteri nec ulla ratione comminui nec augmentum unquam sumere sed immutabilia rerum semina in aeternum manere, et inde fieri ut rerum species perpetuo conserventur.'

[23] *Works of the Honoruable Robert Boyle*, ed. Thomas Birch (London, 1744), vol. II. Boyle states in his prefacee that he had formed his opinions before reading Gassendi. See also Walter Charleton, *Physiologia: Epicuro-Gassendo-Charltoniana: or a Fabrick of Science Natural, upon the Hypothesis of Atoms* (London, 1654).
[24] Boyle, *op. cit.*, note 22, vol. II, 474.
[25] *Ibid.*, 460.
[26] *Ibid.*, 460.
[27] *Ibid.*, 465.
[28] *Ibid.*, 466.
[29] *Ibid.*, vol. IV, 312.
[30] See Chapter 3, for a discussion of this definition and its implications.
[31] Italics mine. Sir Isaac Newton, *Opticks or a Treatise of the Reflections, Refractions, Inflections & Colours of Light* (based on the fourth edition, London, 1730, Dover Publications, 1952), 400.
[32] See Chapter 5.
[33] *Opticks*, 394.
[34] *Ibid.*, 352.
[35] *Ibid.*, 254–269.
[36] *Ibid.*, 255.
[37] *Ibid.*, 262. Henry Power expresses the same conception: 'The Doctrine of Effluxions, their penetrating Natures, their invisible paths, and unsuspected effects, are very considerable, for . . . several Effusions there may be from diverse other Bodies, which invisibly act their parts at any time, and perhaps through any *Medium*: a part of Philosophy but yet in discovery, and will, I fear, prove the Last Leaf to be turned over in the Book of Nature.' (*Experimental Philosophy*, 58.) Power's last sentence is taken directly from Sir Thomas Browne, *Pseudodoxia Epidemica*, edited by Charles Saye (London, 1904), 218–219. It must be noted, however, that like many of their contemporaries, Power and Browne were never entirely unambiguous about observing atoms in principle.
[38] *Opticks*, 267.
[39] *Ibid.*, 340.
[40] This phrase is employed by Newton repeatedly in his work. The greatest and most perceptive of Newton's disciples. Colin Maclaurin, in his *An Account of Sir Isaac Newton's Philosophical Discoveries*, referring to this Newtonian conception as used in the *Opticks* with respect to the 'more hidden parts of nature' observes: 'he endeavours to judge of them [causes], by *analogy*, from what he had found in the greater motions of the system; a way of reasoning that is agreeable to the harmony of things, and to the old maxim ascribed by *Hermes*, and approved by the observation and judgement of the best philosophers. "That what passes in the heavens above is similar and analogous to what passes on earth below"'; *op. cit.*, note 9, 22. As we shall see, this observation accurately reflects the spirit of the third rule.
[41] Joseph Edleston, *Correspondence of Sir Isaac Newton and Professor Cotes* (London, 1850), 155. The Scholium Generale of the second edition of the *Principia Mathematica Philosophia Naturalis* (Cambridge, 1713) reads: 'In hac Philosophia Propositiones deducuntur ex Phaenomenis, & redduntur generales per Inductionem' (484).
[42] He read Johannes Magirus, *Physiologiae Peripateticae Libri Sex cum Commentaris* (Cantabrigiae 1642), and Daniel Stahl, *Axiomata Philosophica sub Titulis XX Comprehensa* (London, 1652). See J. E. McGuire and Martin Tamny, *Certain Philosophical Questions: Newton's Trinity Notebook* (Cambridge University Press, 1983), 3–25, and Richard S. Westfall, 'The Foundations of Newton's Philosophy of Nature', *British Journal for the History of Science* **1** (1962), 172. Newton probably also read Robert Sanderson's *Logic*. See G. L. Huxley, 'Two Newtonian Studies', *Harvard Library Bulletin* **13** (1959), 348–361. In any event coaching in traditional logic was an essential part of the curriculum in Newton's day.

[43] For contexts in which Newton uses this term, see letter to Cotes, *op. cit.*, note 41, 155 and 156, and the Scholium Generale, *Principia* (1713), 484.

[44] *Philosophy, Science and Sense Perception*, 77. Thinkers such as Newton and Descarts, though their philosophies differ in many respects, were concerned more with the way principles are established, and less with 'verifying' them or ascertaining their predictive powers. The Newtonians, however, pointed out that Newton's gravitational theory predicted things not known before, and that its explanatory power covered phenomena hitherto unrelated.

[45] *Principia, op. cit.*, note 39, 484. In the preface to the first edition Newton uses this expression. Newton's use of this verb will be discussed fully in section III below.

[46] *Correspondence of Isaac Newton*, ed. H. W. Turnbull (London, 1961), vol. I, 209 (letter dated 6 July 1672). For Cotes, *Correspondence of Newton and Cotes*, 155 and 156.

[47] Isaac Newton, *The Mathematical Principles of Natural Philosophy*, translated by Benjamin Motte (London, 1729), vol. II, 204.

[48] *Op. cit.*, note 30.

[49] For a full discussion of these early drafts of the Rule, see I. Bernard Cohen's 'Hypotheses in Newton's Philosophy', *Physis Rivista Internazionale di Storia della Scienza* **8** (1966), 163–184, and Chapters 6 and 7.

[50] Vincent Wing, *Astronomia Britannia* (London, 1669). Trinity College Library, N. Q. 18.36.

[51] U. L. C. Add. 4004, Folio 15$^{r}$: 'Ax: 100. Every thing doth *naturally* perserve in y$^{t}$ state in w$^{ch}$ it is unlesse it bee interrupted by some *externall* cause, hence axiome I$^{st}$, & 2$^{d}$, & 3$^{d}$. A body once moved will always keep y$^{e}$ same celerity, quantity & determination of its motion.'

[52] John Herivel, *The Background to Newton's PRINCIPIA: A Study of Newton's Dynamical Researches in the years 1664–84* (Oxford, 1965), 55–64.

[53] *Ibid.*, 9–13. In the early propositions of Book I, *Principia*, Newton operates mathematically with the tangential and inertial components as if they were the same, though he clearly recognized that they are conceptually distinct.

[54] For a somewhat different approach to the problems just discussed see Derek T. Whiteside, 'Newton's Early Thoughts on Planetary Motion: A Fresh Look', *The British Journal for the History of Science* **2** (1964), 117–137, and John Herivel's reply, 'Newton's First Solution to the Problem of Kepler Motion', *BJHS* **2** (1964–65), 350–351. Both scholars tend to underestimate the ontological presuppositions characterized above, which provided the framework for Newton's mathematical analysis of motion.

[55] N. Q. 16.200, Trinity College Library. For an analysis of this draft see Chapter 6; see also section III of the present chapter.

[56] John Passmore, *Philosophical Reasoning* (London, 1961), 38.

[57] Frank E. Manuel, *Isaac Newton, Historian* (Cambridge, 1963), 44.

[58] For a discussion of these thinkers, see A. A. Luce, *Berkeley and Malebranche* (London, 1934); George Berkeley, *Philosophical Commentaries*, ed. A. A. Luce, note to entry 358, 388; Richard H. Popkin, 'Berkeley and Pyrrhonism', *The Review of Metaphysics* **5** (1951), 223–246; Richard H. Popkin, 'Leibniz and the French Sceptics', *Revue Internationale de Philosophie*, nos. 76–77 (1966), 228–247; Richard A. Watson, 'The Breakdown of Cartesian Metaphysics', *Journal of the History of Philosophy* **I** (1963), 177–197; Harry M. Bracken, 'Berkeley and Malebranche on Ideas', *The Modern Schoolman* **41** (1963), 1–15; Anita Dunlevy Fritz, 'Malebranche and the Immaterialism of Berkeley', *Review of Metaphysics* **3** (1949–50), 59–80; Richard A. Watson, *The Downfall of Cartesianism 1673–1717: a Study of Epistemological Issues in Late 17th-Century Cartesianism* (Martinus Nijhoff, The Hague, 1966). The main early anti-Cartesian writers in France were: Simon Foucher, *Critique de la vérité où l'on examine en même-terms une partie des principes de M$^{r}$* Descartes (Paris, 1675); P. D. Hueti, . . . *Censura Philosophiae Cartesianae* (Editio Quarta aucta et emendata, Paris, 1694); J. B. du Hamel, *Astronomia Physicae* . . . (Paris, 1660).

[59] Nicolas Malebranche, *De la Recherche de la Vérité*, ed. Genevieve Lewis (Paris, 1945),

Eclaircissement vi[e]: 'Qu'il est très difficile de prouver qu'il y a des Corps', 24–33. Nicolas Malebranche, *Dialogues on Metaphysics and on Religion*, translated by Morris Ginsberg (London, 1923), 71–75, 167–168. Beatrice K. Rome, *The Philosophy of Malebranche* (Chicago, 1963), chapter VI.

[60] René Descartes, *Philosophical Writings*, translated and edited by G. E. M. Anscombe and P. T. Geach (London, 1966), 202. *The Principles of Philosophy*, from Latin text of 1644. See also (*Œuvres de Descartes*, edited by Charles Adam and Paul Tannery, vol. VIII, Pars Secunda, CCI, 324.

[61] From Descartes, *Philosophical Writings*, 202.

[62] For an interesting critique of Descartes's position see Anthony Kenny, *Descartes: A Study of Philosophy* (Random House, 1968), 214.

[63] See Henry More, *A Collection of Several Philosophical Writings* (London, 1712), 61–113.

[64] Descartes, *Philosophical Writings*, 235.

[65] *Ibid.*, 236. Descartes is quit unambiguous about the unity of matter: 'I do not see how any reasonable man can deny the great advantage of forming ideas about microscopic events (*in minutis corpusculis*), which elude our senses by their mere minuteness, on the pattern of sensibly observed macroscopic events (*in magnis corporibus*), instead of bringing into our explanation some new conception of things wholly dissimilar to sensible objects' (236).

[66] Alan Gewirth, 'Clarness and Distinctness in Descartes' in *Descartes: a Collection of Critical Essays*, ed. Willis Doney (Notre Dame, 1968), 264–277, and Richard Watson, *op. cit.*, note 58 'The Breakdown of Cartesian Metaphysics', 181–184.

[67] Descartes, *Philosophical Writings*, 208.

[68] *Acta Eruditorum*, April 1695, 145–157. Gottfried William Leibniz, *Philosophical Papers and Letters*, translated and edited by Leroy E. Loemker (Chicago, 1956), vol. II, 845, letter to Burcher de Volder (1643–1709) of 23 June 1699. De Volder, among other things, forced Leibniz to formulate some of his clearest statements on substance and the distinction between body and non-extended forces of which it was composed. See also Leibniz's *Specimen Dynamicum* of 1695 (*Philosophical Papers and Letters*, vol. II, 711–738) and *On Nature Itself (Ibid.*, 808–825).

[69] For a discussion of this point with reference to the concept of *phaenomena bene fundata* see H. W. B. Joseph, *Lectures on the Philosophy of Leibniz* (Oxford, 1949), chapter 5. The point is clearly made by Leibniz in the following passage: 'It is to be observed that matter, taken as a complete being is nothing but a collection or what results from it, and that every real collection presupposes simple substances or real unities, and when we consider further what belongs to the nature of these real unities, i.e. perception and its consequences, we are transferred, so to speak, into another world, that is, into the intelligible world of substances, whereas before we were only among the phenomena of the senses'; *New Essays concerning Human Understanding*, translated by A. G. Langley (New York and London, 1896), 428. Also the following passage: 'But strictly speaking, matter is nothing but a phenomenon founded in things, like the rainbow or the parhelion, and all reality belongs only to unities'; C. J. Gerhardt, *Die Philosophischen Schriften von G. W. Leibniz* (Berlin, 1875–90), vol. II, 268.

[70] For a classic discussion of this aspect of Leibniz's thought, see *Leibniz: the Monadology*, translated by Robert Latta (Oxford, 1898), 21–39.

[71] *Philosophical Papers and Letters*, vol. II, 845.

[72] *Ibid.*, p. 838. For a discussion of Leibniz's theory of matter, see Bertrand Russell, *A Critical Exposition of the Philosophy of Leibniz* (Cambridge, 1900), chapters VII and VIII.

[73] *Philosophical Papers and Letters*, vol. II, 843.

[74] I have used the Torino edition of 1958 edited by Paolo Broinghiari, with a preface by Adriano Carugo and Ludovico Geymonat. The *Discorsi* was first published in 1638. See also *Opere, ed. naz.*, vol. VIII, 86.

[75] Galileo seems to mean that various agents can break matter into chemical solutions of a liquid sort. The reference to 'acque Forti' seems puzzling, since it does not affect gold. Perhaps Galileo was thinking of *aqua regis*.

[76] The translation is mainly based on *Dialogues concerning Two New Sciences*, translated by Henry Crew and Alfonso De Salvo (New York, Dover Publications, n. d.), 40–41. I have also consulted Thomas Salisbury's translation, *Mathematical Discourese and Demonstrations touching Two New Sciences* (London, 1665), 33. I am grateful to James MacLachlan of the University of Toronto for calling my attention to this passage.

[77] There is a discussion of this by Ernest Goldbeck: 'Galileis Atomistik und ihre Quellen', *Biblioteca Mathematica Zeitschrift für Geschichte der mathematischen Wissenschaften*, series **3** (Leipzig, 1902), 84–102. The Galilean discussion of 'atoms' began in his *Discorso intorso alle case che stanno in su l'acqua o che in quella si muovono* (Florence, 1612). See *Opere*, vol. IV. The contribution of matter is discussed there almost as an afterthought.

[78] It is difficult to understand how atoms can be infinite in number, finite in size, and countable.

[79] N. Q. 16.200 Trinity College Library; see Draft I in appendix.

[80] *Ibid.*

[81] U. L. C. Add. 4005.15 Folio 81[r–v]. For an analysis of this draft see Chapter 6. See Draft 2 in appendix for Latin.

[82] U. L. C. Adv. b. 39I.402; Draft 3 in appendix.

[83] *Ibid.*, p. 402[a]. The distinction between a 'quality' and a 'property' probably reflects viewing phenomena 'philosophically' as opposed to 'physically' or 'chemically'.

[84] Alexander Koyré, 'Pour une Edition Critique des Œuvres de Newton', *Revue d'Histoire des Sciences* **8** (1955), 19–37.

[85] In his *Wiles Lectures*, delivered in Belfast, May 1965.

[86] *Philosophiae Naturalis Principia Mathematica* (London, 1687), p. 411. Hypothesis II was changed slightly in the third edition; see Alexander Koyré, *Newtonian Studies* (London, 1965), 265–266.

[87] Isaac Newton, *Philosophiae Naturalis Principia Mathematics* (London, 1726), 389.

[88] A. Rupert Hall and Marie Boas Hall, *Unpublished Scientific Papers of Isaac Newton* (Cambridge, 1962), 293.

[89] 'Regulam voco Propositionem ominem quae ex Phaenomenis per argumentum Inductionis (stabilitur) colligitur & cum ysdem congruit.' U. L. C. Add. 3965.13 Folio 420[r].

[90] *Correspondence*, vol. I, 209: 'by deriving it [the theory of light] from Experiments concluding positively & directly'.

[91] *Correspondence of Newton and Cotes*, 155. Newton uses 'infer' and 'deduce' interchangeably in such contexts. As with the logicians of his period the psychological connotations of 'infer' were not distinguished from the 'formal' character of 'deduce'.

[92] *Opticks* (1730), 205.

[93] Isaac Newton, *Optice: sive de Reflexionibus, Refractionibus, Inflexionibus & Coloribus Lucis* (London, 1706), 171. A similar distinction between, and a use of, these verbs is found in Descartes's *Regulae* which Newton probably knew. In Rule XII we find him saying while discussing the properties of magnets: 'Sed insuper ut quaestio sit perfecta, volumus illam omnimo determinari, adeo ut nihil amplius quaeratur, quam id quod *deduci potest ex datis*.' (Italics mine.) *Œuvres de Descartes*, edited by Charles Adam and Paul Tannery (Paris, 1908), vol. X, 431. Also in Rule XII we find: '. . . primo diligenter *colligit* illa omnia quae de hoc lapide habere potest experimenta, ex quibus deinde *deducere* conatur qualis necessaria sit naturam simplicium mixtura ad omnes illos . . .'. There is no doubt that there are abundant scholastic precedents for this usage. Like Newton, Descartes is clearly not using 'deduce' in the sense of 'formally deduce'. It is not unlikely that Newton was influenced by Descartes, since both were more concerned with the mode of establishing principles rather than with their explicit verification. A similar observation has been made of Galileo with respect to establishing principles by E. McMullin. See his Introduction in *Galileo, Man of Science* (Basic Books, New York, 1967), 13.

[94] *Opticks* (1730), 211.

[95] *Optice* (1706), 176.

[96] *Correspondence*, vol. I, 209.

[97] U. L. C. Add. 3970, Folio 479$^{r-v}$.

[98] Newton, *Mathematical Principles* (1729), 392.

[99] *Op. cit.*, note 50, 27. This was written just prior to the first version of Book I of the *Principia*. In February 1692–93 he used the same form of words in writing to Bentley in English: 'That Gravity should be innate, inherent and essential to Matter . . .'. There is no evidence that Newton ever reconsidered his view that *vis insita* was contrary to gravity in this respect. Quotation from the Bentley letter is from I. Bernard Cohen, *Isaac Newton's Papers and Letters on Natural Philosophy* (Cambridge, 1958), 302.

[100] U. L. C. Add. 4004, Folio 15$^{r}$.

[101] *Correspondence of Newton and Cotes*, 154–155.

[102] U. L. C. Add. 3970.13, Folio 420$^{r}$. 'Hypothesin voco opinionem omnem quae nec phaenomenon est, nec demonstratur, nec ex Phaenomenis per Argumentum Inductionis deducitur.' On the justification for Newton's tougher line on hypothesis, see I. Bernard Cohen, 'Hypotheses in Newton's Philosophy', *Physis Rivista Internationale di Storia della Scienza* **8** (1966), 165–184. Also see Chapter 7.

[103] U. L. C. Add. 3965.6, Folio 256$^{r}$. 'Hypoth III. Qualitates Corporum quae intendi et remitti nequent quaeque corporibus omnibus competunt in quibus experimenta instituere licet sunt qualitates corporum universorum. Idem intelligendum est de qualitatibus corporum omnium ejusdem generis. Fundamentum esse videtur Philosophiae totius. Neque enim aliter qualitates corporum insensibilium a qualitatibus sensibilium derivare licet.

'Hypoth IV. Pro novis rerum generibus habenda non sunt quae cum alijs rerum generibus quoad omnes suas notas qualitates congruunt. ['speciebus' is scored out twice and replaced by 'generibus'].

'Hypoth IV. Corpus omne in alternius cujuscunque generis corpos transformari posse, & qualitatum gradus omnes intermedios successive induere.' This was labelled 'Hypothesis III' in the first edition of the *Principia*.

[104] R. Boyle, *Works* (1772), vol. IV, 72.

[105] For two excellent discussions of Boyle and Descartes on analogy, transduction, and hypotheticalism, see Laurens Laudan, 'The Clock Metaphor and Probabilism: The Impact of Descartes on English Methodolgoical Thought, 1650–65', *Annals of Science* **22** (1966), 81–97, and Maurice Mandelbaum, *op. cit.*, note 3, chapter II.

[106] Henry Power, *Experimental Philosophy, In Three Books: Containing New Experiments Microscopical, Mercurial, Magnetical, . . . in avouchment and Illustration of the now famous Atomical Hypothesis* (Introduction by Marie Boas Hall, Johnson Reprint of 1664 Edition, New York and London, 1966). Newton had Power's book in his library.

[107] Ralph Cudworth, *The True Intellectual System of the Universe* (London, 1678), 858.

[108] *Ibid.*, 856.

[109] *Ibid.*, 858–859. Similar arguments are to be found in Henry More's *An Antidote against Atheism*, in *A Collection of Several Philosophical Writings* (London: 1712), 53 and 97.

[110] Hall and Hall, *op. cit.*, note 88, 130–146. I have partially developed this theme in Newton in Chapters 3 and 5.

[111] *Op. cit.*, note 32, and Francis Oakley, 'Christian Theology and the Newtonian Science: The Rise of the Concept of the Laws of Nature', *Church History* **30** (1961), 435–436.

[112] *Dictionary of National Biography* (London, 1888), vol. 18, 271.

[113] Joseph Glanvill, *Sadducismus Triumphatus: or a full and plain Evidence concerning Witches and Apparitions*, fourth edition (London, 1726). The 1681 edition was a re-edition of *Philosophical Considerations touching Witches* of 1666, the fourth edition of which in 1668 was entitled *A Blow at Modern Sadducism. . . .* The quotations used in this paper are unchanged throughout these editions.

[114] *Ibid.*, 24. I do not wish to imply that the phrase 'analogy of Nature' is original with either Glanvill or Newton. But Glanvill's is the earliest English use of the phrase known to me. It can no doubt be located in the countless scholastic manuals used in the seventeenth century.
[115] *Ibid.*, 5.
[116] John Locke, *An Essay concerning Human Understanding* (London, 1690). The quotations used remain the same in editions published later in Locke's lifetime.
[117] John W. Yolton, *Locke and the Compass of Human Understanding: A Selective Commentary on the 'Essay'* (Cambridge University Press, Cambridge, 1970).
[118] Locke, *Essay*, 339.
[119] *Ibid.*, 3.6.12, p. 213. Surprisingly, Locke scholars have scarcely commented on his adherence to the chain of being. A. S. Pringle-Pattison has a short footnote reference to it in his edition of *An Essay concerning Human Understanding* (Oxford, 1924), 341–342. What he says follows almost entirely A. C. Fraser's footnote comment in his edition reproduced by Dover (New York, 1959), vol. II, 380–381. Both fail to see how these ideas provide part of the general framework of Locke's *Essay*, and content themselves with saying that they are 'more comprehensive speculation than is usual with him', Fraser, *op. cit.*, 380.
[120] *Ibid.*, 2.1.15, p. 42.
[121] *Ibid.*, 4.16.12, pp. 339–340.
[122] *Ibid.*, 3.6.12, pp. 213–214.
[123] *Ibid.*, 3.6.12, pp. 213–214. For a different interpretation of Locke's use of analogy-based bypotheses, which places emphasis on his 'hypotheticalism', see Laurens Laudan, 'The Nature and Sources of Locke's Views on Hypotheses', *Journal of the History of Ideas*, **27** (1967), 221–223.
[124] George Cheyne, *Philosophical Principles of Religion: Natural and Revealed* (in two parts, London, 1715), Part I, 50.
[125] See Chapter 5.
[126] Cheyne, *Philosophical Principles*, part II, 46. See also 38–39: 'A Wise Man performs all his Works in Number, Weight, and Measure, and sure *infinite Wisdom, Simplicity*, and *Unity*, must accomplish all it's Works and with the most *Consumate Harmony, Proposition* and *Regularity*, and this in the following parts of this Treatise . . . we shall call the *ANALOGY OF THINGS*'.

These analogical schema were extended into political and social realms in the 1690s. There is no surer indication of the entrenchment of such reasoning in connection with providentialism as a means of justifying political and social events than the poems of Thomas Shadwell: *A Congratulatory Poem on his Highness the Prince of Orange, his coming into England* (London, 1689); and John Theophilus Desaguliers: *The Newtonian System of the World the Best Model of Government: an Allegorical Poem* (London, 1728). Nehemiah Grew in his *Cosmologia Sacra: or a Discourse of the Universe as it is the Creature and Kingdom of God* (London, 1701) also connects nature, religion, government and society by the use of analogy and Providentialism. The New England Divine Jonathan Edwards writing early in the eighteenth century, in developing his theory of reality from Newton's doctrine of gravitation and the Cambridge Platonist doctrines of God and space, makes analogical connections which echo Cheyne. See his *Dissertation on the End For which God Created the World, in Works* (New York, 1881), vol. II, 201f. In *Images and Shadows of Divine Things* (Yale, 1948), the moral and religious worlds are reflected in the physical world. He holds, like Cheyne, that just as the planets are attracted to the sun, so minds are attracted to God by 'moral gravity' or love.
[127] Chyne, *Philosophical Principles*, part I, 42.
[128] *Ibid.*, part II, A3.
[129] *Ibid.*, part I, 36.
[130] *Ibid.*, part II, 103–104. For evidence of Cheyne's cautious attitude towards analogical reasoning see *An Essay on Regimen* (London, 1740), 226. Analogy provides only 'dark hints'

and by it 'we may very readily err in the details, and in the Eliciting and Application of a particular case; but not so readily in the general Law'. The chain of being is discussed by Maclaurin, *Account*, 15–19, in Lockean fashion as background to his discussion of Newton's experimental philosophy: '. . . For tho' all sorts and degrees are equally the objects of philosophical speculation; yet it is from those which are proportional to sense that a philosopher must set out in his enquiries, ascending or descending afterwards as his pursuits may require' (18). The most elaborate use of the doctrine is found in Berkeley's *Siris in The Works of George Berkeley, Bishop of Cloyne*, ed. A. A. Luce and T. E. Jessop (Edinburgh, 1953), vol. V.

131 Sir Isaac Newton, *The Mathematical Principles of Natural Philosophy*, trans. Benjamin Motte (London, 1729), vol. II, 203.

132 *Correspondence*, vol. I, 376.

133 *Opticks* (1730), 125.

134 *Ibid.*, Query 31, 376.

135 *Mathematical Principles* (1729), vol. II, 202. Rule I of the third Book which states: 'We are to admit no more causes of natural things than such as are both true and sufficient to explain their appearances.'

136 See Chapter 7. See also David Kubrin, 'Newton and the Cyclical Cosmos: Providence and the Mechanical Philosophy,' *J. H. I.* **27** (1967), 326–346.

137 Hall and Hall, *op. cit.*, note 86, 341. This passage is from an intended conclusion for the first edition of the *Principia*. It is by no means isolated; a similar passage was added to the closing paragraph of the third Book of the 1713 *Principia*. For an analysis of this and related materials see Chapter 7. In *Les Météores*, Discours VII, *Œuvres*, vol. VVI, 321, Descartes, beginning with exhalations, arrives by a series of transmutations at 'du sang, ou les sauterelles ou choses semblables'.

138 See Chapter 5. Though Newton had a notion of the conservation of matter, with respect to momentum it was not independent of active principles. The former had to be periodically 'recruited'. The sensibility of Newton's thought regarding conservation seems to hark back to the 'biological' thought of such thinkers as Power and Glisson who moved from the material to the immaterial, partly because they had no conscious notion of a 'state' of matter. With respect to both matter and motion Robert Hooke had a relatively clear notion of conservation: 'Both these Powers I take to be the immediate Product of the Omnipotent Creator, and immutable in themselves, without a like command of the same Power; and always to act in a regular and Uniform Geometrical or Mechanical Method; which Method by diligent Observation and curious Scrutiny may by natural and artifical Means be discovered, and, as I conceive, reduced under certain Rules, and Geometrically demonstrated.' Robert Hooke, *Postumous Works* (London, 1705), 73. Hooke derived his ideas on conservation from *Genesis*, pp. 97, 172, 175. For him matter and motion are appearances of one 'Power'.

139 Johs. Lohne, 'Newton's Proof of the Sine Law and his Mathematical Principles of Colors', *Archive for the History of Exact Sciences* **I** (1961), 393.

140 *Correspondence*, vol. I, 366.

141 Lohne, *op. cit.*, 393.

142 See Chapter 7.

143 Is. Newton, *Philosophiae Naturalis Principia Mathematica* (London, 1687), 411. Translation mine.

144 As it stands the significance of the sentence is not easily established; however, in conjunction with the first 'Hypoth IV', namely, 'Things which agree as to all their known qualities with other kinds of things are not to be considered as new kinds of things', we may be able to clarify what Newton was attempting to say. Both sentences seem to refer to analogical reasoning. Thus the 'known qualities' of matter, among others, includes extension, impenetrability, etc., which comprise the known positive analogy. But imperceptible particles operate on the

sensory organs, *i.e.* sight, smells, sounds, etc.: thus they are analogous with perceptible bodies in that they are solid. Given the causal theory of perception generally accepted in his time, through this analogy of effect, Newton could claim to have good grounds for supposing that imperceptible particles have at least other known essential qualities of gross bodies. Therefore they are 'not to be considered as new kinds of things'. This sort of reasoning was employed by Newton in his general argument for the universalizing of gravitation. Similar presuppositions for the argument by analogy from perceptible things is found in Lucretius's *De Rerum Natura*, Book I, lines 599–631: see translation by Cyril Bailey (Oxford, 1924), 46–48. This was a source that Newton knew well. Thus this line of reasoning places Newton's thought regarding transcategorical inference closer to that of Epicurus and Gassendi: see section I above. Nevertheless, this particular type of analogical reasoning would ultimately be justified by Newton through the general metaphysical of analogy described in the last two sections of this paper.

145 Hall and Hall, *op. cit.*, note 86, 138–140, and Newton's intended letter for Hartsoeker of May 1712, U. L. C. Add. 3965.17 fol. $257^{r-v}$. Part of this is published in Chapter 5. The similarity of Newton's views regarding 'immaterial' matter as an entity dependent upon God is remarkably similar to those of Malebranche, *op. cit.*, note 57. For a different view of the unity and inter-dependence of natural objects in relation to Providence, see Locke's *Essay*, 4.6.11.

146 Gowin Knight, *An attempt to demonstrate that all the Phaenomena in Nature may be explained by two simple active Principles Attraction and Repulsion* (London, 1748), Coroll. II, Prop. VII, p. 7. Corollary I states that: 'the Existence, Extension, Impenetrability, Mobility, and *vis inertiae* of matter, are apparently the immediate Effects of God's will' (7).

147 *Philosophy, Science and Sense Perception*, 78. Italics in original.

148 *Ibid.*, 87. Italics in original.

149 *Ibid.*, 88.

150 Alexander Koyré and I. Bernard Cohen, 'Newton and the Leibniz-Clarke Correspondence', *Archives Internationales d'Histoire des Sciences* **15** (1962), 113.

151 *Mathematical Principles* (1729), vol. II, 204.

150 *Principia* (1713), *praefecto*.

153 *Correspondence of Newton and Cotes*, 158–159.

154 Edward Stillingfleet, *Origenes Sacrae, or a Rational Account of the Grounds of Christian Faith* (London, 1680), 425. (The first edition was published in 1662).

155 *Ibid.*, 425.

156 See Chapter 5.

157 But he also conceived micro-particles as having properties different from gross bodies. There is also a strong possibility that Newton may have conceived the particles of his later aethereal medium to be noninertial. In all three editions of the *Principia*, definition I, which is concerned with the notion of the quantity of matter (mass), seems to contrast ponderable bodies with the aether; for Newton says that he has 'no regard in this place to a medium, if any such there is, that freely pervades the interstices between the parts of bodies'. This conjecture is strengthened by Newton's observation in the intended letter to Hartsoeker (*op. cit.*, note 138) of 'a power seated in (an immaterial) a substance in wch bodies move & flote without resistence & which has therefore no *vis inertiae* . . .'. There is furthermore a curious inconsistency in Newton's thought, displayed not only in the difference between these speculations and the third rule, but in the rule itself. In the remarks to the rule he speaks of 'The extension, hardness, impenetrability, and *vis inertiae* of the whole', resulting 'from the extension, hardness, impenetrability, mobility and *vis inertiae* of the parts: and thence we conclude the least particles of all bodies to be all extended, and hard, and impenetrable, and moveable, and endow'd with their *vires inertiae*'. Not only does this commit the fallacy of composition, but the reasoning is evidently circular since the qualities of bodies are held to be known through experience alone. In any event, the possibility of micro-entities different from gross bodies occurred to Newton as early as 1663 in his *Questiones quaedam Philosophiae* (U. L. C. Add. 3996). He states

there four possible positons 'Off y[t] first mater', namely: 'Whither it be mathematicall points: or Mathematicall points & parts: or a simple entity before division indistinct: or individualls i.e. Attomes' (Folio 88[r]). Newton rejects the other positions and argues for Gassendist-like atoms, which he associates with Henry More's notion of 'indiscerpible' parts (Folio 89).

158 U. L. C. Add. 3970. Fol. 338[r].

159 This phrase is not used to indicate that the chain of being and related doctrines are specifically Elizabethan; clearly they are not. These doctrines were vitally transformed in England at the end of the sixteenth century, however, and as I have tried to show were still part of the educated Englishman's view of reality at the end of the seventeenth century.

CHAPTER 3

# BODY AND VOID AND NEWTON'S *DE MUNDI SYSTEMATE*: SOME NEW SOURCES

> Omnis ut est igitur per se natura duabus constitit in rebus: nam corpora sunt et inane.
>
> *Lucretius* Book I

## INTRODUCTION

The manuscripts in the Portsmouth collection amply testify to the fact that Newton from 1686 to the end of his life in 1727 was concerned to strengthen the basic principles of his Natural Philosophy. As is well known both the *Principia* and the *Opticks* went through three editions during this period: each successive edition containing more reflections on many of the leading philosophical and theological questions of the time. Still, many of Newton's speculations never reached the press and have remained in manuscript to be analysed and related to the published material. Some of these documents have been ably discussed by Professor Cohen & Koyré.[1] They have convincingly established that Newton was actively involved, behind the scenes, in the Leibniz-Clarke controversy, and that he also produced many private arguments to rebut the leading Leibnizian charges against his philosophy as developed in the *Opticks* and the *Principia*.

The documents to be presented show, it will be argued, that Leibnizian criticism also caused Newton to contemplate substantial and extensive changes in the introductory section to Book III of the *Principia*. Some of these were to take the form of definitions and discussions of the notions of void space and sensible matter, which together embody a doctrine fundamental to Newton's natural philosophy, and which were intended to be incorporated into the Third Edition. In other documents he was chiefly concerned to clarify the term 'phenomenon'.[2] The main purpose of this article is to put these documents, with a complete analysis, into the public domain. However, Sections VI and VII will be concerned with the nature of the doctrine of the void, as formulated and defended in these definitions, in relation to some of the deeper assumptions of Newton's philosophy of nature.

Since the definitions contained in these documents were intended to be prefatory to Book III I shall, in this section, discuss the main features of the doctrine of void space as developed in the *Principia* and the *Opticks*, taking note of any substantial changes in successive editions. This will serve to present a body of doctrine which, from 1715, Leibniz began to attack, and which Netwon contemplated defending in the form of definitions.

These remarks to Book III begin with a plain reference to the vacuum. Referring to the mathematical principles developed in the First Book Newton says:

These principles are the laws and conditions of certain motions and powers or forces, which chiefly have respect to philosophy; but, lest they should have appeared of themselves dry and barren, I have illustrated them here and there with some philosophical scholiums, giving an account of such things as are of more general nature, and which philosophy seems chiefly to be founded on; such as the density and resistance of bodies, space void of all bodies, and the motion of light and sounds. It remains that, from the same principles, I now demonstrate the frame of the System of the World.[3]

Newton intends, then, to demonstrate the structure of the world using mathematical principles. The demonstrations may be mathematical, but the cardinal ideas are indeed physical ones, 'which philosophy seems chiefly to be founded on'. Thus the conceptual structure of this Book is concerned with the visible world of gross bodies; and the central conception is the mutual interaction of these in an unresisting void. As we shall see the composition of the sub-sensible world is also held to be characterised by these antithetical notions.

The doctrine of the void is mainly discussed and defended in Propositions VI and X. Proposition VI which states

That all bodies gravitate toward every planet; and that the weight of bodies towards any one planet at equal distances from the centre of the planet, are proportional to the quantities of matter which they severally contained – [4]

is discussed extensively by Newton. In the body of his remarks to this proposition he shows, by reference to Corollaries I and II of Proposition XXIV of Book II, where the theory of pendulous oscillations in a vacuum is developed, that "all sorts of heavy bodies descend to the earth from equal heights and at equal times".[5] Newton goes on to state that he tried the experiments with "gold, silver, lead, glass, sand, common salt, wood, water, and wheat" and concluded that:

. . . the quantity of matter in god was to the quantity of matter in wood as the action of the motive force upon all the gold to the action of the same upon the wood: that is, as the weight of one to the weight of the other; and the like happened in other bodies.

He then proceeds to show, by an imaginative use of analogical reasoning that "the nature of gravity towards the planets is the same as towards the earth".

Although the body of this proposition is based on the experimental evidence for the vacuum, the corollaries are directly concerned with the doctrine itself. In the First Edition, Proposition VI has four corollaries; in the Second and Third another is added. In all three Editions Corollary III is concerned with the interstitial vacuum[6] in sensible matter. In the First Edition it reads:

Therefore a vacuum is necessarily granted. For if all spaces were full, the specific gravity of the fluid which fills the region of the air on account of the extreme density of the matter, would fall nothing short of the specific gravity of quicksilver or gold or any other of the most dense bodies. And on that account neither gold, nor any other body, could descend in air; for bodies do not descend in fluids, unless they are specifically heavier than the fluids.[7]

Now in the Second Edition, after an exchange of six letters with Cotes which implicitly criticised the formulation of his theory of the void,[8] Newton changed the first sentence to read "All spaces are not equally full", adding the following sentence at the end "but if the quantity of matter in a given space, can, by any rarefaction, be diminished, what should hinder a diminuition to infinity?".[9] And in order to clarify this, a further corollary numbered IV was added:

If all the solid particles of all bodies are of the same density, and cannot be rarefied without pores then a vacuum is granted. By bodies of the same density, I mean those whose inertia are as their bulks.

Taken together these corollaries are concerned with three distinct but related doctrines. The first is to be found on Corollary III. The argument is not as clearly expressed as it might have been, and the passage would seem to support two alternative interpretations. In the first place, it can be taken to argue from the hypothetical assumption that all spaces whatsoever are entirely filled with a fluid, such that the world becomes an absolute plenum. Then, any given body would be as dense as any other and "the fluid which fills the region of the air" would be equally dense. Thus, since everything would have the same specific gravity, or since the notion of specific gravity would have no application in such a world, it would be impossible for bodies to move: this is patently contrary to experience.[10] The passage cannot be interpreted otherwise. For if, in the second place, it is taken to mean[11] that only the spaces external to bodies are entirely filled, these themselves having void interstices, the argument fails to make sense. According to this interpretation the phrase "would fall nothing short of the specific gravity of Quick-Silver" would imply that the fluid filling external places would be as dense as dense bodies; whereas, it must be denser. On the first interpretation the phrase makes a minimal reference to the assumption that everything is equally dense, which is not only contrary to experience of dense bodies, but is also inconsistent with the doctrine which Newton holds to explain their nature, namely, that difference in density is a result of the comparative ratio between bodies of pores to solid parts. On the second interpretation the phrase would make non-sense of the doctrine itself; for, given that bodies have interstices, the fluid external to them, being destitute of pores, could not be as dense as they without bankrupting the explanation of relative densities embodied in the doctrine.

That the first reading is correct can be seen from two passages in the *Opticks*. In Query 22 of the 1717 edition Newton, arguing against an aethereal fluid of a Cartesian or Leibnizian sort in favour of an aethereal medium, says:

May not Planets and Comets, and all gross Bodies, perform their Motions more freely, and with less resistance in this Aethereal Medium than in any Fluid, which fills all Space adequately without leaving any Pores, and by consequence is much denser than Quick-Silver or Gold?[12]

And in Query 28 where the argument is, in part, concerned to show that resistance is primarily a function of the inertia of the parts of a fluid, he says of the heavens:

If as dense as Quick-Silver, they would not have much less Resistance than Quick-Silver; if absolutely dense, or full of Matter without any *Vacuum*, let the matter be never so subtile and fluid, they would have a greater resistance than Quick-Silver.[13]

Thus we have the contention that any matter whatsoever, devoid of pores, is entirely dense or in the words of Whiston, "the greatest that can be, and the most perfect and absolute, or rather infinite".[14] This provides a basis for a hierarchy of relative densities of different sorts of matter, such as gold and quicksilver, and including the aethereal medium of the last two editions of the *Opticks* in so far as it contains particles of matter.

The second doctrine of the Corollaries of Proposition VI is that the quantity of matter in the world is far less than that usually taken to be the case. This would seem to be the import of the passage added to the end of Corollary III, as a result of the criticism of Cotes. This very view was expressed much earlier by Newton in a document which, as the Halls indicate, was probably intended to follow this corollary.[15] And it would seem likely that Newton had this in mind when he made the change under discussion. It can definitely be assigned to the period of the early 1690's. The document forms part of an elaborate, but unimplemented, revision of the first edition of the *Principia* undertaken in the same period. There exists one earlier draft variant in the Portsmouth Collection, as well as five other sheets containing closely related material.[16] There is not space in this article to discuss these. But they clearly indicate that the corollaries to Proposition VI were intended to have been greater in number and rather different in content from any of the printed versions.[17] They also indicate that Newton was concerned, not many years after the publication of the *Principia*, to work out the full implications of this doctrine.[18]

In the document published by the Halls Newton argues, on the basis of the motions of the planets and comets, that celestial space contains very little mater and "That bodies are more rare by far than is usually believed". This conclusion is supported, for example, in a number of ways: the comparative specific gravity of various sorts of heavy bodies, the fact that liquids, even very dense ones such as Quick-Silver, permeate the densest bodies, and that magnetic effluvia readily propagate through them. Newton then concludes:

One must have recourse to a certain wonderful and exceedingly artificial texture of the particles of bodies by which all bodies, like networks, allow magnetic effluvia and rays of light to pass through them in all directions and offer them a very free passage: and by such an hypothesis the rarity of bodies may be increased at will.

This view of the nature of matter is strikingly similar to that propounded by Fatio De Duillier. Indeed, Newton mentions his "mechanical" explanation of gravity, since it too must be founded on the existence of void space.[19] Although Newton's ideas on matter and void as expressed in this document were not available to Leibniz, those of a similar type were to be found in the 1704 English and 1706 Latin editions of the *Opticks*. For instance the Latin edition, which he certainly read, says of the texture of bodies in Proposition VIII, Book II, Part III: "Atq; hinc quidem id insuper intelligi potest; Corpora multo esse rariora, multoq; plures intra se Meatus habere, quam vulgo existimatur".[20] The supporting arguments are also the same as those of the document.

The third doctrine of the Corollaries to Proposition VI, Book III, asserts that the "Primigenial" particles are of the same density. Newton cannot be taken to include the ultimate particles, when he speculates that "a dimunition (of matter) to infinity" may be possible. Rarefaction and condensation can only be explained, in his view, by supposing that forces operate on these particles causing them to recede from or to approach one another; and hence diminishing or increasing their numbers, and the resulting agglomerations, which have interstices, present in any given void space. As Corollary IV states, the lack of pores in the ultimate particles demands that a vacuum must be granted so as to account for such phenomena. Thus, it is the quantity of sensible and sub-sensible matter that may be diminished, not the ultimate particles which "are of the same density". The definition of inertia embodied in the corollary is the same as that to be found in Definition XV of *De Gratiatione*. There Newton says: "Bodies are denser when their inertia is more intense, and rarer than it is more remiss".[21] Accordingly, from about 1670 inertia becomes the primitive concept in terms of which density and quantity of matter are to be explicated,[22] and in modern terminology Newton believes that the ratio of inertia to bulk is a constant for all "elementary" particles.

The debate over the vacuum was not restricted to the interstitial vacuum. Leibniz also took issue with Newton's assertion of the vacuity of celestial space. Newton's published statements on this are not univocal. These will be examined and the apparent discrepancies explained, in general, by chronology and shift of reference. Proposition X Book III of the Principia is straight forward, stating:

That the motion of the planets in the heavens can be conserved for an exceedingly long time.[23]

The remarks are plainly concerned to show that celestial spaces are void of sensible matter. We find in the Third Edition a paragraph added which appeared in neither of the two earlier editions. Making use of the mathematical machinery of the Scholium to Proposition XXII of Book II, Newton argues that the air two hundred miles above the earth is very rare in comparison with that near the surface. He then says:

In the spaces near the earth the resistance is produced only by the air, exhalations and vapours. When these are carefully exhausted by the air pump from under the receiver heavy bodies fall within the receiver with perfect freedom and without the least sensible resistance. And therefore the celestial regions being perfectly void of air and exhalations, the planets and comets meeting no sensible resistance in those spaces will continue their motions for a very great length of time (diutissime movebuntur).[24]

Essentially the same argument is to be found in Query 20 of the Latin edition of the *Opticks* and in the 1717 English edition. In the English edition it is as follows:

Now if the Resistance in a Vessel well emptied of Air, was but a hundred times less than in the open Air, it would be about a million of times less than in Quick-Silver. But it seems to be much less in such a Vessel, and still much less in the Heavens, at the height of three or four hundred miles from the Earth, or above. For Mr. *Boyle* has shew'd that Air may be rarified above ten thousand times in Vessels of Glass; and the Heavens are much emptier of Air than any Vacuum we can make below.[25]

Newton then computes the rarity of the air up to a height of 228 miles above the earth from the relation of its density to the atmospheric force compressing it. Notice that he claims in the passage from the *Principia* that the celestial regions are "perfectly void of air and exhalations", and in the Query that regions without atmosphere are possible. This seems to imply that the extreme outer regions are very probably absolutely void. However, already in the Second Edition of 1713 Newton had begun to put some matter into space. Reference is made to Proposition XL of Book II at the beginning of the basic Proposition X of the Third Edition. There, in an addition to a long Scholium, made in the Second Edition and repeated in the Third, we find:

And therefore the celestial spaces, through which the globes of the planets and comets are continually passing towards all parts, with the utmost freedom, and without the least sensible diminuition of their motions must be utterly void of any corporeal fluid, excepting perhaps, some extremely rare vapours and the rays of light.[26]

Also in Query 28 of the 1717 Edition of the *Opticks* we find:

And therefore to make way for the regular and lasting Motions of the Planets and Comets, it is necessary to empty the Heavens of all Matter, except perhaps some thin Vapours, Steams, or Effluvia, arising from the Atmospheres of the Earth, Planets, and Comets, and from such an exceedingly rare Aethereal Medium as we described above.[27]

Thus we have an apparent incompatibility between the perfect celestial vacuum of the addition to Proposition X in the Third Edition of the *Principia* and of the first passage cited from Query 28, and the imperfect vacuum of Proposition XL and of Query 28 with respect to the second quotation from it. There is the further difficulty that the addition to Proposition X concerning the perfect vacuum occurs some time after Newton had begun to entertain the possibility of an aethereal medium of differential density in celestial spaces.

These questions are best answered if we first consider, in general outline, the development of Newton's thought on the void and the aether in the early years of the Eighteenth Century. There is no decisive evidence to indicate whether or not he accepted some sort of aethereal hypotheses in the period from 1692 to 1713. In this period he seems to have seen the world as being comprised of matter and void, the former depending for its existence and motion on the ultimate efficacy of God's will. In the years prior to the republishing of the *Opticks* in English,[28] he began again to consider a "mechanical" aether: hence, the Queries of the 1706 edition had to be changed, in part, to accommodate this development in his thought. Accordingly Query 20 of that edition was enriched with a mention of "such an exceedingly rare Aethereal Medium . . .". And if spaces can contain a "rare" Aethereal Medium, it is not surprising that they should also contain vapours and exhalations.

The restriction of "vapours and exhalations" to the *solar* system can finally remove the apparent incompatibility. Newton is not clear in his use of the term *celestial*, applying it indifferently to regions within and without the solar system. From the passage quoted from Query 28 it is clear that he is supposing the existence of vapours, exhalations and effluvia, in the atmosphere of the *Solar system*, 'arising from . . . the Earth, Planets, and Comets'. Thus, since Proposition X and the first quotation from Query 28 state that the farther above the earth the rarer becomes the amount of matter, by implication, eventually regions far beyond it and entirely void will be reached.[29] So it seems that only the spaces of the solar system contain some matter. And the aethereal medium being 'elastick', very 'rare' and presumably diffusible, by virtue of its very 'subtile' particles, through the great distances of this inner system is supposed by Newton to be of little hindrance to the passage of dense bodies. Only fluid media of the Leibnizian variety would produce such an effect. Still, as I shall show, the Newtonian Aethereal Medium *is* incompatible with the possibility of a perfect vacuum in the outer celestial regions, and involves other difficulties as well.

Thus it is clear that in both the *Principia* and the *Opticks* Newton accepts the interstitial and celestial vacuum. Acceptance of this latter doctrine was restricted to Newton and his immediate circle of disciples. Even Huygens, an adherent of the atomic philosophy, did not postulate a *Vacuum Separatum*. Criticising Newton's aether for not being able in his terms to account for the action of gravity or of light he says in the *Addition* to the 1690 edition of the *Discours de la Cause de la Pesanteur*:

Pour examiner donc ce point, je dis que la matiere etherée peut etre censée rare de deux manieres, sçavoir ou que ses particles soient distantes entre elles, avec beaucoup de vuide entre deux; ou qu'elles se touchent, mais que le tissu de chaunne soit rare, & entre-meslé de beaucoup de petits espaces vuides. Pour ce qui est du vuide, je l'admets sans difficulté, & meme je le crois necessaire pour le mouvement des petits corpuscules entre eux. N'estant point du sentiment de Mr. Des Cartes, qui veut que la seule étendu fasse l'essence du corps; mais y adjoutant encore la dureté parfaite, qui le rende impenetrable, & incapable d'estre rompu ni écorné. Cependant à confiderer la rareté de la premiere, je ne vois pas comment alors on pourroit rendre raison de la Pesanteur: & quant à la Lumiere, il me semble entierement impossible, avec de tels vuides, d'expliquer sa prodigieuse vitesse, . . . C'est pourquoy je tiens qu'une telle rareté ne sçauroit convenir aux espaces celestes.[30]

Thus only the existence of tiny interstices within sensible bodies is acceptable to Huygens.

## II

The various facets of Newton's doctrine of the void as expressed in his major published writings have been discussed. It remains to consider the nature of Leibniz's opposition to this doctrine as developed in the controversy with Clarke.

As early as 1689 in the *Tentamen de Motuum Celestium Causis*[31] Leibniz implicitly attacked the void. Probably provoked by the sustained attack against the vortex theory in Book II of the *Principia*,[32] he attempted to show that Keplerian concepts are applicable in a world full of matter. Thus, the *Tentamen* can be seen as an attempt to sustain the plenum against void spaces. From 1710 his general criticisms of Newton's natural philosophy, as shown in the *Théodicée*, become explicit. Still, at this stage there is no definite attack on the vacuum as such. The first public attacks on the doctrine occur in an exchange of three letters between N. Hartsoeker and leibniz in the *Memoires de Trevoux* for February 1711, and in a letter published by P. Rémond in August 1713.[33] Sustained criticism, however, only begins in the *Leibniz-Clarke Correspondence*,[34] and falls into three categories. In the first place there are arguments of a theological nature based on the "principle of the best". Those of a philosophical character are based on the principle of sufficient reason, and the view that void space constitutes, if it exists, an attribute without a subject. Leibniz also holds, in general, that the atomic philosophy embodies a shallow and incomplete metaphysics. Lastly, there are arguments of a "quasi-experimental" nature such as the view that there is matter void of heaviness and that viscosity is the important factor in the resistance of fluids.

It is only in his second letter to Clarke towards the end of November 1715 that Leibniz launches his first attack on the vacuum. It is here that he introduces the principle of sufficient reason, the implications of which had been fully discussed earlier in the *Théodicée*.[35] Leibniz claims that from this principle (together with the principle of contradiction) the whole of theology and natural philosophy can be justified. The following passage plainly shows that he interprets the Newtonian doctrine as holding that the world contains a great paucity of matter:

. . . LA MATIERE EST LA PARTIE LA MOINS CONSIDERABLE DE L'UNIVERS: c'est qu'il admet outre la Matiere, un espace vuide, et que selon luy la matiere n'occupe qu'une trés petite partie de l'espace. Mais Democrite et Epicure ont soutenu la même chose, excepté qu'ils differoient en cela de M. Newton du plus au moins; et que peut étre selon eux, il y avoit plus de matiere dans le monde, que selon M. Newton. En quoy je crois qu'ils étoient preferables; car plus il y a de la matiere, plus y-a-t-il de l'occasion à Dieu d'exercer sa sagesse et sa puissance.[36]

The argument is strictly a theological one, grounded on the principle of "la perfection de Dieu", and not on the principle of sufficient reason if that is taken to mean that God must always have a motive for acting in one way rather than in another. For the objection to the vacuum is couched in terms of God's wisdom and power; if there is a vacuum in nature there is less matter and less reality, and hence there is a limitation in God's power of creation and this in turn constitutes a violation of his infinite perfection.

In his second reply of December 1715 Clarke rebutted this by pointing out that:

How small soever the QUANTITY of matter be, God has not at all the LESS SUBJECT to exercise his WISDOM and POWER upon: For, Other things, as well as MATTER, are equally subjects on which God exercises his POWER & WISDOM.[37]

Leibniz in his next letter of the 25th February 1716 simply denies that there are things other than matter on which God can exercise his power; "car je tiens que toute substance cree est accompagnée de matiere".[38] On the 15th May Clarke replies that the "determinate quantity of matter" now in the world was the "most convenient" for the "present state of things".[39] To this Leibniz replies that there is no possible reason to limit the quantity of matter for "il faudra tousjours y adjouter quelque chose, pour agir suivant le principe de la perfection des operations Divines".[40] Thus the present constitution of things cannot be the best possible. Now in his fourth reply of the 26th June, Clarke changed the character of the exchange on the vacuum by introducing experimental evidence. This was probably in response to a letter which Leibniz sent to Caroline on June 2nd and in which he says:

Je ne crois pas qu'il ya ait aucun espace sans matiere exclus des experiences qu'on appelle du vuide, n'excluant qu'une matiere grossiere qu'on tire de la cavité du cerre par le poids du vif argent avec Torricelli, et par la pompe avec M. Guerike. Car les rayons de la lumiere, qui ne sont point sans quaelque matiere subtile passent à travers de verre.[41]

Clarke points out that though rays of light and other matter in small quantities may be in a receiver, "the WANT of RESISTANCE plainly shows that the GREATEST PART of THAT SPACE is void of MATTER. For SUBTLENESS or FINENESS of MATTER, cannot be the cause of WANT of RESISTANCE".[42] He also argues that quicksilver is as subtle as water but yet makes "more than ten times the resistance". Therefore resistance arises from the quantity of matter not its grossness.

Leibniz's reply to this, on August 18th, attempts to argue that in the exhausted receiver there are beams of light, magnetic effluvia and other matter "non pesante et qui ne resiste point sensiblement".[43] To this Clarke replies on October 29th that the notion of matter void of gravity is absurd and that resistance is proportional to the quantity of matter of inertia. To Leibniz's claim that it is not the density of matter but the difficulty of "giving place" of the liquid which causes resistance to moving bodies, Clarke replies that with such fluids as water and quicksilver (the resistance of which is in question) there is little tenacity, and hence the only important factor is their quantity of matter.[44]

The exchange concerning the void ends with Clarke's defence in October 1716. In his edition of this *Correspondence* published in 1717,[45] he added as an appendix to Leibniz's fourth paper a postscript to a letter which Leibniz had written to Caroline on the 12th May 1716. It is almost entirely devoted to repudiating the vacuum. Leibniz begins by saying:

Tous ceux qui sont pour le Vuide, se laissent plus mener par l'imagination que par la raison. Quand j'étois jeune garçon, je donnay aussi dans le vuide et dans les Atomes; ⟨mais la raison me ramena⟩.[46]

He then argues that to admit the vacuum "c'est attributer a Dieu une Production tres imparfaite", pointing out that as a principle every perfection that God could create in the natural order, "sans deroger aux autres imperfections", has been created. "Or figurons-nous un Espace vuide, Dieu y pouvoit mettre quelque matiere sans deroger en rien à toutes les autre choses. Donc il l'y a mise. Donc il n'y a point de d'Espace entirement vuide: Donc tout est plein". Moreover, he claims that the same argument can prove that there are no atoms. Leibniz next outlines a specific argument against the void based on the principle of sufficient reason and the notion that matter is more perfect than the void; namely, that if there were a vacuum there would not be a sufficient reason for determining the proportion of matter to void space; space being homogeneous, why should God create matter here rather than there?

Such, then, are the Leibnizian objections to the vacuum. It might be thought that he is entirely misguided in attempting to confute a purely empirical statement – the existence of a vacuum – with predominantly a priori statements about the world being a plenum. Yet, as will become clear, he had some justification for regarding the Newtonian expression of the atomic theory as a metaphysical system. And it is clear that Newton did so as well. For, as we shall see, he recognised an irreducibly metaphysical substratum supporting his world-view at its deepest level. This he chose to minimize, mainly for polemical reasons, in the Definitions to be discussed.

## III

The Definitions will now be given in the order in which they were probably written. In the Appendix the Latin transcriptions will be found. These successive draft variants show clearly the development of Newton's ideas. They also indicate that he debated the manner and form of their presentation. His practice of using square brackets to mark off material to be omitted when the document would be copied will be followed; angle brackets will be employed to indicate my additions, interpolations or glosses to the text in the interest of clarity; and finally words or phrases scored out or interlineated by Newton will be italicised.

*Draft No. 1* (1 p. Fol. 644r). ULC. Add 3965. 17.[47]

Judging from its lack of organisation and its incompleteness this draft probably represents an early, perhaps the earliest, attempt on the part of Newton to formulate his ideas on the notion of void and body. There are no definitions of these terms as there are in later draft variants. It is a catalogue of ideas, developed in a rather random fashion, which is given shape in the later documents. The page is heavily scored and has many interlineations. It would seem that the material follows on from a separate sheet which is no longer to be found in the manuscript remains.

Mathematical solids are not perceived by touching nor cause a resistance, nor are they usually called bodies . . . The Quintessence is different from the four elements and is subject to none of the senses nor can it numbered among phenomena. Prime matter which is neither a thing nor possessed of quality nor a thing that can be measured is not a phenomenon. The subtle matter by which the heavens are filled ⟨an unintelligible place is interpolated here⟩ is not a phenomenon. And things which are not phenomena, have no place in experimental philosophy. Inductive argument taken from experiments *taken* and the observations of sensible things on which experimental philosophy is based, cannot be applied to hypothetical or metaphysical entities *except by means of hypotheses it ought not to be extended* in as much as they are phenomena; therefore the things which are said in this book concerning bodies *by the power of induction* bears no reference to entities of this sort. Here we are concerned only with *phenomena* sensible things *and their parts, the other things are more properly treated of in metaphysics and hypothetical philosophy*. ⟨Above this last phrase is written another which is heavily scored out and quite unintelligible.⟩ *Because in these things only the argument of induction has place.* The other things which cannot be perceived, but are termed bodies hypothetically by some people, these things are more properly treated of in metaphysics and hypothetical philosophy. We must begin from phenomena. Experimental philosophy exists in treating of such things. From this philosophy to *efficient and final causes and* to hypothetical philosophy men must proceed. The solid orbs in which the planets inhere *are not phenomena*. The subtle matter in which flow bodies *move without resistance* is not a phenomenon. *And things which are not phenomena* ⟨The rest of the sentence, which is not complete, is too heavily scored out to be intelligible.⟩

And just as geometers define a line as that which has length without breadth so that *their propositions* concerning lines of this sort are in this way understood and *however a line having breadth* in mechanics and other sciences *however* a line *having breadth* has a place: thus body *and vacuum* are here defined [*not in order that we deny that other bodies exist but in order that we may show in what sense these words are to be understood in what follows. The propositions which follow are understood of bodies of this kind. About other bodies let authors in other sciences dispute.*] So that these words may be understood in the sense *defined* in what follows. About other sorts of bodies and other sorts of void let authors in other sciences dispute.

## *Draft No. 2* (1 p. Fol. 437v) ULC. Add. 3965. 13

In this draft the ideas take on more definite shape. Definitions of the terms body and vacuum are given. The interlineations and phrases struck out indicate that Newton's views are still developing; nor is he sure, yet, of their final order of presentation. Here a list of things is given that are taken to exemplify the notion of a body. In the discussion of this term there are to be found some new ideas not in Draft I. The remarks to the definition of vacuum on this sheet are interesting and entirely different from what is to be found in later drafts. The notion of exhalations and vapours is here introduced. The sheet is badly torn in the left-hand bottom corner so that some of the words supplied are necessarily conjectural. These will be indicated in the Latin appendix.

*Definition I*

Body I call everything tangible in which there is a resistance to tangible things, and whose action *resistance*, if it is great enough, can be perceived.

It is indeed in this sense that the common people always accept the word body. And of this sort are *The Earth, Planets, Comets*, metals, stones, sand, clay, mud, *earth*, salts, wood, bones, flesh, water, oil, milk, blood, air, wind, smoke, exhalations, flames, and whatever can be included under the four elements, and flow from them by exhalations and return to them by condensation: *I add the stars and the heavenly bodies. The Planets and Comets act* ⟨written above this last phrase is the following.⟩ These emit and reflect light and are weighed down by their constituent parts *and are numbered among phenomena and in their motions observe the laws of bodies*. [Mathematical solids are not perceived by touching nor cause a resistance nor are they usually called bodies.] Vapours and exhalations on account of Their rarity lose almost all perceptible resistance, and in the common acceptance often lose even the name of bodies and are called spirits. And yet they can be called bodies in so far as *they are the effluvia of bodies and* have a resistance proportional to density. [But if the effluvia of bodies were to change thus in respect of their forms so that they are to lose *all* power of resisting, *and cease to be numbered among the phenomena*, these I would no longer call bodies: for I speak with the common people.]

*Definition II*

Vacuum I call every place in which a body is able to move without resistance.

Thus it is called by the common people. If any man should contend that bodies are given, which by touch are neither felt nor cause a resistance, this man is now disputing the grammatical significance of the word by calling things bodies which are generally not called such by people: and I should prefer to go with the crowd *who at least are capable of giving things names*. Things which by touching press together and by pressing together act on other things, and things which do not press together and, since they do not press together, do not act on other things, thus they can be mutually distinguished: so that they may truly be called bodies;[48] but let these other bodies be called intangible things or things ⟨rest of this phrase is missing⟩ *or insensible matter or* let them be designated by another name, and the common type of body let it rejoice in a third name such as substance, or being, or that which acts, or ⟨rest of this phrase is missing.⟩ Thus that ambiguity of words will be avoided from ⟨a phrase is missing⟩ which ever arguments of ideas are accustomed to arise and since it is disputed about bodies ⟨a phrase is missing⟩ as phenomena but ⟨because of a torn corner the rest of the remarks cannot be made out.⟩

## *Draft No. 3* (1 p. Fol. 422r) ULC. Add. 3965. 13

On this sheet the definitions of body and vacuum are substantially the same as in Draft 2: however, they are now numbered "*Definito II*" and "*Defintio III*". On the verso side of the sheet there is a definition of the term 'phenomenon' marked "Definitio I". The significance of this will be discussed later. In the top left-hand corner of the sheet we find written in Newton's hand "pag. 359". This is the pagination of the Second Edition of the *Principia*. It is plain then that these definitions were intended to occupy a place in the Third Book just after the *Regulae*, displacing Phenomena I and II to the following page. This draft appears to be a more polished version of Draft I; moreover, some of the ideas are more fully developed. For instance, the comments on the definition of body are more extensive than in either Draft I or 2. For this reason and for the following it is a probably a later version. In later drafts the stars, planets and comets no longer appear as examples of material bodies. Again, the sentence "I add the heavenly bodies", and the criteria of materiality *viz.*, bodies emitting and reflecting light, and their parts coming together by the force of gravity, is interlineated in Draft 2 and is not present at all in Draft I. But a more important reason is this. In Draft 1 Newton does not wish to deny that there are other sorts of bodies; but he is concerned to indicate the sense that the word will have within the framework of his *Systemate Mundi.* In Draft 2 he develops this notion by attempting to formulate criteria for distinguishing other sorts of bodies from material bodies. Also implicit in Draft 1 is the view that body and void are correlatives, one being tangible and the other not. This view is explicitly developed in Draft 3 and in later draft variants; whereas the criteria for distinguishing types of bodies found in Draft 2 is not. Therefore Draft 3 is probably a later variant than either Draft 1 or 2.

*Definition II*

Body I call everything which can be moved and touched, in which there is resistance to tangible things, and its resistance, if it is great enough, can be perceived.

It is indeed in this sense that the common people always accept the word body. And of this sort are metals, stones, sand, clay, mud, earth, salts, wood, bones, flesh, water, oil, milk, blood, air, wind, smoke, exhalations, flames, and whatever can be included under the four elements or flow from them by exhalation and return to them by means of condensation.[49] I add the heavenly bodies. These emit and reflect light and are numbered among phenomena, and are weighed down by their parts pressing heavily upon them, and assume a round shape, and in their motions observe the laws of bodies. Vapours and exhalations on account of their rarity lose almost all perceptible resistance and in the common acceptance often lose even the name of bodies and are called spirits. And yet they can be called bodies in so far as they are effluvia of bodies and *have a force* of resistance *have* proportional to their density. Mathematical solids are not *perceived* do not move by touching *nor cause a resistance*, nor are they usually called bodies. [The quintessence is different from the four elements, and is subject[50] to none of the senses nor is it numbered among phenomena.[51] The prime matter which is neither a thing, nor 'possessed of quality, nor a thing which can be measured, is not a phenomenon. The solid orbits in which the planets inhere are not phenomena. The subtle matter in which the *planets* float, and *in which* bodies move without resistance is not a phenomenon.[52] And what are not phenomena, and subject to none of the senses, have no place in experimental philosophy. Inductive argument taken from experiments and the observations of sensible things on which experimental philosophy is based cannot be applied except by means of hypotheses to hypothetical or metaphysical entities. Therefore, the things which are said in this book about bodies through the power of induction bear no reference to entities of this sort. Here we are concerned only with sensible things and their parts, for it is in these things alone that the inductive argument has its place. The other things which cannot be perceived, but yet are hypothetically termed bodies by some people, these things are more properly treated of in metaphysics and hypothetical philosophy. *We must begin from phenomena* ⟨The following phrase is written above the last.⟩ *Philosophy begins with phenomena*. Experimental philosophy consists is treating of such things. One must pass from *experimental* philosophy to the efficient and final causes of *things* and *from all these to the nature of imperceptible things and finally* to hypothetical philosophy. At the beginning of the first book I have defined the quantity of matter so that it may be treated in mathematical terms; here I have defined body composed of such matter in order that it may be treated in physical terms.

*Definition III*

Vacuum I call every place in which bodies are able to move without resistance.

It is in this sense that the common people understand the significance of the word. But it follows from the first Definition *that vacuum is that which is not a tangible thing and does not impede the motion of bodies*. For just as geometers define a line that has length without breadth, so that their propositions concerning lines of this sort are only understood, and in mechanics, however, and other sciences the line having breadth has a place; thus body and vacuum are here defined so that these words may be understood in the sense defined in what follows. About other sorts of bodies and other sorts of void let authors in other sciences dispute.

## *Draft No. 4* (1 p. Fol. 430r) ULC. Add. 3965. 13

There is plain indication in this draft that Newton considered presenting his views in another form: the sheet is entitled "Scholium" and was intended as a redaction of Page 360 of the Second Edition which contains "*Phaenomena*" III and IV. This material is interesting. For it connects the new ideas intended for Book III with basic concepts, such as inerta and mass, from the First Book. Thus we have in one document a terse philosophical presentation of Newton's leading doctrines concerning the nature of bodies in the physical world. This draft was probably written after No. 3. Newton has purposely left a space in it for the list of material bodies and the criteria of materiality to be filled in when the document would be copied. The material which follows on after this indication is the same in form and content as the remark supporting the definition of vacuum in Draft 3. It is bracketed and a line drawn through it. This usually meant that such material was not to be incorporated into the final copy. In this case, however, it would seem that Newton intended that the material not be suppressed – for it appears in the final draft – but that it should follow on after the next section which summarises the remainder of the remarks on the term body of Draft 3.

*Scholium*

In the first definition of the first book I *defined I described what is* the quantity of matter in any given body, and said that throughout I understood this quantity by the *name* body or mass, and that in the meanwhile I have no account here of the medium, if there is any such thing, freely pervading the interstices of the parts. And in the third Definition *I said* that the force of inertia was proportional to the body and was innate and essential, and that this force is the power of resistance by which every body *attempts to stay preserves* in its state of rest, or moving uniformly in a right line. By body I understand everything tangible, in which there is a resistance to bodies touching it, and whose resistance, if it is sufficiently great, can be perceived. For it is in this sense that the common people always understand the word body. And of this type[53] . . . *especially because* [and finally we must pass to *philosophy* hypotheses. Therefore I understand the word body *in this proposition* in no other sense than the one which I have defined. And vacuum I call all space which is destitute of bodies of this kind. Just as geometers define a line as that which has length but no breadth, so that their propositions concerning lines of this sort are only understood; and in mechanics, however, and in other sciences, the line has breadth: so body and vacuum are here defined so that they may be understood in the sense defined in *this proposition and in the following of this book*. But let others who, wander beyond the limits of experimental philosophy, *authors in other sciences* discuss other bodies and other sorts of void][54] ⟨there is a large space left in the manuscript here: Newton begins again with a dash⟩ – Because the inductive argument has place in these things alone. Therefore *the things which are here affirmed* by the power of induction about the universal gravity of bodies, *here affirmed* have no bearing on the quintessence, and first matter, and solid orbs, and subtle matter, which are not phenomena. [Here I am speaking of bodies which are *truly touched and* truly have the *power* of inertia and by means of this power truly cause resistance to tangible things.][55] *And these are not the bodies of which I am speaking here*. These things which are not perceived, but are hypothetically named bodies by some people, must be more properly dealt with in metaphysics and hypothetical philosophy. Philosophy begins with phenomena through the inductive argument. Experimental philosophy consists in dealing with these things. It is necessary to pass from experimental philosophy to the *final and* efficient causes of things, and from all these to the nature of imperceptible things and finally to hypothetical philosophy.

## *Draft No. 5* (1 p. 540r) Add. 3965. 13

This sheet contains Definitions of body and vacuum which are the same as those of the earlier drafts. Accordingly I shall not give a translation, but the Latin will be found in the Appendix. The amplificatory remarks appended to the definition of body are considerably shorter than those of Draft 3, not including any of the ideas which come after the reference to 'mathematical solids' in that Draft. The definition of vacuum and the supporting remarks are, however, the same as in Drafts 3 and 4. Thus Newton has formulated his ideas in final form. At the top of the sheet is written 'Corrigenda et addenda in Lib. III Princip.' The Definitions were to occupy a place between Regula I and Regula II on the revised Page 357 of the Second Edition. In any event there is a line drawn through both Definitions on the sheet: and they do not reappear in subsequent revision sequences of the Third Book.

We are now in a position to attempt to date these documents. The most obvious fact is that the pagination on some of the manuscript sheets matches that of the Second Edition: thus, the intended emendations were made after 1713. Yet, we cannot assume without question that they were written sometime just prior to the publication of the Third Edition in 1726. For it can be established, I think, that Newton had ample motivation for formulating these Definitions earlier.

Since Koyré & Cohen have shown that the extent of Newton's involvement in the development of the Leibniz-Clarke controversy was considerable, and since the doctrine of the void is fundamental to his thought, it is reasonable to assume that he would have been especially aware of any Leibnizian criticism of this theory.

If successive exchanges on the void are compared, it becomes apparent that they converge to a critical point, the most detailed and extensive criticism from Leibniz occurring in the fifth letter of August 18th 1716.[56] Here he argues, as we have seen, from what he takes to be empirical evidence. There can be little doubt that Newton thought these criticisms to be wide of the mark, the result of insufficient physical insight. But Leibniz twice repeats[57] in his letter the same criticism that he had made earlier in the "Apostille" to Conti; namely, that the notion of atoms and the void is an *hypothesis* based on an "easy" and "superficial" philosophy, the result of "limited" metaphysical views.

This "Apostille" was sent with a letter to Conti about the 6th December 1715, and embodies a general attack on Newton's natural philosophy.[58] The Leibnizian letter to Caroline, which is similar in character, was sent at the same time, and Clarke's response began the controversy on natural philosophy. In preparing a reply to Leibniz's Apostille to be sent via Conti, Newton wrote no less than twelve separate drafts, in two of which he mentions that Leibniz denies the conclusion that the vacuum exists without rebutting the premises from which he (Newton) argues.[59] Thus the implication is that the charge of having a shallow metaphysics is irrelevant criticism. The letter actually sent to Leibniz on the 26th February 1716 was shorter than most of these drafts. It strongly states, however, that any doctrine Newton would defend in natural philosophy was based on "arguments of induction drawn from experiments".[60]

Yet, not only were the same "philosophical" arguments used against the doctrine of the void in the *Correspondence*, but, as we have seen, in the "postscript" of Leibniz's letter to Caroline on the 12th May 1716, – wherein he made the arrogant assertion that he accepted atoms and the void as a young man – the Newtonian void is again strongly criticised on philosophical grounds. We have manuscript evidence to show that Newton received this postscript from Caroline and that he copied it out in full.[61] Its tone, style and character must have annoyed him beyond measure.

Thus this philosophical thrust along with the nature of the criticism of atoms and void in the fifth letter, probably caused Newton to draft the Definitions some time towards the end of 1716, during the last stage of the controversy or just after it. As we shall see this interpretation is supported by the nature of the remarks to the Definitions themselves, and by other related documents which record Newton's opinion of Leibniz's theory of matter.

## IV

It is not time to discuss the Definitions themselves attempting, where possible, to relate them to the body of published writings. The first thing to notice, on considering the various formulations, is that the terms body and void are interdefined: they are, in fact, treated as correlative terms. What one is, the other is not. Body is tangible, impenetrable, movable, that which acts and offers resistance; whereas, void space has contrary qualities: intangibility, penetrability, immovability, passivity, and hence offers no resistance.

There are a great number of treatises on natural philosophy written in England during the seventeenth century which, as a result of accepting one or other of the many expressions of the mechanical philosophy, either define or use body and void as correlative terms. The origin of this formulation is, of course, to be found in such ancient writers Democritus, Epicurus and later Lucretius, all of whom had throughout this period a decided influence on the character and development of English natural philosophy. Thus neither Newton's definitional treatment of the terms, nor the philosophical view which lies behind it is original. It was he, however, who first translated this view, in part, into an effective natural philosophy of the perceptible world resting on an impressive basis in experience. It will be important, then, to consider historically some of the influences that may have led Newton to this commitment.

Many of his problems and characteristic doctrines were formulated in his early years towards the end of, and just after his undergraduate days, when his creative intellect burned intensely. This is certainly so with regard to his theory of sensible matter, which then remained essentially the same throughtout successive refinements of his philosophical position. It is therefore, to this early period in his intellectual development that we must turn in order to establish influences on the formulation of his views. Apart from its intrinsic interest this investigation will provide part of the necessary background for an analysis of the Definitions themselves.

There are two key texts; the *Quaestiones quaedam Philosophicae*, entries which were made about 1664, Newton's last year as an undergraduate; and *De Gravitatione*, written no later than 1670.[62] These works are by no means equivalent: the setting and purpose of each is different. But they do deal with many of the same questions and in the same spirit. Professor Westfall has argued, convincingly, that the Gassendist influence on Newton in the *Quaestiones* is more decisive than the Cartesian.[63] This influence was probably through the mediation of Charleton's *Physiologia*,[64] a work saturated with Gassendist ideas, there being no direct evidence that Newton had read either Gassendi's *Syntagma Philosophicum* or the earlier *Philosophicae Epicurae Syntagma* as an undergraduate.[65] The nature and character of the theory of void space developed in *De Gravitatione*, however, strongly suggests that Newton by this time had read the *Syntagma* and that he had also been influenced by the writings of Henry More.

In both these early texts it is evident that Newton accepts the existence of void spaces and the antithetical character of the terms 'body' and 'void'. In each case he uses the traditional arguments against the plenum as found in Charleton and Gassendi.[66] There are, furthermore, terminological similarities in the expression of these arguments. In the *Quaestiones* we find reference to the interstitial vacuum made by such phrases as 'vacuities interspersed' and bodies 'divided by interspersed inanities'.[67] Also phrases such 'vacuum disseminatum', 'vacuo inter . . . corpuscula disseminato' and 'extramundana spatia' support the view that Gassendi influenced Newton's discussion, in *De Gravitatione*, of motion in non-resisting media. It must be noted, however, that by this time Newton is using these phrases primarily as a means of describing his own theories and, unlike Gassendi, is in no way interested to formulate the distinction between the *vacuum separatum*, the *vacuum disseminatum*, and the *vacuum coacervatum* as an end in itself.

There is, however, an important difference in the two works with respect to the Newtonian commitment to the void. In the *Quaestiones* he only affirms the existence of void interstices in matter. There is no evidence that he had come to accept a complete celestial void.[68] In *De Gravitatione*, arguing against the Cartesian identification of matter with space, he holds space entirely devoid of matter to be a distinct physical possibility; moreover, with regard to movements in the heavens he says:

> If there were any aerial or aethereal space of such a kind that it yielded without resistance to the motion of comets or any other projectiles I should believe that it was utterly void.[69]

Newton is well on his way to regarding the celestial void as a fact of nature.

In both works the contrariety of void and body is accepted, though in *De Gravitatione* it is more explicit than in the earlier work. There it emerges from newton's discussion of the motions of bodies, the assumption being that that in which they move must be opposite in nature.[70] Charleton, discussing 'inanity' in the context of motion, says:

> Epicurus names it a *region, or space*, and *a Nature that cannot be touched:* thereby intimating the direct contrariety betwixt the essential notion of Corporiety and Inanity: which antithesis *Lucretius* plainly expresseth in that verse, *Tactus Corporibus cunctis intactus Inani.*[71]

Since some of the discussions of motion in the *Quaestiones* display an anti-Aristotelian bias, it is possible that passages such as the above remained fixed in Newton's mind from his reading of Charleton's *Physiologia.*

With *De Gravitatione* the discussion of space and matter becomes at once more philosophical and theological. One of the central aims is to bring out the categorical difference between these notions. Such passages as the following from the *Syntagma* may well have influenced Newton's view regarding the difference between perceptible matter and void space:

: . . . cum circa Reum divisionem memoravimus Epicurum, & cum ipso Lucretium Universum seu naturam rerum in Corpus, Inanéque partiri, Exinde enim habetur corpus ita ab Inani distingui, ut cum Inane sit molis expers, intactile, incapax actionis, passionis, resistentiae; Corpus sit mole praeditum, tactile, actionis, passionis, resisteniae capax.

And at the end of the same paragraph Gassendi says:

ex horque est illa apud Lucretium Antithesis, qua convenire dicitur *Tactus corporibus cunctis, intactus Inani.*[72]

Unmistakably, then, Gassendi holds the view that body and void are correlative terms. There is an elaborate development of this in Books II and III of the *Syntagma.* Throughout, Gassendi's analysis is set within a broad sceptical framework. We cannot have certain knowledge of matter; its real nature is beyond understanding. Experience alone gives us the properties of matter, and also shows that it is essentially different from space. If matter is finite, there must exist something greater than itself which it occupies; this is void space infinite in extent.

While in *De Gravitatione* Newton largely accepts the sceptical epistemology of Gassendi regarding the phenomenal world,[73] it is More who most probably influenced the general principles of his account of the creation of matter. In terms of an imaginative hypothesis Newton suggests that it is possible that God by an act of will could have endowed empty spaces with the sensible properties of matter such as impenetrability.[74] In appearance it would be no different from what it is. The essential point so far as Newton is concerned, however, is its contingent nature. Its existence is utterly dependent on the will of God. This is made clear from the following passage on body:

For certainly whatever cannot exist independently of God cannot be truly understood independently of the Idea of God. God is no less present in his creatures than they are present in the accidents, so that created substance, whether you consider its degree of dependence or its degree of reality, is of an intermediate nature between God and accident. And hence the idea of it no less involves the concept of God than the idea of accident involves the concept of created substance.[75]

Thus in comparison with the necessary existence of Space and God, matter is an entirely impoverished being embodying the lowest degree of reality and perfection. Hence the 'downgrading' of matter is clearly a stated assumption of Newton's position and plays an important role in his discussion. As

More puts it, "From the highest life viz, the Deity, there does result that which has no life or sense at all, to wit, the stupid matter".[76] These themes recur often in the *Immortality of the Soul*, a work which Newton certainly read.[77] That he still accorded matter the lowest degree of reality and perfection in his system some four decades later, just prior to the *Correspondence*, is attested by a passage from a letter of Conti to Rémond, extracts from which were sent to Leibniz. In the course of a summary of Newton's philosophy, based on actual conversation, Conti says:

> L'Espace n'a aucum changement comme les corps; et comme il contient les corps, qu'il est le lieu des corps, qu'il est eternel, qu'il est immense, il a bien plus de perfections que les corps. Si vous demandez ce qu'il y a dans les espaces vuides de matiere, les newtonistes vous repondent, que comme l'espace est une proprieté de la divinité, il y a tout ce qui accompagne la divinité.[78]

Thus Newton's account of the creation of matter vividly dramatises the lowly status that it has in his natural philosophy. Moreover, the later stress on its paucity in favour of the void on the part of Newton and his disciples, is yet another expression at a different level of this theologically orientated world view originating in the writings of More and the Cambridge platonists. A still more fundamental expression of this attitude is involved in the doctrine of matter's passivity. There is sufficient evidence to show that Newton plumped for 'infinitely' hard bodies as the basic constituents of the actual world. If matter is inert and hard in the Newtonian sense,[79] this entails that it has the property of non-elasticity. Immediately this precludes the Leibnizian solution of dynamical activity in the world, since this view entails that the defining characteristic of ultimate matter is *elasticity*. Clearly this theory allows matter from the theological point of view to be more independent of God, activity being in some sense inherent in it. Thus not only would a force inherent in bodies and conceived as *vis viva* open the door for atheism in Newton's view,[80] but it would also grant matter more reality and perfection.[81] Both consequences are inimical to his theology and ontology; and probably explain in a general way, apart from technical objections, his unsympathetic attitude toward Leibniz's dynamics.

The major intellectual forces which probably led Newton, by 1670, to an acceptance of void space and its antithesis to matter have been indicated. It will now be helpful to consider briefly this commitment as it develops from the time of the First Edition of the *Principia* to the time of the Definitions. From what has been said it is plain that Newton's correlative use of the terms void and body presupposed a definite view of the world, and thus was not merely a matter of definitional convenience. Hence an attempt to combine the scattered elements of the conceptual background relating to these Definitions, will help to characterise the deeper substrata supporting the natural philosophy of the Third Book. Also such a reconstruction may indicate the inter-connections which these elements had for Newton when he drafted the Definitions.

The question that now needs clarification is this: in what sense does Newton think that body and void are correlative terms? I must first indicate by example two basically different senses in which terms may be described as correlatives. The terms 'husband' and 'wife' illustrate the strongest possible relation. If the term 'husband' has significant application, then it necessarily follows that the term 'wife' has as well, and conversely. With the terms 'hot' and 'cold' the relation is weaker. If it is significant to say that something is hot, then the term cold has significance and conversely. This does not entail, however, that nothing can be hot but that something else is cold; but only that if the term 'hot' has application then it is possible that the term 'cold' has as well, and conversely.[82] Thus it is clear that the stronger relation entails the weaker: the converse, however, does not hold. Some important controversies in the history of natural philosophy have centred on the nature of correlative terms. For example, the Peripatetic philosophy treated gravity – levity' as correlatives in the stronger sense. The 'corpuscularian' opposition of the seventeenth century admitted the symmetry of significance, but denied the existence of 'levity' altogether.

I shall argue that in the Third Book, Newton uses the terms 'body' and 'void', as applied to the sensible world, in a sense which admits a symmetrical relation of significance, namely, the weaker relation. At a deeper level, however, his view of the connection between void space and matter does not accord with the way the terms in either example have application. My procedure will be to consider some characteristic physical doctrines which use these concepts, to analyse the philosophical assumptions which lie behind them, and to indicate some of the empirical facts which the entire 'theoretical' framework was devised to explain.

We have seen that the doctrine that differences in weight and density are to be explained by the ratio of interstices to solid parts in matter, and the view that planetary motions are impossible in fluid media, vortically structured, both presuppose that as a matter of fact vacuities exist in nature: thus no matter without void. This position is brought out more forcibly by the following two physical doctrines.

For Newton it is an indisputable fact that solar bodies move along their orbits as a result of mutual gravitation towards one another and toward the sun. Dynamically it is equally incontestable that this action and reaction is analysable into an inertial and a centripetal component. Clarke concluding a summary of this view, says of bodies which satisfy the above conditions:

> . . that the space betwixt them is void, that is, hath nothing in it which sensibly resists the Motion of Bodies passing transversally through: All this, is nothing but Phaenomenon or actual matter of fact, found by experience.[83]

But Newton also holds it to be an 'actual matter of fact' that if the solar bodies move according to Keplerian laws which in turn follow demonstrably from the law of gravity as it mathematically describes the effects of such a force, this force:

> . . . operates not according to the quantity of the surfaces of the particles upon which it acts (as mechanical causes used to do) but according to the gravity of the solid matter which they contain . . .

and thus

> penetrates to the very centres of the sun and planets, without suffering the least diminuition of its force.[84]

Newton certainly thought in the early 1690's that it was possible to give a mechanical explanation of gravity along the lines of Fatio's hypothesis. This involved the postulation of a great number of vacuities in matter.[85] But it seems that the mechanical aether of the 1717 edition of the *Opticks*, in order to give an account of the operation of gravity within bodies, is also committed to this hypothesis. There is, however, a grave difficulty, which probably accounts in part for the fact that Newton abandoned Fatio's hypothesis, and that he may not have taken his own aethereal medium seriously. For if gravity is as the total quantity of matter of any solid body, large or small, then it must be possible for its action to pass entirely through such bodies. But it is plain that such action cannot be explained by the hypothesis that bodies are extremely rare, since it is the operation of gravity on *solid* matter that is in question. Nevertheless this seems to be the only hypothesis available to Newton as a means of explaining how his mechanical and material aether is able to act on bodies. Thus in 1716 when he drafted the Definition of void space and was entertaining an aethereal medium as a means of explaining gravity, he may still have thought that such an hypothesis was plausible in principle. Accordingly some hypothesis involving the existence of voids *in* bodies is to be granted so as to account for, in terms of an aethereal medium, the action of gravity upon them.

Again from the demonstrations of Book II and III it is plain that only in spaces void of resistance would planetary bodies falling under Newton's laws move in elliptical orbits. These laws were formulated in terms of precise and well attested Newtonian concepts: the proportionality of inertia to gravity and of both to the quantity of matter. Thus the total conception of discrete bodies or particles interacting with one another according to a definite function of the distance separating them, presupposes the near vacuity of the intervening spaces. Any vortical fluid, such as the Cartesian or Leibnizian, as Newton shows in Section IX[86] of Book II, would be incompatible with the only system that accounted for them dynamically – the Newtonian. Newton in his criticism of the *Tentamen* which draws on this Section of the *Principia*, puts the matter unambiguously.

Quae de Vorticibus dicuntur sunt merè hypothetica, & cum motu Cometarum conciliari non possunt, neque quadrant cum Planetarum temporibus periodicis quae sunt in ratione sesquiplicata distantiarum ab orbium centro communi.[87]

So much for the triumph of Newtonian dynamics with regard to the perceptible world. But Newton's own tentative explanation of the cause of gravitational attraction, in 1717, by means of an 'aethereal medium' was bedevilled by conceptual tension. Detailed discussion of this issue is beyond the scope of the present paper, but since the cosmological implications of this aether affect the doctrine of celestial voids, the following difficulties must be mentioned. Given the Newtonian account of differential densities discussed in Section I, and given that the aether is material in nature, how can the theory that the aether is "much rarer within the dense bodies of the Sun, Stars, Planets, and Comets, than in the empty celestial spaces between them? And in passing from them to great distances", grows "denser and denser perpetually?"[88] be reconciled with the view that matter in these spaces becomes rarer and rarer as distances increase? Secondly, so as not to hinder the passage of planets Newton must also hold that the gross aether has an extremely small density. How then can it possibly transmit gravitational influences, especially when other sorts of vibrations, such as sound, must be 'propagated through the solid, pellucid and uniform Capillamenta' of the nerves?[89]

Leaving these special difficulties aside, we can find other theoretical considerations which will throw light on Newton's position regarding these correlatives. The first is his theory of force. Newton often states that gravity is a real physical force in nature:[90] and he denies, moreover, that it is inherent in matter. It is therefore something of an independent nature which operates on bodies, its properties in no way being reducible to them. For matter can only react to such a force, never originate it. So, at least from the time of the *Principia* Newton began to think of the world in terms of a richer ontology than that of matter in motion.[91]

This can be further clarified in contrast to the Cartesian view of force, Descartes tended to assimilate the concept of force to the description of relative changes of position between bits of matter. Such changes for Newton could be no more than effects, the manner in which force are acting, which can be treated mathematically in terms of magnitude, direction, and quantity of matter. Although he would agree with Descartes that forces can be dealt with only in so far as their effects are observable, this in no way reflects, as it did for Descartes, what he meant by the term. In short, Newton's was not a positivistic notion of force, but a realistic one.

What implications does such a view of force have for his theory of void space and matter? Since forces are real existents, they must be physically present somewhere. Being of a different ontological type from matter they are neither assimilatible to it nor one of its essential attributes, nor in any sense inherent in it The only alternative for Newton was to suppose that they operate on inert matter in void space. Thus the doctrine of a hierarchy of forces ranging from the subsensible to the sensible world requires void space as a physically real existent.

The same point is made, at a more philosophical level, with the theory of absolute space. I shall not deal with the doctrine as presented in the *Principia* but turn to *De Gravitatione* where the essential features of the position are to be seen clearly. There, as we have noted, Newton strongly objects to the Cartesian theory of matter. Now this position takes the occupancy of space, that is voluminousness or geometrical solidity, to be the essential attribute of matter. Thus Descartes regarded empty space without any matter in it as a demonstrable impossibility. For being spatial was a logically sufficient as well as necessary condition of being material.[92] Plainly, then, matter and space are identical: equally plainly Newton repudiates this and the theory of knowledge which supports it. For he argues in the manner of Gassendi,[93] Charleton and More,[94] that neither space nor time can be comprehended under the traditional categories of substance and attribute. As Charleton puts it: "We say *more general than these two* because as well all substances or accidents whatever, have both their existence in some place and their duration in some time."[95] Thus in the Newtonian view void space, far from being a mere conceptual entity, has the most general mode of existence: "For it has its own manner of existence which fits neither substances nor accidents".[96]

Newton also holds that void space, being an unconditioned entity,[97] is ontologically more basic than its material content. Consider the following passages from *De Gravitatione*: "But it is usually believed that these spaces are nothing: yet indeed they are true spaces. Although space may be empty of body, nevertheless it is not in itself void; and *something* is there, because spaces are there, although nothing more than that."[98] And a few pages earlier we find the following:

> Moreover, since we can clearly conceive extension existing any subject, as when we imagined spaces outside the world or places empty of body, and we believe (extension) to exist wherever we imagine there are no bodies, and we cannot believe that it would perish with body if God should annihilate a body, it follows that (extension) does not exist as an accident inherent in some subject.[99]

Thus Newton following Gassendi, Charleton and More is arguing that void space, which is incorporated, is a physical reality. Moreover space, being independent of body, could exist even if body were destroyed. The latter, however, essentially depends for its existence on the former. Therefore void space is a fundamental ontological category, a necessary condition for the existence of bodies *qua* extended things.

From the foregoing analysis it is clear that Newton could not have regarded the terms body and void as correlatives in the strong sense but only in the weak sense in which the significance of terms is mutually presupposed. This means that just as something can be hot without something being cold, so it is possible that something could fall under the concept of void space without something falling under the concept of matter. There is an important difference in the two cases however. In the first case something may be hot without something being cold and conversely; but according to the basic principles of Newton's ontology this symmetry does not hold with respect to the void and matter. For something could always fall under the concept of void space whether or not matter existed.

Thus at this level, since void space is the most general and basic category, it cannot be for Newton a correlative of matter. He allows, of course, that the concept cannot be formed independently of experiencing matter. Since in the *actual* contingent world Newton holds that certain empirical facts can only be explained by the physical doctrines discussed, the terms are correlatives in the weak sense as applied to this world by means of these doctrines. Thus with respect to the composition of concrete things it is contingently the case that void does not exist part from matter. Matter is matter only with reference to some void, and void is void only with reference to some matter.

## V

"About other sorts of bodies and other sorts of void let authors in other sciences dispute." Who are these writers from whose views on matter Newton wishes to disassociated himself? A necessary preliminary to answering this question is a consideration of his conception of sensible matter with respect to its causal properties. An integral part of this conception is the view that bodies act upon one another and by implication on the sensory organs.[100] It is clear from the various formulations in the Definitions that his aim is to stress the perceptible and tangible aspects of matter. Furthermore, he is concerned with a small but central sub-class of tactile qualities, namely, those of hardness and resistance to touch. These qualities have as an objective counterpart in the physical world solid and tangible bodies which can interact with one another.

There is sufficient evidence to show that Newton formulated these conceptions as an undergraduate at Cambridge. For example, in the *Quaestiones* he says:

> The nature of things is more securely & naturally deduced from their operations once upon another and upon yr senses. And when by ye former experiments we have found ye nature of bodys, by ye latter wee may more clearely find ye nature of yr senses.[101]

Later in *De Gravitatione* the same view is developed in a more epistemological context. Again it is emphasied that bodies are phenomena, only in so far as they act upon one another and upon the senses. Substance is "an entity that can act upon things".[102] Mere extension cannot do this since contrary to the opinions of Descartes, solidity and the power of resistance are not derivable from it.

It can thus be readily seen that Newton's views remained essentially the same from his student days to the early period of his more mature thought; moreover, as the Definitions testify, it remained unchanged throughout his life. The authors whom he read as a student, and who probably influenced him in this respect, held essentially the same doctrines. For example Digby, to whom he refers in the *Quaestiones*, says in the opening page of his *Two Treatises*:

> . . . when a man would distinguish a corporeal substance from a spiritual one (which is accounted indivisible) he naturally pitcheth upon our apprehension of its having bulk, and being solid, tangible, and apt to make impression upon our outward senses: according to that expression of Lucretius, who studying nature . . . telleth us *Tangere enim & Tangi, nisi Corpus nulla potest res.*[103]

Similar passages are to be found in the writings of More, Charleton and Gasssendi. And Gassendi and Charleton also single out this Lucretian phrase as epitomising their view of sensible matter.[104] Although it is nowhere quoted by Newton it nicely characterises his point of view.

There are four separate pieces of evidence which indicate that Leibniz was primarily the person whose views of matter Newton repudiated. In nearly all the draft variants to the letter sent to Conti in 1716 he charges Leibniz with changing the meanings of words so as to criticise his views on gravity, the nature of the soul, on hypotheses, and on miracles. In a letter actually sent to Conti he says: "As to philosophy, he colludes in a signification of words, calling those things miracles which create no wonder; and those things occult qualities, whose causes are occult".[105] Now in the document under discussion Newton is concerned that it be perfectly understood in what sense he understands the terms body and void. In Definition II of Draft No. 2 he puts the matter plainly by stating: "If any man should contend that bodies are given which by touch are neither felt nor cause a resistance, this man is now disputing the grammatical significance of the word . . .". Newton's aim is to ensure "that ambiguity of words will be avoided from whichever arguments of ideas are accustomed to arise". Since the character of the criticism here is the same as that made against Leibniz in the letter to Conti, it would seem likely that the Leibnizian view of matter is the primary target in these passages.

This suspicion is confirmed. For in one of the variants of the letter to Conti Newton says: "He sayeth that I have not demonstrated a vacuum nor universal gravity nor atoms". Later in the same document we find:

> He of opinion that space void of tangible body may be full of a corporeal intangible fluid whereas the Ancients believe that all things intangible were incorporeal. I understand tangibility not in a mathematical sense but in a physical sense, such a tangibility as by some resistance (can) affects the sense of touching.[106]

This is precisely the doctrine that Newton is defending in the Definitions and consequently rejecting the notion that an intangible but corporeal fluid is a sort of matter. And he does so in much the same manner as in the latter part of the above passage.

From two further sources there is evidence of a more conclusive sort, that Newton has in mind not only the Cartesians, but also Leibniz and Rémond, a follower of Leibniz. First there is an unpublished set of notes by Newton defending his natural philosophy against criticism by Rémond.[107] This attack, originally a letter to N. Bernoulli, was printed in the Second Edition of Rémond's *Essai d' Analyse sur les Jeux de Hazard* in 1713.[108] There a rebuttal of the Newtonian philosophy is directed against the recently published Second Edition of the *Principia*. Among other things Rémond is concerned with method, the theory of attraction, and the doctrine of the void, wishing to show that all phenomena can be explained in terms of matter and motion alone.

In these elaborate notes which were probably written in 1714, and the basis of an unpublished letter by Keill,[109] Newton takes issue with Rémond on twelve separate points numbered as such. The sixth point is concerned to contest the assertion that Cartesianism is a 'Metaphysicam sanam' and that everything can be explained in terms of moving bodies: the latter view is also attributed to Leibniz earlier in the draft. However in an earlier set of notes[110] for the letter, it is clear that Newton intended to examine Leibniz's metaphysics in greater detail. In these he reiterates time and again that experimental philosophy, being based on phenomena, is to be distinguished from metaphysics. He then proceeds to ask what is to be understood by metaphysics:

> By a healthy metaphysics he (Rémond) understands the Cartesian in so far as it proceeds from innate ideas to the existence of things. But let us hear what Leibniz himself has to say about true metaphysics.[111]

Than Leibniz's *De Primae Philosophiae Emendatione & de Notione Substantiae* which appeared in the *Acta Eruditorum*[112] for March 1694, is given its full title followed by a quotation from the article which would have comprised nearly half its published length. Newton begins his criticism of both metaphysical views by saying:

> Metaphysics certainly is founded on ideas, ideas and all true philosophy are founded on phenomena & Newton begins from phenomena. Momortus contends that ideas must not be deduced from phenomena but formed with closed eyes *and from dreams of this kind.* He contends further that we must proceed from ideas to phenomena the eyes being opened again so that we may by the help of ideas see the phenomena clearly and distinctly.[113]

Thus in so far as we are doing natural philosophy, says Newton ironically, we must begin from the phenomena with our eyes open. This is what both Descartes and Leibniz failed to do in Newton's view.

But it is the content of the article, from which Newton intended to quote, that is important. There Leibniz makes a plea for greater exactness in the defining and use of metaphysical notions. One of the main motives is to underline Descartes' shortcomings as a metaphysician: he failed to attain certainty, and adhered to absurd doctrines such as "putting the nature of corporeal substances in extension". Leibniz then outlines his own theory of force and the essence of matter. This is activity which "seems not to have been perceived clearly by those who have found the essence of bodies to be in extension, alone or together with the addition of impenetrability".[114]

If Newton had had to choose between the Leibnizian or Cartesian metaphysics of matter, it is probable that he would have found the latter more congenial. After all the Cartesian system fitted together in a coherent way; whereas the Leibnizian dynamics, especially since it depended for its complete meaningfullness on a deep and difficult framework of metaphysical ideas not easy to grasp systematically, seemed to Newton, at least, to have no validity for natural philosophy. Although Leibniz agreed with Newton in rejecting the Cartesian theory of matter, he replaced it by non-extended substances, characterised by substantial forms and internal principles of change. Ultimately all of this was embedded in a system of 'windowless' monads in pre-established harmony, a doctrine Newton disparages on many occasions.[115]

Since both sets of notes were written about the same period as the Definitions and for the same reason, a defence of his natural philosophy, it would seem clear that Leibniz, his followers, and the Cartesians are those authors who speak of "other bodies". Moreover Leibniz's view in *de Primae Philosophiae*, that metaphysical concepts are "hidden from popular understanding" and must be defined clearly would constitute, for Newton, another example that he "colludes in the signification of words".

The other manuscript source has been published by Cohen & Koyré.[116] It comprises two fragments criticising Leibniz's doctrine of motion and force. Newton shows himself to be highly familiar with Leibnizian writings on the subject. So as to comment on the views that he is criticising, he refers to a series of articles published by Leibniz in the *Acta*: "The grounds of his error is that he confounds Acta Erudit: ad Ann. 1686 pag. 162; & ad Ann. 1690 pag. 234; & ad Ann. 1691 pag. 439; & ad Ann. 1695 pag. 155 . . .".[117] The last reference is the most important for the problem under discussion. For in 1695 Leibniz published the first part of his *Specimen Dynamicum*[118] which is the most mature and extensive discussion of his theories of force and matter. Again Descartes is criticised, and, by implication, the view of matter that Newton holds: "I found a proof that something more than magnitude and impenetrability must be assumed in body, from which an interpretation may arise". Thus in the light of the above evidence it is plain that, apart from what was said in the *Correspondence*, Newton had detailed and first-hand knowledge of Leibniz's theory of matter, and that it was he whom he mainly had in mind as someone who merely disputed the significance of words and who held a 'metaphysical' theory of matter.

VI

The following two questions now demand consideration. We require to know what Newton understands by the term 'phenomenon', and in the light of this to decide how far he is advancing in his remarks criteria whereby various intellectual disciplines may be classified. As was pointed out above, ordinary terrestrial bodies are phenomena because they have certain causal properties, offer resistance, and can be described in terms of the four elements in connection with the operations of various agencies. But the heavenly bodies are phenomena because their parts come together by the power of gravity into spherical masses, they either emit or reflect light, and they are subsumable under the laws of motion. In a short unfinished treatise entitled *Cosmography* Newton says:

The Universe consists of three sorts of great bodies, Fixed Stars, Planets and Comets, and all these have a gravitating power tending towards them by which their parts fall down to each of them after the same manner as stones and other parts of the earth do here towards the earth and by means of this gravity it is that they are all spherical.[119]

He goes on to point out that the fixed stars and the sun shine "strongly with their own heat"; whereas the earth & planets shine "by his light reflected from them". Furthermore all bodies move according to the laws of motion except the fixed stars. Thus we have in this passage the same criteria of materiality applied to heavenly bodies as in the Definitions. And Newton may well have drawn on this treatise when drafting them.

Furthermore we have in the remarks to the Definitions a classification of matter in the visible world into terrestrial and heavenly bodies. It must not be supposed, however, that Newton is maintaining an ontological distinction in the manner of the Aristotelian tradition; rather, he is concerned to emphasise the different ways by which we come to know through the senses the properties of two classes of physical things. Terrestrial bodies can be handled, felt to offer resistance, and be experimented with. Heavenly bodies, on the other hand, are not amendable to such procedures. Both sorts of bodies, however, fall under the other criteria. Indeed Newton's whole assumption is that in terms of these criteria we are justified in analogically transferring concepts which apply to terrestrial bodies to the heavenly ones.

Although it is patently obvious from Newton's writings that he held phenomena to be known only through sensory experience, nevertheless, because of what he took to be persistent misunderstandings of his position, he felt constrained to define the notion. This he does on three separate manuscript sheets.[120] Since the other formulations do not differ substantially from the one to be discussed, which is on the reverse side of the sheet containing the Definitions of Draft No. 3, special mention of them will not be made. As these Definitions were an intended revision of Page 359 of the Second Edition, the Definition of Phenomena, if the plan had been carried out, would probably have occupied Page 360. It is as follows:

*Definition I*

Phenomena I call *whatever can be seen and is perceptible* whatever things can be perceived, either things external which become known by the five senses, or things internal which we contemplate in our minds by thinking. As fire is hot and water is wet, *and gold is heavy, and sun is light,* I am and I think. All these are sensible things and can be called phenomena in a wide sense; but those things are *properly called* phenomena which can be seen, but I understand the word in a wider sense.[121]

Apart from the fact that Definition I makes no reference to the uncertainty of hypotheses, and mentions, as example, only one sort of mental activity presupposing personal existence, it is the same in content as the suppressed Fifth Rule of reasoning published by Professor Koyré.[122] The most striking thing about this Definition is the fact that it indicates an acceptance of the Cartesian 'Cogito' argument. In few other places in Newton's writings do we find such positive Cartesian influence. Since Rule V similarly employs this argument it is surprising that Professor Koyré makes no mention of it in his analysis. He is right, however, to suggest that it is a manifestation of 'Lockean' empiricism against the *apriorism* of the Continent. The striking similarity between Definition I and Rule V, argues that they were probably written in the same period; and having the same polemical tone, they would seem to be closely related to the Definitions of body and void and Rule IV of the Third Edition. Thus whether they concern Definitions, Rules or Scholia, *all* of these documents were written with the intention of presenting a systematic reply, in the *Principia*, to Leibniz's current attacks on his natural philosophy. Thus the Fifth Rule, which was only one of the forms into which Newton considered casting his ideas on phenomena, is not primarily anti-Cartesian in spirit as Koyré suggests.

The remarks to the Definitions[123] constitute one of the clearest expressions of what Newton considered to be the task and nature of experimental philosophy. If he had decided to incorporate them in the Third Edition along with the projected Fifth Rule they would have set it and the Fourth Rule into clearer relief. It is noticeable that he repeatedly states that his definition of sensible matter is in accord with the views of the 'common people'. Clearly this is a way of reaffirming that the subject matter of the Third Book is the material objects of everyday experience and, by implication, that the *Principia* is not concerned with metaphysics. It is pain, nevertheless, that he is not ruling out metaphysics, but claiming that such considerations do not find a proper place in the realm of experimental philosophy as he conceives it. Neither is he claiming that purely philosophical speculation is to be discredited. For Newton could not deny that such speculation has a role to play in the consideration of ultimate questions. Rather he is at pains to point out that such speculation should not be thought to constitute arguments of the same sort as apply to phenomena, or be thought to repudiate their results.

Consequently the remarks are also concerned to distinguish both the subject matter and types of argument to be found in such disciplines as experimental philosophy, mathematics, and metaphysics. It would seem that Newton is labouring obvious distinctions in pointing out that the arguments of experimental science are basically inductive in character and, as such, have no rightful place in either mathematics or metaphysics. Moreover, he emphasizes it by referring to the fact that different subject matter demands different arguments. Still these remarks have some weight if considered in a wider context.

In an interesting passage from one of the draft variants of the *Scholium Generale* we read:

> From phenomena we know the properties of things, and from the properties we infer that the things themselves exist and we call them substances: but we do not have any idea of substances. We see but shapes and colours of bodies, we hear but sounds, we touch but external surfaces, we smell odours and taste flavours; but we know the substances or essences themselves by no sense, by no reflex action and therefore we have no more idea of them than a blind man has of colours. And when it is said that we have an idea of God or an idea of body, nothing other is to be understood than that we have an idea of the properties or attributes of God or an idea of the properties by which bodies are distinguished from God or from each other.[124]

This passage in conjunction with the remarks on the Definition of body shows that Newton, within the framework of his natural philosophy, is distinguishing perceptible and imperceptible physical bodies from such metaphysical entities as substance. In the passage just quoted he is concerned with the proper object of each sense. Colours, sounds, odours, and tastes are the objects of immediate perception; and it is by abstracting from these phenomenal properties that experimental philosophy begins.

In contrast to perceptible bodies, however, the primordial particles are clearly not phenomena: nonetheless Newton never doubted their existence. Still the problem – as the Third Rule plainly illustrates – was to establish the nature of such bodies. "Without the Third Rule it cannot be proved that all matter is impenetrable".[125] There was no other course open to him but to employ analogical arguments. Such arguments, needless to say, were anchored in direct experience, basing themselves on perceptible characteristics. On the other hand, though from the properties of phenomena we might construct metaphysical notions congruent with them, this would provide no clear grounds that they stood for anything in nature. Similarly from the natural order the existence of God can be inferred. But for Newton his complete nature is only revealed, if at all, in the scriptures by exegetical argument and the testimony of the prophets.[126]

Yet another version of the Fourth Rule, quite different from those discussed by Professor Koyré, is specifically concerned with the nature of arguments in different disciplines.

In experimental philosophy *objections* Hypotheses ought not to be allowed to oppose arguments gathered from experiments by induction, in order that *this is to say* experimental philosophy may not be confused with hypothetical. *Geometrical demonstrations are universal.* Arguments based on induction *they are stronger than hypotheses* are not demonstrations, yet they are stronger than hypotheses; and ought to be reckoned general, except in so far as exceptions gathered from experiment may be met with. Accordingly when no objections of this kind are met with they ought to be stated as of general application.[127]

This draft is less polemical than both the published Rule and the other manuscript redactions. Newton's aim is to characterise the nature of inductive arguments: they need not be general but, in so far as there are no exceptions, they can be held to be so. Plainly they are not to be confused with hypothetical arguments – those of the metaphysics and the natural philosophy of Leibniz – or with mathematical demonstrations: the former are a weaker type of reasoning, the latter are a stronger.

Thus various disciplines have distinctive subject matter and consequently develop arguments appropriate to their needs. Experimental philosophy is no exception. In Newton's view, however, his critics had failed to appreciate this. For this reason he is concerned to say that 'metaphysical and hypothetical entities' are 'more properly treated in metaphysics and hypothetical philosophy'.

In the closing sentences of some of the above documents in Section III, Newton suggests that speculative thought, whether it is metaphysical or theoretical, ought to base itself on a sure foundation of experimental philosophy. This is in character with many passages in which he sees his views as providing a basis for natural theology. For instance in Query 28 he says:

Whereas the main Business of Natural Philosophy is to argue from Phenomena without feigning Hypotheses, and to deduce Causes from Effects, till we come to the very first Cause which certainly is not mechanical . . .[128]

We shall see however, that these classification in no way reflect the actual development of Newton's thought at various levels, and that they are, in part, expedient moves in defense of his position.

## VII

In the last two sections the empirical and experimental character of the Definitions of void and body and related documents have been analysed. Quite a different Newton is revealed in Section IV. There the character and interconnections of void space and matter are seen within the context of a theological metaphysics. It must not be supposed that this view is absent from the *Principia*. It is clearly present in the *General Scholium* of the Second Edition. And while in the *Scholium* on space and time, vacuity is not explicitly cited as being one of the attributes of absolute space, nevertheless, it is clearly implied by Newton's discussion. To be sure he attempts to secure the measure of absolute entities by experimental investigation of rotational motion. Nonetheless, the basic presuppositions of his analysis are still those of the earlier treatises, especially *De Gravitatione* discussed in Section IV.

This leads to a consideration of the following question. Is it necessary to reconcile Newton's "operational" statements of the Definitions with the metaphysical and theological approach of the *Scholium*? Especially when the latter presupposes the assumptions discussed in Section IV? The question begins to have point when it is noticed that Newton tells us at the end of this *Scholium* that he is concerned to show "how we are to obtain the true motions from their causes, effects, and apparent differences and the converse", and that this will "be explained more at large in the following treatise. For to this end it was that I composed it". Now it is the Third Book that deals preeminently with the motion[129] of physical bodies: the Definitions, however, make no reference to the absolute character of such motion. Rather, they tend to given the impression that the Third Book is concerned with purely relative motion. Again, the question can be put into relief if we compare the status Newton accords to absolute space and motion in the *Scholium*, with what he says of the motions of bodies in an unresisting void in the Definitions. In the *Scholium* we find the following passage:

> I do not define time, space, place, and motion, as being well known at all. Only I must observe, that the common people conceive those quantities under no other notions but from the relation they bear to sensible objects. And thence arise certain prejudices . . .[130]

But in the remarks to the Definitions, Newton repeatedly refers to the common people, as a means of suggesting that he is not doing anything like metaphysics in the Third Book, but simply defining the nature of void space and matter in a way which accords with everyday experience. In the context of the controversy with Leibniz, this appeal to the common people is also a way of saying that the Leibnizian criticisms of the doctrine of the void are irrelevant so far as the *Principia* is concerned.

There are a number of factors to be considered in answering this general question. First, in so far as the Third Book is concerned with the movements of gross bodies, the Definitions are not unreasonable as a reply to Leibniz. But as we have seen from the discussion of the corollaries to Proposition VI, the doctrine of void spaces is also held, in part, to characterise the deep substructure of the physical world. It is plain that the Definitions as they stand do not cover the doctrine of the interstitial vacuum. Nor does the Third "Rule of Reasoning", which is concerned with justifying the existence and character of imperceptible particles, but not with the existence of innumerable and actual vacuities in the constitution of matter.[131] Thus even within the framework of the Third Book the Definitions would seem to be unsatisfactory. For with the doctrine of the interstitial vacuum and its implications, Newton goes far beyond the world of everyday experience. And the evidence which he adduces for it is not to be found in any of the editions of the *Principia*, but only in suppressed material such as the draft addition to Proposition VI of the Third Book,[132] or in the *Opticks* and their Queries. Moreover, his discussion of such evidence always makes it plain that he thought it did little more than 'illustrate' the existence of these sub-sensible vacuities. Yet the doctrine was essential, as pointed out in Sections I and IV, for his explanation of a large range of phenomena. Thus when faced with Leibniz's criticisms of his doctrine, and profoundly disliking any sort of controversy, Newton decided to make a minimal claim about the way he was using the terms vacuum and body in the Third Book. 'Other authors' could then dispute about other notions as they pleased. Accordingly, by means of this manoeuvre, he intended to give the impression that nothing more was involved than what the Definitions stated, even though their ostensive 'operational' purpose was out of keeping with the *Scholium* on absolute space and motion.

It is also stated in the remarks to the Definitions that any metaphysical system must be developed in the light of phenomena as understood by experimental philosophy. Since the *Principia* is a technical treatise, is it not the case that with respect to the Definitions Newton is justified in discounting the metaphysical and theological setting of his philosophy of nature? However, it is clear from Section IV that his commitment to a metaphysical world view was prior to many of his effective scientific results, and that the refinement of his position and the ability to do science within such a framework developed simultaneously. Indeed it is difficult to see how this could have been otherwise. Thus we must see the assertion that metaphysics comes after experimental philosophy as another polemical move in a bitter controversy.

Again it is clear from the Clarke-Leibniz *Correspondence* that the Newtonians were finally forced by Leibniz to fight at the most fundamental level. As a result of the work of Cohen & Koyré we are now in a position to say for the first time with certainty, that the philosophy expressed by Clarke was that of Newton himself. Indeed, it is almost certain that one of the reasons why Clarke was asked to perform this task was the fact that Newton did not feel entirely at home with philosophical matters, as is evident from his early treatises. Hence Clarke performed this philosophical assignment with Newton's continuing approval, defending not only the doctrines of the Queries, but those of the *Principia* as well, but means of philosophical arguments.

In the light of this was Leibniz justified in regarding matter and void as constituting the bare bones of a metaphysical theory? There can be little doubt. It was not merely that there was insufficient experimental evidence for a conclusive demonstration of the theory of level upon level of actual vacuities in matter and its consequent paucity, nor was it entirely the philosophical and theological framework within which Newton's theory of matter was ultimately embedded: more important, perhaps, it was the fact that Leibniz perceived matter and void to be the 'scientific' cutting edge of a systematic way of seeing a very wide range of diverse phenomena. It was not merely a way of seeing empirical facts, but it also involved distinctive religious and moral consequences.[133]

Thus Leibniz saw that Newtonianism embodied the generality, the all-embracing characteristics of a metaphysical system. But for him it was a shallow system for which he had a deep intellectual distaste. As a young man he came to feel 'cramped' in its categories. Now, at the end of his years, he saw this basically ancient way of seeing things becoming ascendant. Fundamentally this is what the *Correspondence* is all about: the serious confrontation of two radically different philosophies concerning the nature of God, man, and the physical world. Newton and Clarke were acutely aware of this fact. The latter did public battle, while the former developed many private arguments, some of which were certainly seen by Clarke. It is equally plain from the *Correspondence* and from Newton's deepest thoughts concerning the significance of void space in his system, that should this doctrine be successfully contraverted by Leibnizians, the whole Newtonian position would be seriously threatened.

Thus we have three facts. In the first place the Definitions are not adequate to justify Newton's appeal to the interstitial void in the Third Book and, connected with this, from a general point of view, they are not in character with the ideas of the *Scholium* on absolute motion in absolute space. Secondly, the Baconian contention that metaphysics, though it has a different subject matter, comes after arguments based on induction, is simply not borne out by Newton's own intellectual development. And lastly, it is clear that he supported Clarke's 'philosophical' defence of his own position which is tantamount to admitting, contrary to the views expressed in the remarks concerning the nature of various disciplines, that the arguments formulated in the *Correspondence* are relevant to his experimental philosophy.

We can now consider the initial problem mentioned at the beginning of this section. In the context of these Definition was Newton entirely justified in disregarding the metaphysical and theological orientation of his general philosophy of nature, when the presuppositions underlying this orientation support the *Scholium* on Space and Time, and indeed, certain aspects of Book III? He was certainly entitled to define his terms as he pleased; and in fact he did. But as we have seen, his Definitions did not reflect his total use of the term 'vacuous space' in the *Principia*, and especially in the Third Book. Thus it seems that he was to some extent being disingenuous, by implying that he never meant anything more in his application of the term. Furthermore, it seems that if he had not been formulating these Definitions in defence of his position against Leibniz, he might well have been less prudential in his formulation, allowing himself the liberty of stating or implying the true significance that the doctrine of void space held for him. Since the Definitions were intended to come prior to the Third Book itself, such a liberty would have been in keeping with the policy of the *Scholium* on space and time and the *Generale Scholium.*

The Definitions analysed thus show Newton on the defensive, attempting by "experiment" and "commonsense" to bury the true metaphysics of his system in a characteristically Newtonian response to the penetrating criticisms of his Continental rival.

## APPENDIX

Add 3965 sec 17 Draft No. 1 Folio 641

Solida mathemathica non sentiuntur tangendo nec resistentiam creant, neque corpora dici solent. ⟨At this point Newton has drawn eight lines across the sheet.⟩

Quintessentia ab Elementis quatuor diversa nullis est sensibus obnoxia, nec inter phaenomena numerari potest. Materia prima quae nec quid nec quale nec quantum est non est Phaenomenon. Materia subtiles quae caeli omnes impliantur ⟨a phrase id interpolated here that cannot be made out⟩ non est phaenomenon. Et quae Phaenomena non sunt ea in Philosophia experimentali locum non haebent Argumentum Inductionis ab experimentis *desumptum* & sensibilium observationibus desumptum, in quo Philosophia experimentalis fundatur, ad entia vel hypothetica vel metaphysica quae Phaenomena non sunt, *extendi non debet: applicari non potest nisi per hypothesin:* ideoque quae de corporibus *vi Inductionis* in hoc Libro dicuntur ad ejusmodi entia nil spectant. De solis *Phaenomenis* sensibilibus *et earum partibus* hic agitur *Reliqua* ⟨a phrase is interplated here that is too heavily scored to be intelligible⟩ *in Metaphysica & Philosophia hypothetica tractanda* ⟨a word is illegible⟩ *sunt* ⟨the following phrase is written above the line⟩ *eo quod Argumentum Inductionis in ÿs solis locum habeat.* Reliqua quae non sentiuntur sed per hypothesin ⟨illegible word crossed out⟩ *tamen a nonnullis* corpora nominantur, in Metaphysica et Philosophia hypothetica rectius tractanda sunt. A phaenomenis incipiendum est. In hic tractandis Philosophia experimentalis consistit. Ab Hac Philosophia ad *causas efficientes & causas Finales* & Philosophiam hypotheticam pergendum est. Orbes solidi in quibus Planetae inhaerant *non sunt Phaenomena.* Materia subtilis & *sine resistentia* ⟨the next word is unclear; but reading 'movement' makes sense⟩ *movent* in qua innatent, non est ⟨Newton had first written 'sunt' and then crossed it out⟩ Phaenomenon, *Et quae Phaenomena* ⟨two or three words are unclear⟩ *in hoc fluida sine resistentia.*

Et quemadmodum Geometrae lineam definiunt quae longa est sine latitudine ut *eorum Propositiones* de ejusmodi lineis solummodo intelligantur & *lata tamen linea* in Mechanica *tamen* & alÿs scientÿs *linea lata* locum habeat: sic corpus & *vaccum* hic definiuntur [*non ut alia esse corpora negemus sed ut sequuntur in quo sensu voces accipiantur in sequentibus ostendamus*

*Propositionis quae sequuntur de ejusmodi corporibus hic definitis intelliguntur De alÿs corporibus disputent* Authores in alÿs scientÿs] ut voces in hoc sensu *definito* accipiantur in sequentibus. De alÿs corporibus et alÿs vacuo disputent authores in alÿs scientÿs.

Add 3965. sec 13 (53–94) Draft No. 2 Folio 437$^{v}$

Definitio I

Corpus voco rem omnen tangibilem qua tangentibus resistitur, et cujus actio ⟨written above the line⟩ *resistentia*, si satis magna sit, sentiri potest.

Hoc enim sensu vulgus vocem corporis semper accipit. Et hujus generis sunt ⟨the word 'Tellus' is heavily crossed out⟩ *Tellus, Planetae, Cometae* metalla, lapides, arena, argilla, lutum, *terra*, salia, ligna, ossa, carnes, aqua, oleum, lac, sanguis, aer, ventus, fumus, exhalatio, flamma, & quicquid sub elementis quatuor comprehendi potest, vel ab his exhalando manare & in haec per condensationem redire: Addo *Tellurem & Corpora caelestia. Haec emittunt & reflectunt* ⟨this phrase is written above the following⟩ *Planetae & Cometae* agunt – Lucem. & *inter* ⟨there are three words here quite illegible⟩ *Phaenomena numerantur* ⟨there is a word written below 'numerantur' which is too heavily scored out to be legible⟩ & *motibus suis observant leges corporum* & a partibus suis incumbentibus premuntur. [Solida mathematica non sentiuntur tangendo, nec resistentiam creant, neque corpora dici solent] Vapores et exhalationes ob raritatem suam amittunt resistentiam prope omnem sensibilem, & apud vulgus saepe amittunt etiam nomen corpora & spiritus vocantur. Corpora tamen vocari possunt quatenus *sunt effluvia Corporum* & resistentiam habent densitati proportionalem [Quod si effluvia corporum ita formis mutarentur ut vim *omnem* resistendi amitterent *et inter Phaenomena numerari cessarent* haec non amplius ⟨an illegible word⟩ vocarem corpora: nam loquor cum vuglo]

Definitio II

Vacuum voco locum omnem in quo corpus sine resistentia movetur.

Sic enim vulgus loqui solet. Siquis contendat corpora dari quae tangendo nec sentiri nec resistentiam creare queant, hic jam disputat ⟨there is an illegible phrase interpolated here⟩ de vocis significatione Grammatica, corpora nominando quae vulgus corpora non vocat: et malim *loqui* cum vulgo *loqui, cujus utique est vis et nomina loquendi.* Quae tangendo (premunt) et premendo agunt in alia, quaeque non premunt & (premendo) non agunt in alia, ita ab invicem distingui possunt; ut (vera)[134] vocentur corpora; altera dicantur res intangibiles, vel res ⟨phrase is missing⟩ *vel materia insensibilum vel* alio nomine designentur; & horum genus commune tertio (nomine) gaudeat, quale est Substantia, vel Ens, vel Agens, vel ⟨there is a phrase missing⟩ Sic enim evitabitur ambiguitas illa verborum ex ⟨a phrase is missing⟩ idearum & contentiones oriri solent. Et cum de corpori ⟨part of the sentence is missing⟩ tanquam Phaenomenis hic disputetur, spatium ⟨part of the sentence is missing⟩

ut Vacuum considero. De natura rerum ⟨part of the sentence is missing⟩ ut Vacuum considero. De natura rerum ⟨rest of sentence is missing⟩ destitutis, disputent par ne licet. ⟨the remaining sentences can not be made out⟩

Add. 3965 sec 13 [53–94] Draft No. 3 Folio 422$^r$

Definitiones

Definitio II

Corpus voco rem omnem *Mobilem* & tangibilem qua tangentibus resistitur, & cujus resistentia, si satis magna sit, sentiri potest.

Hoc enim sensu vulgus vocem corporis semper accipit. Et hujus generis sunt metalla, lapides, arena, argilla, lutum, terra, salia, ligna, ossa, carnes, aqua, oleum, lac, sanguis, aer, ventus, fumus, exhalatio, flamma, & quicquid sub elementis quatuor comprehendi potest, vel ab his exhalando manare, & in haec per condensationem redire. Addo corpora caelestia. Haec emittunt et reflectunt lucem et inter phaenomena numerantur, a partibus suis incumbentibus premuntur, figuram rotundam induunt & in motibus suis observant leges corporum. Vapores & exhalationes ob raritatem suam amittunt resistentiam prope omnem sensibilem, & apud vulgus saepe amittunt etiam nomen corporum & spiritus vocantur. Corpora tamen vocari possunt quatenus sunt effluvia corporum & *vim habent* resistentiam *habent* densitati proportionalem. Solida mathematica non *sentiuntur* agunt tangendo *nec resistentiam creant*, neque corpora dici solent. [Quintessentia ab elementis quatuor diversa nullis est sensibus obnoxia nec inter phaenomena numerari potest. Material prima quae nec quid nc quale nec quantum est, non est phaenomenon. Orbes solidi in quibus Planetae inhaereant non sunt phaenomena. Materia subtilis in qua *Planetae* innatent et *in quibus* corpora sine resistentia moveantur, non est phaenomenon. Et quae phaenomena non sunt nec ullis sensibus obnoxia, ea in Philosophia experimentali locum non habent. Argumentum Inductionis ab experimentis et sensibilium observationibus desumptum, in quo Philosophia experimentalis funatur, ad entia vel hypothetica vel metaphysica quae phaenomena non sunt, applicari non potest nisi per hypothesin, ideoque quae de corporibus vi Inductionis in hoc libro dicuntur, ad ejusmodi entia nil spectant. De solis sensibilibus et eorum partibus hic agitur propterea quod argumentum Inductionis in ÿs solis locum habeat. Reliqua quae non sentiuntur sed per hypothesin tamen a nonnullis corpora nominantur, in Metaphysica et Philosophia hypothetica rectius tractanda sunt. A Phaenomenis Philosophia incipit *A phaenomenes incipiendum est.* In his tractandis Philosophia experimentalis consistit. Ab Philosophia *experimentali ad rerum ad* causas efficientes & finales, & *ab his omnibus ad naturam rerum insensibilum & ultimo* ad Philosophiam hypotheticam transeundum est: Initio Libri primi quantitatem materiae definivi ut tractaretur *physice* mathematice: hic corpus extali materia constans definio ut tracetur physice.

Definitio III

Vacuum voco locum omnem in quo corpora sine resistentia moventur.

Sic enim vulgus solet; et haec vocis significatio ex definitione prima consequitur *Vacuum est quod rebus tangibilibus et contactu sua motum corporum impedientibus vacat.* Quemadmodum vero Geometrae lineam definiunt quae longa est sine latitudine ut eorum Propositiones de ejusmodi lineis solummodo intelliguntur, et in Mechanica tamen et alÿs scientÿs Linea lata locum habet; sic corpus et vacuum hic definiuntur ut voces in sensu definito accipiantur in sequentibus. De alÿs corporibus et alio vacuo disputent anthores in alÿs scientÿs.

Add 3965 sec 13 [53–94] Draft No. 4 Folio 430r

Scholium

In Definione prima Libri primi *definitur dixi quid sit* quantitas Materiae in Corpore quovis & me hanc quantitatem per nomine Corporis vel Massae passim intelligere et Medÿ siquod fuerit interstitia partium libere pervadentis hic nullam rationem habere. Et in Definitione tertia *dixi* vim Inertiae corpori insitam proportionalem *insitam* seu essentialem esse et hanc vim esse potentiam resistendi dequa corpus *omne conatur permanere perseverat* in statu suo quiescendi vel movendi uniformiter in directum. Per corpus utique rem omnem intelligo tangibilem *intelligo* qua tangentibus resistitur, et cujus resistentia, si satis magna sit, sentiri potest. Hoc enim sensus volgus vocem corporis semper accipit. Et hujus generis *proptera quod* [et ultimo ad *Philosopiam* hypothesis *Philo* ⟨the word was not completed⟩ transeundum est. Vocem corporis igitur *in hac Propositione* non latus accipio quam in sensu hic definito. Et *vacuum voco* spatium omne hujusmodi corporibus destitutum. Quemadmodum vero Geometrae Lineam definiunt quae longa est sine latitudine, ut eorum Propositiones de hujusmodi lineis solummodo intelligantur; & in Mechanica tamen & alÿs scientÿs Linea lata est: sic corpus et vacuum hie definiuntur ut in sensu definito accipiantur in *hac Propositionem et sequentibus his Libres*. De alÿs Corporibus et alio Vacuo disputent *authores in alÿs scientÿs* alÿ qui extra limites Philosophiae experimentalis vagantur] ⟨There is a large space in the manuscript. Newton begins again with a dash⟩ propterea quod argumentum Inductionis in ÿs solis locum habeat. Ideoque quae *hic affirmantur* vi Inductionis de universali corporum gravitate *hic affirmantur*, and Quintessentias, & Materias primas, et Orbes solidos, et Materias subtiles quae phaenomena non sunt, nil spectant. [Hic loquor de corporibus quae *vere tanguntur* & vere habent *vim* inertiae et per hanc vim tangentibus resistentiam vere inferent] *Ea non sunt corpora de quibus hic agitur* Quae non sentiuntur sed per hypothesin tamen a nonnullis corpora nominantur, in Metaphysica et Philosophia hypothetica rectius tractanda sunt. A phaenomenis *Philosophia* per argumentum Inductionis incipit. In his tractandis Philosophia experimentalis consistit. A philosophia experimentali ad rerum causas *finales et* efficientes, & ab his omnibus ad naturam rerum insensibilium, & ultimo ad Philosophiam hypotheticam transeundum est.

Add 3965 sec 13 [126–149] Draft No. 5 Folio 504r

Definitio I

Corpus voco rem omnem quae moveri eet tangi potest et qua tangentibus resistitur.

Hoc sensu vulgus vocem corporis semper accipit. Et hujus generis sunt metalla, lapides, arena, argilla, lutum, terra, salia, ligna, ossa, carnes, aqua, oleum, lac, sanguis, aer, ventus, fumus, exhalatio, flamma, et quicquid sub elementis quatuor comprehendi potest, vel ab his exhalando manare & in haec per condensationem redire. Adde corpora caelestia. Haec *moventur* emittunt et reflectunt lucem & inter phaenomena numerantur; a partibus suis incumbentibus premuntur, figuram rotundam induunt, & in motibus suis observant Leges corporum. Vapores et exhalationes ob raritatem suam amittunt resistentiam prepe omnem sensibilem, et apud vulgus saepe amittunt etiam nomen corporum et spiritus vocantur. Sed sunt spiritus corporei, *et corporei*, et corpora vocari possunt quatenus *sunt* effluvia corporum & vim habent resistendi densitati *suae* proportionalem. Solida mathematica non aqunt tangendo neque corpora dici solent.

Definito II

Vacuum voco spatium omne per quod corpus sine resistentia movetur.

Sic enim vulgus loqui solet. Et haec vocis significatio ex Definitione prima consequitur.Vacuum est quod rebus tangibilibus et contactu suo motum corporem impedientibus vacat. Quemadmodum vero Geometrae Lineam definiunt quae longa est sine latitudine ut *eorum* de hujusmodi Lineis solummodo intelligantur, et in Mechanica tamen et alÿs scientÿs Linea lata locum habet: sic Corpus et Vacuum hic definiuntur ut voces in sensu definito accipiantur in sequentibus. De alÿs corporibus et alio vacuo disputent authores alibi.[135]

NOTES

1 Alexandre Koyré & I. Bernard Cohen, "Newton and the Leibni-Clarke Correspondence", *Arch. Inter. d'Hist. des Sc. Quinzieme Annee*, Nos. 58–59, 1962, pp. 63–126.

2 These are to be found in ULC. Add. 3965 in sections 13 and 17. Explicit details are given below in conjunction with the manuscripts. It will be shown in Section VI that these definitions are essentially connected with the Fourth Rule, and the suppressed Fifth Rule of reasoning in Philosophy, found in this section.

3 Isaac Newton, *Mathematical Principles of Natural Philosophy*. Edited by F. Cajori, Berkeley, California, 1960 p. 397.

4 *Ibid.*, Book III, pp. 411–414.

5 *Ibid.*, Book III, p. 411.

[6] In the sense that the Corollary is primarily concerned with the specific gravity of bodies, which for Newton presupposes pores in their composition.

[7] *Principia*. London, 1687, Book III, p. 411.

[8] J. Edleston, *Correspondence of Sir Isaac Newton and Professor Cotes*, London, 1850, pp. 65–80. Cotes's criticisms and Newton's replies are contained in a series of important exchanges from February to March 1711–12. His main concern is with the Newtonian definition of inertia. He argues that Newton has not shown in Corollary III, either *a priori* or experimentally, that two bodies which occupy equal volumes, and are entirely filled with matter, necessarily have the same density or inertia. That this is a necessary consequence, is implied in the Corollary as formulated in the First Edition.

[9] Cajori, *Mathematical Principles, op. cit.* p. 414. Newton is taking the idea of infinite rarefaction seriously; in the limit there is empty space, apart from the entirely dense primordial particles.

Furthermore, Newton would seem to have the following argument in mind for the vacuum. All matter from its *vis inertiae*, or resistance to change its place, must give resistance to motion in proportion to its quantity. And as we find different degrees of resistance to bodies moving in various mediums, there must be different quantities of matter in equal spaces. Consequently there is just so much more of *vacuum*, or absence of matter in one of the spaces, as there is *less* of resistance. When there is no resistance, there will be *no matter*: thus a vacuum.

[10] Essentially the same argument, often similarly expressed, is to be found in the writings of the Newtonians. For example: J. Keill, *An Introduction to Natural Philosophy*, London, 1745 (translated from the last Latin Edition). These lectures were given in 1700, Lecture X, p. 117. W. Whiston, *Sir Isaac Newton's Mathematick Philosophy*, London, 1716. Lecture XXXIV, April 26, 1708, p. 340. R. Bentley, *Confutation of Atheism* reprinted in *Isaac Newton's Papers and Letters on Natural Philosophy*, ed. I. Bernard Cohen, Cambridge, 1958, p. 321. And S. Clarke's note on p. 96 of *Rohault's System of Natural Philosophy*, vol. II. London, 1728.

[11] Because of the reference to the "region of the air" which tends to give the impression that Newton is speaking of external spaces.

[12] Isaac Newton, *Opticks*, New York, 1952, p. 352.

[13] *Ibid*, p. 368.

[14] *Vide*, note 10, item Whiston, p. 341.

[15] A. Rupert Hall & Marie Boas Hall, *Unpublished Scientific Papers of Isaac Newton*, Cambridge, 1962, p. 312.

[16] The earlier variant of the manuscript published by the Halls is to be found in ULC. Add. 3965. 6. F. $266^{v}$. The published variant is in Add. 4005, Fols. 28–29. The five related sheets, which are really alternate versions of the Corollaries of Proposition VI, are to be found in Add. 3965. 6. Fols. $311^{r-v}$ and $310^{r-v}$.

[17] For instance, in Add. 3965. 6. F. $266^{v}$, we find the following intended corollaries. "Corollary 4. Without atoms gravity cannot be explained mechanically. Corollary 5. Without void gravity cannot mechanically be explained. Corollary 6. Void is granted Corollary 7. Bodies are far rarer than is commonly believed. Corollary 8. Atoms are granted. Corollary 9. Granted is an infinite and omnipresent spirit, in which matter is set in motion according to mathematical laws." Also in Add. 3965. 6. F. $267^{r}$, Newton postulated, as an alternative to the void as a condition of mechanically explaining gravity, a hypothesis involving non-gravitational particles. He says ". . . and the other kind of less dense particles which have to be the cause of the gravity of the denser ones but themselves have no gravity, lest their gravity might have to be explained by a third kind, and that again by a fourth and so on to infinity". Newton then attempts to secure acceptance of a vacuum as a means of mechanically explaining gravity by indicating the absurdity in the notion of non-gravitational matter.

[18] He also intended in this unimplemented edition to support Propositions IV to IX of the Third Book by references to the opinion of the ancients regarding gravity, the void and motion.

Proposition VI was to have been supplemented by a scholium containing four quotations from Books I and II of Lucretius' *De Rerum Natura* regarding matter and void.

[19] *Cf.* Bernard Gagnebin, "De la Cause de la Pesanteur. Memoire de Nicolas Fatio de Duillier Presenté à la Royal Society le 26 Fevrier 1690", *Notes and Records of the Royal Society*, **6**, 1949, p. 117. Fatio says: "Je suppose que les diferens Epaces du Monde sont presque entierement vuides de Matiere. Plusieurs Phenomenes que nous observons dans la Nature, etablissent cette extreme Rarété des Corps. Elle est necessaire, dans mon Hypothese, afin que les Particules de la Matiere, qui cause la Pesanteur, aient leurs Mouvemens, en Lignes droites, extrement Libres." p. 127.

[20] Isaac Newton, *OPTICE sive de Reflexionibus, Refractionibus, Inflexionibus Coloribus LUCIS*, London, 1706, p. 229. There is no substantial difference in the English and Latin editions. In a letter to Leibniz in 1694. Fatio states that Newton believes there is much more 'de vuide que de plein'. However, no arguments are given. *The Correspondence of Isaac Newton*, Vol. III, p. 308.

[21] Hall & Hall, *op. cit.*, p. 115.

[22] Density is put proportional to inertia when volume is constant. It is interesting to note that inertia is an intensive quality rather than an extensive one. However, in so far as Newton asserts that the proportionality of matter to weight is proved experimentally (Def. 1 of the *Principia*), his definition of mass is circular since this involves density being proportional to gravity of matter and volume conjointly.

[23] Cajori, *Op. cit.*, Bk. III, p. 418.

[24] *Ibid.*, Bk. III, p. 419.

[25] *Opticks, op. cit.* Query 28, p. 367.

[26] Cajori, *op. cit.*, Bk. II, p. 366.

[27] *Opticks*, *op. cit.*, Query 28, p. 368.

[28] Henry Guerlaac, 'Newton et Epicure', Conférence donnée au Palais de la Dêcouverte, Paris, 1963. Professor Guerlac has gone some way to establishing this view by a comparative analysis of the difference in the Queries of the 1706 edition from those of the 1717 edition. There is much manuscript evidence from the period under discussion to support this; for instance, *Vide*. note 17, F. 266$^{v}$.

[29] Bentley, using the same argument, is quite explicit. "And yet the higher you ascend above that region (Orb of Saturn), the Rarefaction still gradually increases without stop or limit: so that, in a word, the whole Concave of the Firmament, except the Sun and Planets and their Atmospheres, may be considered as a mere Void." *Vide*. Note 10, item *Papers and Letters*, p. 325.

[30] *Oeuvres completes*, The Hague, Vol. XXI, p. 473.

[31] J. Gerhardt, ed., *Leibnizens Mathematische Schriften*, Halle, 1860 VI, p. 144 sq. 161 sq. Originally in *Acta Eruditorum* 1689, 81–96.

[32] It is not certain that Leibniz had read the *Principia* prior to composing his *Tentamen*. He himself tells us that he had only read an account of the *Principia* in the *Acta Eruditorum* for June 1688, pp. 304–315. *Vide* C. I. Gerhardt. *Leibnizens Mathematische Schriften*, Halle, 1859, IV, pp. 87ff. Also Huygens, remarks in a letter to Leibniz on 8 February 1690: "You have only seen an extract of his book and not the book itself." *Oeuvres Completes* **IX**, p. 367.

[33] These three letters form only a part of the correspondence which began in 1706. The void is criticised implicitly by Leibniz. The letters were republished in the *Memories of Literature* in 1712 and Cotes drew Newton's attention to them. *Vide*. Note 8, p. 158. Rémond's criticisms will be discussed below.

[34] Throughout, references and quotations are from Robinet's edition of the Correspondence, unless otherwise indicated. A. Robinet, "Correspondence Leibniz-Clarke", Paris, 1957.

[35] Ed. J. E. Erdmann, *Opera Philosophica, pars prior*. Berlin, 1840, Preface, pp. 468–479.

[36] Robinet, *Op. cit.*, p. 36.

[37] *Ibid*., p. 48.
[38] *Ibid*., p. 55.
[39] *Ibid*., p. 70.
[40] *Ibid*., p. 91 29th May 1716. It is in this letter that Leibniz makes the charge that void space is an attribute without support: "Si l'espace est un realité absolue bien loin d'etre une proprieté ou accidentalité opposée á la substance, il sera plus subsistant que les substances." Clarke replies in Fourth Letter "Void Spacce is not an *Attribute without a subject;* because by *void space*, we never mean *Space void of everything*, but void of Body only." *Vide*. Note 1, Koyre & Cohen pp. 90–91, where there is a discussion of this point.
[41] *Ibid*., p. 78.
[42] *Ibid*., p. 109. Clarke also stresses in this passage that "EXTRA-MUNDANE SPACE, (if the Material World be Finite in Dimensions), is not IMAGINARY, but REAL."
[43] *Ibid*., p. 138.
[44] Ibid., p. 192. Also see *Opticks*, Query 28 p. 365, the Scholium to Proposition XIX, Bk. II of the *Principia*, and Cotes' Preface to the Second Edition.
[45] *A collection of papers which passed between the late learned Mr. Leibnitz and Dr. Clarke, in the years 1715 and 1716. Relating to the principles of natural philosophy and religion*, London, 1717.
[46] Robinet, *Op cit*., p. 76.
[47] Where ideas are common to the five variants, Note remarks will be given in Draft No 3. In some cases I have added puctuation. The latin, however, is presented in the Appendix as it was written. This type of sub-heading will be used for all manuscripts in both the text and the Latin Appendix.
[48] The sentence structure leaves much to be desired. The meaning, however, is clear. Newton is indicating a criterion of materiality. Bodies act on one another in virtue of their solid, impenetrable content. Mathematical solids (*viz*. above) cannot act; for they pass through one another. The same criterion is implied in Defintion II of Draft No. 3.
[49] The reference to the four elements is puzzling. It is probably a compressed way of referring to the theory of transmutation which is elaborately developed in the Third Book, Cajori, Op. cit. pp. 522, 526–530 and p. 542. The passage at p. 542 was only added in the Third Edition. The Comets continually supply a "vapourous spirit" which considerably rarefies and dilates in the heavens. By the power of gravity this vapour is drawn slowly to the earth and, condensing into water, augments the dwindling supply of nourishing fluids necessary to sustain life: whereupon the various transmuting agents produce physical stuffs such as mud, clay, sand etc. In the *Opticks* there are many passages which support a theory of terrestrial transmutation, i.e. Query 30, pp. 374–375 and also the Draft *Conclusio* to the *Principia*, Hall & Hall, *Op. cit*., p. 341. In each case the process begins with water. All phenomena, accordingly, arise from and return to this common stuff, which is itself replenished by the "vapours" of the Comets' tails. Thus, the phrase under discussion, coming as it does directly after the list of physical bodies, is a convenient way of referring to the contents of the perceptible world, the theory which explains their transmutations, and the occurrence of such intermediary states as vapours and exhalations. The influence of alchemical theories is beyond the scope of this paper.
[50] Newton did not close this bracket. It is clear however that he intended all the material after the bracket to be deleted when the passage would be copied; for this is precisely the material left out when the Definitions became part of a revision sequence. *Vide* Draft No. 5 of Latin appendix.
[51] This would appear to be a reference to the chemical use of the term "Quintessence" in the seventeenth century: that is, something which is approximated to, but never reached in experiment. It is certain that the term had alchemical significance for Newton.
[52] This is plainly a rejection of the Fluid aether of the Cartesians and the Leibnizians. For instance in one of the variants of the letter sent to Leibniz via Conti in February 1715/16

Newton says of Leibniz: "He pretends that all places not filled with tangible bodies may be filled with intangible corporeal Fluid." Koyré & Cohen, *Op. cit.*, p. 109.

53 Newton has marked a space here with three dashes. As suggested above, he probably intended the list of bodies to be copied in here from one of these drafts.

54 There is a line drawn through the bracketed material. See my remarks above.

55 This bracketed sentence has two lines drawn through it.

56 The Fourth letter contains a fair number of criticisms of the void; but these are consequences of the general discussion of space.

57 Robinet, *Op. cit.*, pp. 135–136, 181. Where Leibniz says that neglect of the principle of sufficient reason leads to "l'origine des chimeries, comme par example d'un temps ou, d'un espace absolu réel, du vuide, des atomes . . ."

58 *Ibid.* p. 41.

59 Koyré & Cohen, *Op. cit.*, pp. 73, 110. All twelve drafts are published in this substantial article.

60 H. G. Alexander, *The Leibniz-Clarke Correspondence*, Manchester, 1956, p. 187.

61 ULC. Add. 3968. 36. Fol. 517 at the bottom of the sheet on which the post-script is copied in French by Newton, we find, "Received of ye Princess May 7th 1716. *Vide* Koyré & Cohen, *op. cit.*, p. 81.

62 The *Quaestiones* are found in ULC. Add. 3996 and *De Gravitatione et Aequipondie Fluidorum* is published by the Hall's. *Vide.* Note 15.

63 Richard S. Westfall, "The Foundations of Newton's Philosophy of Nature", *Brit. J. for the Hist. of Sc.* **1** (1962), 172–182.

64 W. Charleton, *Physiologia Epicuri-Gassendi-Charltoniana*, London, 1654.

65 *Opera Omnia*, Vols. I and II ed. T. Gregory.

66 Charleton, *op. cit.* Chap. IV, pp. 21–50 and Chap. III, p. 19. P. Gassendi, *Syntagma Philosophicum* Vol. 1, *Sect.* 1. *Lib.* 11. *Cap.* 111, pp. 192–196. With respect to the *Quaestiones*, it is more prbable that if Gassendi influenced Newton at this early stage it was the *Philosophia Epicuri Syntagma* Vol. III, *pars secunda-Cap.* 1. pp. 11–12.

67 *Quaestiones* Folios ii$^{v}$, 89$^{r}$, 89$^{v}$, 97$^{r}$, 114$^{r}$. It cannot be maintained that Newton acquired this language exclusively from Charleton or Gassendi. Such expressions were widely used in the plenist-vacuist controversy of the first half of the seventeenth century. see C. Webster, "The Discovery of Boyle's Law and the Concept of the Elasticity of Air in the Seventeenth Century", *Arch. for History of Exact Sciences* **2**, 141–502 (1966). Also, such phrases are used by Stanley in his *History of Philosophy*, London, 1655–62, part 5, 1660, pp. 144, 154, 159 and 178, which was read extensively in this period. However, since many of Newton's doctrines seem to come from Charleton and Gassendi, it is probable that the terminology was, in the first instance, taken from them.

68 This is not surprising, for at this stage Newton still accepted the Cartesian Vortices.

69 Hall & Hall, *op. cit.*, p. 146. On page 147, Newton says 'Since the resistance of the aether is on the contrary so small when compared to the resistance of quick-silver . . . , there is all the more reason for thinking that by far the largest part of the aetherial space is void, scattered between the aethereal particles'.

70 *Quaestiones*. Folio 114$^{r}$.

71 Charleton, *op. cit.*, p. 17. He is, of course, only repeating, in part, Gassendi's elaborate discussion in Bk. III of the *Syntagma*.

72 Gassendi, *Syntagma*, Vol. I p. 231. Similar passages are to be found on pp. 230 to 232 and on pp. 180, 184, 185 of *Lib.* II. In the earlier *Philosophiae Epicuri Syntagma, Pars Philosophiae Secunda, Caput* I, p. 11, the doctrine is expressed in a similar way. The influence of Henry More in this respect cannot be overlooked. The correlative nature of these notions is clearly expressed in several places in his writings. In *A Collection of Several Philosophical Writings of Dr. Henry More*, London, 1712, *Vide* the *Epistola Prima H. More*

*ad R. Cartesium*, p. 62; the *Immortality of the Soul*, Chap. III, pp. 8–9, Chap. VII, pp. 25–28, and Chap. XI, pp. 40–43. In the *Divine Diaglogues*, 1668, More says, with reference to the reality of infinite extension. 'This evidently demonstrates the existence of the ancient *Democritish Vacuum*, and withall that Extension and Matter are not convertible terms, for which Cartesius so much contends.' p. 103.

73 Professor Westfall has urged this view in his 'Newton and Absolute Space'. *Arch. Inter. d'his. des Sc., Numero* 67, Avril–Juin 1964, pp. 121–132. However, as I shall attempt to show, the thesis needs important qualification.

74 Hall & Hall, *op. cit.*, pp. 139–140. Although there is no direct evidence that Newton read More's *Divine Dialogues* the *manner* in which the doctrines of *De Gravitatione* are expressed is mindful of the first Dialogue. This dialogue is More's mature expression of his position, and the whole discussion is anti-Cartesian in a vein similar to Newton's treatise. Moreover the leading doctrines expressed are closer in character to those of *De Gravitatione* than in any of More's earlier treatises. God's omnipresence, his identification with space, the reality of general extension (this is the first time that More argues for this position), the 'vital congruity' of spirit and matter, that everything, including spirits, must exist 'somewhere', *etc.* Thus it is quite probable, that Newton knew this treatise. Accordingly his theory of the manner of matter's creation could have occurred to him on reading the first Dialogue. After the Platonist Barthnous argues that general extension (space) is a necessary existent, utterly distinct from extended *things*, and the only ground for the existence of an infinite Spirit, Cuphophron suggests that 'Extension is the *Capacity* of matter'. Barthnous then asks, 'What do you mean by Capacity, Matter *in potentia*?' p. 108. The idea is not pursued; but in the context of the discussion, where it is repeatedly affirmed that matter is a dependent existent moved by God's will, we surely have the germ of Newton's hypothesis, namely, that space is the potentiality of matter made actual when determinate parts of space are made to manifest sensible appearances. If this suggestion can be substantiated, Westfall's assertion that 'Newton's empty spaces moved about by God are the direct product of Gassendist skepticism', Westfall, *op. cit.*, p. 129, can no longer stand.

75 Hall & Hall, *op. cit.*, p. 144. Clarke makes the same point in different terms '. . . if *Extension* were the Essence of Matter, and so *Matter* the same as *Space* itself: it would follow that Matter is infinite, and necessarily eternal, and could neither have been created, nor reduced to nothing; which is very absurd.' *Vide*, Note 10, item *Rohault*, Vol. I, p. 24. The passage in the text makes it plain that Newton regards matter to have even less reality and perfection than accidents, by which he probably means properties. For the epistemology supporting this, see Section VI below.

76 More, *A Collection. Immortality of the Soul* II:XI:10.

77 *Quaestiones*, Fols. 89, 104, 105, 117 and 132.

78 Robinet, *op. cit.*, p. 19. The letter is dated 18th October.

79 Newton includes both impenetrability and non-deformability in the meaning of the term. But since impenetrability does not exclude deformability, what he means essentially by the term is non-deformability.

80 'Indeed, however we cast about we find almost no other reason for atheism than this notion of bodies having, as it were, a complete and absolute and independent reality in themselves.' Hall & Hall, *op. cit.*, p. 144. This view is commonly found in English Natural Philosophy of the period. With Newton, however, it is a basic principle. In an early draft of Query 31 he says: '. . . matter depends upon a Deity for its laws of motion as well as for its existence. The Cartesians make God the author of all motion and it is as reasonable to make him the author of the laws of motion.' Add 3970, F. 243$^r$. And in a draft version of the *General Scholium*: 'Materia non est aeterna sed originem habuit a voluntate Dei.' Add 3965; 13 [150–182] F 547$^r$. See also Note 75.

81 Notice how this view of matter is diametrically opposed to that of Leibniz, who clearly

states that the more active matter there is in the world, the more reality and perfection. *Vide.* section II and V.

82 In either case there is a logical connection of significance between the terms; but the modalities are different. In the first case the connection is one of necessity; and the second it is one of possibility. Put notationally we have: $A$ is correlative to $B$ if $\Diamond(\exists x)Ax \mid \Diamond(\exists x)Bx$. And in the second case $(\exists x)Ax \mid (\exists x)Bx$.

83 Robinet, *op. cit.*, pp. 209–210.

84 Cajori, *op. cit.*, p. 546.

85 After stating that if the particles of bodies were pressed together 'in a compact mass like grains of sand or heap of stones' the force of gravity would tend less to the inner than the outer ones, thus violating the proportionality of gravity to matter, Newton says, 'Other textures of the bodies must be devised by which their interstices are rendered more ample, and this is the necessary condition of an Hypothesis by which gravity is to be explained mechanically.' Hall & Hall, *op. cit.*, p. 315. It is necessary, however, to be cautions about Newton's complete acceptance of the hypothesis even in the 1690's; for we have Fatio's own testimony to the effect that Newon 'did not scruple to say *that there is but one possible mechanical cause of Gravity, to wit that which I had found out.* Thô he would often seem to incline to think that Gravity had its foundation in the arbitrary will of God . . .' *Vide.* Note 19, *Notes and Records.* p. 117. But on the other hand, notice Newton's comment to Leibniz in 1693 *Correspondence*, Vol. III, p. 287, 'But some very fine matter seems to fill the heavens. And if all the laws of motion can be shown to be compatible with it, Newton says he is 'far from objecting'. Again this is evidence of a state of vacillation.

86 Cajori, *op. cit.* pp. 385–396. Especially important are the corollaries and scholium for Proposition L 11, and Propositions L 111 of Book II. The main charges against the vortex theory are: the radical instability of the system (the corollaries to Prop. L11), the fact that whether the planets move through, or with the vortices (Leibniz's 'harmonic motion') the third law of Kepler is violated, and that the areal law would also have no application. Cotes gives a very clear account of these criticisms in his Preface to the second edition. He especially stresses the irregularity of the motions of the Comets as evidence against vortices. *Vide.* Bk. III.

87 Edleston, *op. cit.* p. 310. *Ex Epistola cujusdam ad amicum.* This manuscript was not published during the controversy. It gives a clear indication of Newton's commitment to the void.

88 *Opticks. op. cit.* p. 350.

89 *Ibid.* p. 353. The real difficulty here is not that of understanding how an aether of small density can *quantitatively* transmit gravitation as L. Rosenfeld suggests 'Further Papers of Isaac Newton.' *Nature* **196** (1962)] but how such a medium can transmit any influence at all. This is especially so in view of the fact that Newton repeatedly states that very dense media are necessary for maintaining vibrations and for propagating them quickly. *viz. Opticks* p. 344. J. Lohne in an important article, 'Newton's theory of colours', *Arch. for the Hist. of Ex. Sc.* **1**, 402 (1961), has suggested that Newton gave up the cosmological aether because it was incompatible with the exact application of the inverse-square law, which indicated a non-resisting void. Again, Rosenfeld has suggested that he restricted its operation to short-range interactions in his theory of light and colours, 'Newton and the Law of Gravitation', *Arch. or Hist. of Ex. Sc.* **2**, 385 (1965). I agree with this. And these difficulties, in conjunction with those indicated above, make it difficult to explain why Newton outlined such an elaborate hypothesis in the Queries of 1717; unless of course to attempt an explanation of the independent existence of Forces. Professor Guerlac's view – apart from the greater merit of his papers – that he did so as a result of Hauksbee's work on electrical and magnetical phenomena is of little help in the light of such conceptual dilemmas: unless, of course, these phenomena supported in Newton's eyes a notion of an immaterial aether. See Chapter 5. See H. Guerlac, 'Francis Hauksbee', experimentateur au profit de Newton', *Arch. Inter, d'his. des Sc.* **16**, 113–128 (1963) and 'Sir Isaac and The Ingenious Mr. Hauksbee' *L'aventure de la Science, Melanges Alexandre Koyré*, Vol. I, Paris 1964, pp. 228–253.

[90] Hall & Hall, *op. cit.*, pp. 305–308 317, 333, 341, and 345. *Opticks*. Query 31, pp. 389–390, 395 and 397, the Preface to the first edition of the *Principia*, and many other references scattered throughout the published and unpublished writings.

[91] In his *Waste Book* Add. 4004, 1664, which contains many dynamical writings, Newton seemed to conceive the programme of the mechanical philosophy as extending the machinery of matter and motion, as treated mathematically in impact phenomena, to *all* natural phenomena. Some time in the 1680s, he came to the view that many phenomena cannot be treated in terms of such a narrow conceptual framework. Thus he enriched the 'classical' mechanical philosophy with the category of force.

[92] Descartes, *Principia Philosophiae, pars secunda,* art. 10, *Ouvres*, Vol. VII: p. 45, ed. C. Adam & P. Tannery, Paris, 1897–1913.

[93] *Syntagma Philosophicum, Opera Omnia*, Vol. I, *Liber Secundus, Cap.* I. p. 179.

[94] More, *A Collection. Epistola Prima H. More ad R. Cartesium*, p. 62.

[95] Charleton, *op. cit.*, p. 66. Chapters VI and VII of the *Physiologia* could be shown to be an important influence on Newton's views of space and time.

[96] Hall & Hall, *op. cit.*, p. 132.

[97] Leibniz points out in the *Correspondence, Robinet*, p. 136. that for Newton space is 'un realité absolué'. Koyré & Cohen agreeing that he is right, clarify the implications by saying that for the Newtonians, 'infinity implies necessity thus eternity'. *op. cit.*, p. 90. This does not get to the heart of the matter. For the sense in which Newton thought space to be infinite must be made clear. It seems that he held it to be an unconditioned existent. Thus no temporal change or condition could possibly affect its existence; not even the annihilation of the physical world. Hence, in the following sense, it is an infinite and necessary existent: namely, the unconditioned condition for the existence of anything in the world. This is plainly a traditional position.

[98] Hall & Hall, *op. cit.*, p. 138.

[99] *Ibid.*, p. 132. See also pp. 137–138.

[100] *Opticks. op. cit.* pp. 353–354. Hall & Hall., *op. cit.*, p. 132. The same view is clearly developed in Newton's first reply to Hooke's criticisms of his theory of colours in 1672. *Vide.* Note 10, item Cohen, pp. 118–119.

[101] *Quaestiones*. Folio 101$^{v}$.

[102] Hall & Hall., *op. cit.*, p. 132.

[103] K. Digby, *Two Treatises*, London, 1658, p. 1.

[104] *Syntagma. Liber Tertius, Cap.* 1, p. 231. Gassendi's reference is corrupt. He quotes '*Tangere enim, tangi sine corpore null potest res*'. It is interesting to notice that Charleton, in the *Physiologia*, quotes it in the same form. Digby, on the other hand, is correct.

[105] Alexander, *op. cit.*, p. 187.

[106] Koyré & Cohen, *op. cit.*, p. 111.

[107] ULC. Add. 3968. 33. Folios 468–470. For details of Pierre Rémond de Montfort see D. Brewster, *Memoirs of the Life, Writings and Discoveries of Sir Isaac Newton*, Vol. II, p. 436, Edinburgh, 1855.

[108] The letter is copied and in Keill's hand. Folios 465–467. Add. 3968. 33.

[109] Folios 471–473. There are corrections in Newton's hand in the margins of Keill's letter.

[110] ULC. Add. 3968. 41. Folios 69$^{r}$–70$^{r}$.

[111] Add. 3968. 41. Folio 69$^{r}$.

[112] Pp. 110–112.

[113] Add. 3968. 41. Folio 69$^{r}$. This passage is translated from the Latin.

[114] *Acta Eruditorum*. 1694, p. 112. My translation.

[115] Cf. the letter to Conti, the Review of the *Commercium Epistolicum. Phil. Trans.* Vol. 29. 1715, p. 224, and the Preface to the 1722 edition of the *Commericum.*

[116] Koyré & Cohen, *op. cit.*, pp. 118–121.

[117] *Ibid.*, p. 120.

[118] *Acta Eruditorum*, April, pp. 145–157.

[119] ULC. Add. 4005. Fols. 21–22. Hall & Hall, *op. cit.*, p. 374. As the Hall's point put, this seems to be a shortish version of '*Phaenomena*', another treatise in the same manuscript bundle: the latter manuscript closely resembles the 'Phaenomena' of the 1713 edition of the *Principia*; cf. Halls p. 378.
[120] ULC. Add. 3965. 13. Fol. $420^r$. This sheet also contains definitions of the terms hypothesis', 'rule', 'body' and 'vacuum'. Folio $421^r$, this sheet contains a definition of body similar to the above; and folio $422^v$.
[121] Folio $422^v$. ULC. Add. 3965. 13. This passage is translated from the Latin.
[122] 'Les regulae philosophandi' *Arch. Inter. d'His. de Sc. Treizieme Annee*, 1960, 14. The sheets published by Koyré are to be found in the same section of the *Principia* box as the definitions of 'phenomena', sec. 13 [53–94] F. $417^r$, $428^r$.
[123] *Vide*. sec. 111.
[124] Hall & Hall, *op. cit.*, p. 361. A similar, but less elaborate passage is to be found in the *Principia*. Cajori, *op. cit.*, p. 546.
[125] Koyré & Cohen, *op. cit.*, p. 113. See Chapter 2.
[126] In the Keynes Collection of Newtonian manuscripts in King's College Library is the following passage under the heading 'Seven Statements on Religion': 'That religion and philosophy are to be preserved distinct. We are not to introduce divine reveletions into Philosophy, nor philosophical opinions into religions.' K. MS. 6.
[127] ULC. Add. 3965. 13. Foli $428^r$. On the same sheet is written yet another definition of body. The passage is translated from the Latin.
[128] *Opticks. op. cit.*, p. 369.
[129] Cajori, *op. cit.*, p. 397. The preface to that Book.
[130] *Ibid.*, p. 6. The same point is made on p. 11.
[131] I am not suggesting that Newton uses the term 'void' in a different *sense* when referring to the interstices of matter; for clearly it is the *same* space that is divided whether by gross bodies or imperceptible particles.
[132] Hall & Hall, *op. cit.*, p. 316. *Vide*. Section 1.
[133] *Opticks. op. cit.*, pp. 405–406.
[134] The words in round brackets are conjectural, since the manuscipt is torn in the left-hand corner.
[135] Immediately following this on the sheet is the new addition to *Regula* II, the addition to the end of *Regula* III, and *Regula* IV; all of which appeared for the first time in the Third Edition. *Vide* Koyré, *Les Regulae Philosophandi* Note 121.

CHAPTER 4

# SPACE, GEOMETRICAL OBJECTS AND INFINITY: NEWTON AND DESCARTES ON EXTENSION

## INTRODUCTION

In an early treatise, now entitled *De Gravitatione et aequipondio fluidorum* (c. 1668), Newton attacks Descartes' doctrine of material extension. The occasion for the attack occurs in a discussion of the nature and existence of space, in which Newton argues that the extension of space is uncreated, eternal, immutable, and endowed with an autonomous geometrical structures.[1] Not surprisingly, the proposition on which Newton concentrates is his positive claim that "Space everywhere extends to infinity. [*Spatium in infinitum usque omnifariam extenditur*]" (*De Grav.* p. 101).

In order to establish this claim against Descartes' position Newton advances five principal theses: (1) that the extension of space is in fact infinite and not merely 'indefinitely large'; (2) that the infinity of space is an object of the understanding but not of the imagination; (3) that we can have a positive conception of the infinity of extension, though language is compelled to express it in a negative mode; (4) that infinity in itself is not a perfection, but merely 'completes' those perfections or imperfections which are intrinsic to natures; and, lastly (5) that geometrical argument can provide a positive basis for grasping the actual infinity of spatial extension.

It is clear that Newton's theses are philosophical in character. Moreover, they have an intrinsic interest which goes beyond a mere rejection of Cartesian extension. In the main body of my discussion I shall critically analyze Newton's views on infinitely extended space, and also compare them with the relevant doctrines of Descartes' philosophy. In what follows, my analysis is restricted to Newton's early views on extended space (c. 1664–1670), and no attempt is made to consider the evolution of his thought. It is fair to say, however, that the general account of space presented in *De Gravitatione* sets the framework for Newton's later thought on the subject. Furthermore, the epistemic and metaphysical questions that I shall examine, comprise an intricate web of concepts, presuppositions and levels of thought; consequently I shall probe similar sorts of questions in various but related contexts.

A brief statement on the methodology of the present study is in order. It is recognized nowadays that Newton's *De Gravitatione* reveals a systematic grasp of Descartes' natural philosophy, as well as an understanding of the metaphysical principles on which it is based. An earlier source, a unpublished pocket book, entitled *Questiones quædem philosophicæ* (c. 1664), further testifies to Newton's careful reading of the Frenchman's *Meditations* and the

*Objections and Replies*.[2] This is important, as it proves textual evidence that Newton was familiar with Descartes' epistemology, and especially with the latter's approach to the representative character of knowledge. When I consider Newton's use of the distinction between imagining and understanding, in the context of conceiving extension, the fact that he knew Cartesian epistemology will become important. But at the outset, this fact allows me to put Newton and Descartes into dialogue. Not only does the device help to dramatise their individual response to various conceptions of extensions and infinity, but it helps in understanding the subtle ways in which Newton and Descartes redefine elements common to Ancient, Scholastic, and Renaissance thought concerning problems of extension and the infinite.

It is not possible, of course, to give a full treatment of the doctrines of infinity, space, and extension which the 17th century inherited from the past, though I shall have something to say about Aristotle. Moreover, I shall also consider ancient atomism as a source for Newton's views on the infinity of extension. For the most part, however, I refer to those thinkers whose doctrines usefully highlight the views held by Newton and Descartes. In this regard, I shall attempt to show that Newton's early views on how we conceive infinite extension rest on an epistemological basis entirely different from the one later proposed by John Locke in his *Essay*. In making this suggestion I intend nothing anachronistic; rather, my argument will be that Newton's claim to understand *that* extension is infinite, cannot be accounted for by the sort of epistemology which Locke will later attempt to justify in Book II of the *Essay*.[3] Indeed, Newton's views emerge from an entirely different tradition. Furthermore, I shall suggest that Newton's epistemic approach to extension *per se*, has significant affinities with Descartes' treatment of extension (*qua* the essence of matter) as found in the *Meditations* and elsewhere in his writings.

In an earlier study I have pursued the question of Henry More's influence on Newton's doctrine of extended space.[4] I conclude that it is minimal in the period from 1664 to 1668, on the ground that Newton's conception of how space relates to the being and existence of God is subtly different from More's Neo-Platonic conception. Moreover, unlike More, Newton does not conceive infinity as being constitutive of divine essence, but as second level attribute which characterizes each of God's perfections (see Section 4). Consequently, though it is true that Newton's commitment to an ontology of natures is Platonic in character, I shall suggest that the primary influence in this respect is Descartes' *Meditations*, rather than the eclecticism of Renaissance Neo-Platonism, of which we find little evidence in *De Gravitatione*.

## 1. SPACE IN INFINITELY EXTENDED

I shall begin by simply stating Newton's position before justifying my interpretation. The *Questiones* provides early evidence of Newton's commitment to the actual infinity of extension. Under the heading 'Of Quantity' he says:

> If extension is indefinite only in greatness and not infinite, then a point is but indefinitely little; and yet we cannot comprehend anything less. To say that extension is but indefinite (I mean all the extensions which exists and not so much only as we can fancy) because we cannot perceive its limits, is as much to say God is but indefinitely perfect because we cannot apprehend his whole perfection (folio $131^{r}$).

There is no doubt that Descartes is the target here: it is he who uses the term 'indefinite' when claiming not to know whether extension is truly infinite. In fact, many entries in the *Questiones* testify to Newton's careful reading of the Frenchman's views on extension from both the *Meditations* and *Principles of Philosophy* (folio $83^{r-v}$), folio $92^{r-v}$, folio $93^{v}$). Newton's main point is straightforward enough: the scope of what is imaginable (i.e., what we can 'fancy') provides no grounds for the conclusion that in reality extension is "indefinite only in greatness and not infinite", any more than our inability to comprehend the fullness of God's perfections in any way affects the reality of divine being. While Newton gives no indication here of how we know the infinity of extension, he clearly rejects what he takes to be a negative thesis, i.e., that if things are epistemologically indeterminate, this provides grounds for claiming that they are indeterminate in reality.

A similar anti-Cartesian argument is made in *De Gravitatione*:

> Nor is it an objection that space is said to be indefinite in relation to us, that is, that we are simply ignorant of its limits and do not positively know that there are none (Part I, Art. 27 [Descartes' *Principia*]). This is because, although we are ignorant beings, God at least understands that there are no limits, not merely indefinitely, but certainly and positively, and because though we negatively imagine it [space] to transcend [*transcendere*] all limits, yet we positively and most certainly understand that it does so (p. 102).

Here Newton rejects what he understands to be the Cartesian view, namely, that since we can only determine negatively that extension lacks limits, this merely provides grounds for claiming that it is "indefinitely great". Moreover, the distinction between what is properly understood by the intellect and what can be grasped by the imagination is made explicit. In Newton's view, knowledge of the infinity of spatial extension falls under the scope of the understanding, and he seems to imply that Descartes conceives extension to be indefinite because he illicitly restricts our conceptions of it to what can be determined by the imagination.

There are a number of interrelated issues here which need to be disentangled and considered in their own right. Let's ask first: what conception of the infinite extension of space does Newton have? He begins his account of space's ontology by defining extension as "the uniform and unlimited stretching out [*distensio*] of space in length, breadth, and depth"; (*De Grav.* p. 100) for "space everywhere extends to infinity" and "we cannot imagine [*imaginari*] any limit anywhere without at the same [*simul*] time understanding [*intelligamus*] that there is space beyond it" (*De Grav.* p. 101). Thus our conception of space is not restricted to negating in imagination any boundary that can be assigned to extension; rather, the understanding grasps *that* extension transcends all limits "not merely indefinitely, but certainly and positively" (*De Grav.* p. 102). It is clear, then, that Newton holds that the understanding not only grasps *that* space is infinite, but also implies that it has a direct and positive conception of the infinity of extension, even though it can never be sure that it has an adequate comprehension of all that is contained in that nature. So while it is impossible that the mind can imagine the infinity of extension, Newton contends that "we can understand it. We can imagine a greater extension, and then a greater one, but we understand [*intelligamus*] that there exists a greater extension than any we can imagine. And here, incidently, the faculty of understanding is clearly distinguished from imagination" (*De Grav.* p. 101).

Newton's reasoning turns on two commitments which he draws on explicitly in the section from which these passages are taken. The first is the epistemological principle that knowing how something lacks with respect to a kind presupposes a positive conception of the genus in question. I shall return to this principle presently. In the second place, Newton holds that any real being "either has limits or not and so is either finite or infinite" (*De Grav.* p. 102). Accordingly, whether or not we are in a position to justify making true statements about extension lacking limits, or able to imagine an obtaining state-of-affairs of that sort, it is the case that extension is in fact either limited or not. And if it is not limited, it is true that it is infinite, and true to say that we can understand *that* it is. When combined, these principles allow Newton to hold that the imagination's capacity for successively negating the assigned limits of extension presupposes that the understanding grasps that extension is an infinite nature. So though it is indefinite how much extension can be imagined, this in itself indicates no ground for denying that extension is infinite in the sense of being bound by no limits.

We are now in a position to reconstruct the conception of infinitely extended space which is implicit in Newton's thought. Though the imagination can't grasp it, the intellect understands that space's extension is in fact infinite in the sense that it is not intrinsically self-delimiting: it is therefore impossible that anything can exist beyond it since it lacks boundaries. This conception is clearly to be distinguished from the fact that beyond the boundaries of a finite being it is possible that something other than itself exist. And since Newton holds the non-Aristotelian principle that that which is unlimited is fully actual, he at once links actually of spatial infinity with the completeness of being. So, instead of the Aristotelian notion that "The infinite is not where there is nothing further, but where there is always something further" (*Physics* III, 6. 207a 1–2), Newton holds that the infinity of space's extension is a completed reality beyond which there can be nothing. The Newtonian position is consequently far removed from Aristotle's claim that the infinite is that in which always more can be taken, such that each taking is non-identical with anything taken before (6. 207a 5–10).

It is evident that Newton's conception of the magnitude of spatial infinity is 'Epicurean' in spirit. As Newton's contemporaries understood the atomistic tradition, an infinite quantity is one whose existence cannot be limited by anything separate from, or other than, itself. In a work which Newton probably read, Thomas Stanley's *History of Philosophy*, Stanley puts the Epicurean claim as follows:

> – that which is finite hath a bound, that which hath a bound, is seen from some other thing, or may be seen from out of an interval beyond, or without it. But the Universe is not seen out of any other things beyond it; for there is no interval or space which it containeth not within itself.[5]

Of such an entity it can be said that it lacks boundaries, and anything that lacks boundaries has no limit; hence it is impossible that anything other than itself exist beyond it; therefore it is unlimited or infinite. The reasoning here is similar to my reconstruction of Newton's position, and it exemplifies a pattern of thought to be found in Epicurus' *Letter to Herodotus* as preserved in Diogenes Laertius, Book X, an edition of which Newton owned and used. "The universe is boundless. For that which is bounded has an extreme point: and the extreme point is seen against something else. So that as it has no extreme point, it has no limit; and as it has no limit it must be boundless and not bounded".[6] As well as these sources, Newton was familiar with Charleton's *Physiologia Epicuro-Gassendo-Charletoniana*, and probably (in this early period) with Lucretius' *De Rerum Natura*, both of which conceive infinite quantity in a similar way.[7] Although these sources argue that the universe is boundless, it is easy to see that Newton could adopt the arguments to the claim that the extension of space is infinite.

We can get a better grasp of Newton's conception of the infinity of spatial extension if we compare it with the leading doctrines concerning quantity common in the period. According to the Aristotelian tradition, things are quantitative just in case they are finite, possess determinate boundaries, and are conceived to have parts which can be increased indefinitely. Given this conception of quantity, the positive existence of an infinitely extended space is contradictory to the notion of quantity itself, since it supposes the actual existence of a quantity than which there could be none greater. But this is a claim which is inconsistent with the view that quantity is essentially capable of increase or decrease. It is clear that Newton must deny that the infinity of extension presupposes quantity thus understood, i.e., that infinity of extension is to be understood in terms of the potentially of endlessly dividing or increasing a determinite quantity such that at any given time it cannot consist of an actual infinity of parts. And he must deny this, because he holds that the extension of space is an unlimited expanse, at once containing an infinite number of simultaneously actualized parts. In other words, it is a completed totality, beyond which it is impossible that anything separate from itself exist.

But can Newton accept the view, common to thinkers like Cusanus, Bruno, and Leibniz, that though the notion of a completed infinity is incompatible with the notion of divisible quantity, there are grounds for speaking of actual infinity? This is the view that when we truly conceive of a positive infinity, we grasp that it pertains to natures which cannot fall under the category of determinate quantity.[8] For given that completed infinity – conceived as a mode dependent on quantity – is a self-contradictory notion, it provides a *reductio ad absurdum* of the view that the notion of a positive infinity pertains to that which is capable of parts. That is to say, according to the tradition which Leibniz represents, the infinite pertains to unities which lack composition, and whose being or essence are one, eternal and indivisible.[9] I shall have more to say about this conception presently.

It is clear that Newton cannot accept the tradition that opposes completed infinity and quantity, since the infinite extension of space, though not a quantity in the Aristotelian sense, is nevertheless quantitative in character: it is a completed and actual totality. Furthermore, Newton holds in *De Gravitatione* that space is an indivisible quantity in the sense that it comprises an infinite number of indivisible units which cannot in any way be actually changed or altered (*De Grav.* p. 103). To the notion that space is an indivisible quantity Newton relates his theory that infinity is a second-level character which characterizes the intrinsic reality of the extended nature of space itself (see Section 4). Thus the grounds on which Newton is willing to legitimize talk of an actual infinity are different from Leibniz's. They are involved in the conception that space is a differentiated being beyond which nothing other than itself can possibly exist, a reality which (as we shall see) embodies in its nature a link which connects its indivisibility and infinity. And in Newton's view, spatial extension is infinite in the straightforward sense that it is without limit.

Early in 1665 Newton defined the notion of an 'infinitely Greate' quantum in a little treatise entitled 'quantity or Muchness'.[10] Euclidean in origin, the treatise deals, among other topics, with proportions, ratios and incommensurables. On folio 10$^{r}$ Newton writes: "That quantum is infinitely Greate or greater $y^{n}$ any finite quantity $w^{ch}$ is not increased or diminished by being joyned to a finite quantum". It is clear that the phrase beginning "$w^{ch}$ is not – ", refers to the terms 'quantum'. Accordingly, Newton is attempting to capture the notion that there is a 'something' which can be expressed in absolute terms in a geometrical context. But what does he mean to convey by the term 'quantum'? We get a clue at the beginning of the treatise. "The subject of Quantity or Muchnesse, *as it is such*, my bee called a Quantum" (folio $9^r$. My italics). If we don't talk of quantity in general, *as such*, but always of this quantity or that quantity, when answering the question how much, Newton's 'quantum', the subject of "Quantity or Muchnesse", is not an appropriate answer, i.e., it is not a quantity. To speak of a thing as a quantity is to refer to it's length or to its area; whereas, to say that something is a quantum is to say that it is quantifiable in some respect or other. Accordingly, when Newton says that a quantum is infinitely great, he is not saying that there is a *quantity* which is non-comparatively or intrinsically great. Rather, he is claiming that the 'measure' of an infinite quantum is in no way altered in comparison with the 'measure' of a finite quantum. Indeed, the point of that comparison is to attempt a definition of an infinite quantum which denies that it is capable of addition and subtraction in the manner of a finite quantum.

It is clear that Newton's definition is couched in negative terms. Nevertheless, it seems to me that he has in mind a Platonic distinction between something's being a nature and something's having a nature, where the nature defines the sort or class, and something is finite in a class if it lacks in an appropriate way what an infinite member of that class has. Accordingly, when Newton says that there is an infinitely great quantum, he is presupposing an infinite nature which defines the category of the quantitative, a nature presupposed by our understanding of the quantitative character of individual things, just in case they have determinate embodiments of the nature. A full justification of this claim must await Section 5, in which I consider infinite extension and its geometrical determinates; at this point let it suffice that the interpretation is at least consistent with Newton's text.

## 2. IMAGINING AND UNDERSTANDING

So much for a statement of Newton's general position; it is now time to justify the interpretation. I begin with the epistemic distinction between imagining and understanding which figures prominently in both Descartes' and Newton's conceptions of extension. Interestingly enough, there is much that they both agree on, though the ways in which they relate this distinction to questions of infinity and extension differ. Admittedly, in what follows I shall have to reconstruct Newton's reasoning on the basis of the evidence provided by the *Questiones* and *De Gravitatione*. Fortunately, the *Questiones* show that Newton, as well as knowing Descartes' scientific works, was familiar with the epistemology of the *Meditations*. He made notes on, and listed extensive page references to the 1656 Amsterdam edition of Descartes' writings. His references pertain to topics such as necessary being, the idea of conceiving infinity and of grasping an infinite nature, the distinction between existence and essence, the mind-body distinction, the conception of a true and immutable nature, and the distinctions between objective, formal, and eminent reality, etc.[11] Besides these and many other references to Descartes's epistemology, Newton wrote an essay on Descartes' theory of objective and formal reality as it is presented in *Meditations Three*, *Five*, and *Six*, and gave his own version of Descartes' approach to the ontological argument (folio 83$^r$). Furthermore, many of Newton's page references pertain to the *Objections and Replies* to the *Meditations*, in which Descartes employs his distinctions between infinity and the indefinite, between understanding and comprehending, and between imagining and understanding. In fact, Newton's page references show his familiarity with Descartes' replies to *Objections One, Two*, and *Four*, replies which contain important statements of Descartes' epistemological principles (folio 83$^r$). Though Descartes and Newton draw on common epistemological traditions, it's my belief that Newton was influenced by Descartes' approach to the ontology of natures, as well as by his use of the distinction between imagining and understanding the nature of extension. After all, it is part of Newton's general strategy in *De Gravitatione* to argue that, by genuinely espousing the same principles as Descartes, conclusions can be established which differ from Descartes'.

Both thinkers draw a distinction between the 'faculties' of imagination and understanding, together with the objects appropriate to these ways of conceiving. Though hedged with many qualifications by different thinkers, the distinction is essentially this: the imagination represents the sensuously based features of a given object, but the understanding is capable of producing a non-sensuous representation of an object' essential nature. On this conception, the mind has the capacity of knowing one and the same object in different ways, and, correspondingly, there are levels of knowing appropriate to each way of conceiving. Moreover, we can be said to have confused as well as distinct representations of things.

Accordingly, the epistemological framework presupposed by Newton and Descartes is this. The mind in understanding or imagining 'grasps' or 'apprehends' natures in so far as they are embodied in the reality of things, and in so far as that reality is present in sensuous experience. Furthermore, like Descartes, Newton accepts the representative character of knowledge. In the epistemological tradition followed by both thinkers, what the mind is said to be immediately aware of is the content of its constructs or ideas, where to conceive of an idea is not necessarily to imagine or to picture. On the contrary, abstract reasoning concerning the properties of ideas and their relations, involves the mind's conceiving ideas as objects of mental vision, in much the way in which things are said to be perceived as objects of sight, but with this difference: when the understanding 'perceives' something, it *sees* through the inessential to the essential nature in question.

There is a further distinction that Descartes explicitly accepts, which provides an ontology for his use of the distinction between imagining and understanding. This is an act/content conception of thought, which is found in Medieval writers and in the works of the Scholastic commentators. A mental act exhibits the mind's power of representing, while the content is what the representation is about. Thus, the content of any mental act has something of which it is about; whether it is an act of imagining or understanding, in each case there 'exists' an intentional object *in* the act. In other words, the object has intentional inexistence, the mode of being an item has when it is thought of *qua* thought; in Descartes' language, the "objective reality" of the idea. Both representings and representings *qua* what they represent (i.e., their contents) are modifications of mind. Furthermore, if the representative content of an act is *true*, it can be said to exist not only *in* the act, but also to possess a ground which exists independently in the world. In other words, everything that is represented by a *true* represented as existing in the world *is* in the world.

While there is not sufficient evidence to show that Newton accepts (as does Descartes) a fully developed act/content ontology of mind, he certainly holds that there is a correspondence between the 'objective' content of true ideas *qua* thought of, and their extra-mental ground in the natures of existing things (*Questiones*, Folio 83$^{r}$). In other words, he accepts the 'mentalization' of inner representations of outer reality. Given these epistemological principles, 'grasping' or 'apprehending' describe the mind's attempt to conceive an independent world through the presumed accuracy of the representative content of its ideas. Thus we see again that there are levels of knowing in as much as representative contents can refer to the sensing, imagining or understanding of a given object through one and the same type of experience.

While Descartes accepts the distinction between conceiving through the imagination and conceiving through the understanding, he denies that the mind is endowed with distinct faculties. His is a dynamic conception of the mind's capacities, in which affirming, denying, willing, conceiving, understanding, and imagining are all *modi cogitandi*, or cognitive powers appropriate to one and the same indivisible mind.[12] Thus each of these activities is an appropriate modification of a mind endowed with the innate capacity to think, but not a distinct power belonging to a separate cognitive faculty.

Descartes' use of the distinction between imagining and understanding occurs at two important places in his *Meditations*: the argument concerning the essential nature of a piece of wax; and the argument concerning the manner in which the mind apprehends figures such as the chiliagon. In both cases, Descartes argues that through imagining alone, the mind attains an incomplete and confused representative content of these respective natures. Moreover, ideas of imagination presuppose changing sensory representations of particular things, the images or 'pictures' of which the mind produces, while the mind's power of understanding forms a purely conceptual representation, in the content of which the thing's essential nature is grasped. Of the wax, for example, Descartes says that the nature of its extension, flexibility and changeability is not captured by the mind's imaginative capacity to picture what is now round as becoming square or triangular, etc. On the contrary, " – I conceive it (the wax) capable of undergoing an infinity [*innumerabilium*] of similar changes, and I cannot compass this infinity in my imagination. Consequently, this conception that I have of the wax is not achieved by the faculty of imagination" (AT. VII. *Meditation* II, 31). Rather, the essential dispositional nature of the wax is conceived by the power of the understanding; for it grasps distinctly and clearly that the wax is an extended, flexible and changeable nature; in contrast, that fact is conceived confusedly in terms of the sensuously-based and occurrent representations, the contents of which are present in the imagination.

It would be wrong to suppose that Descartes denies that the wax is perceived by the senses. The mind grasps the wax through a notion formed by the understanding, a notion that is originally confused by various sensory ideas and images which we mistakenly suppose are a part of what the wax is. However, once the understanding apprehends that the wax is something purely extended, flexible, and changeable, the conception of its nature is clear and distinct. "But what is here important to notice is that perception is not a vision, a touch, nor an imagination, and has never been that, even though it formerly appeared so; but is solely an inspection by the mind, which can be imperfect and confused as it was formerly, or clear and distinct as it is at present, as I attend more or less to the things of which it [the wax] is composed." (AT. VII. *Meditation* 11, 31). Here we have an epistemological argument which has metaphysical consequences, at a stage in the dialectic prior to Descartes' being in a position to show that extended things exist extra-mentally.

To a large extent, the Cartesian understanding of the distinction between the imagination and the understanding is one which Newton accepts, though not necessarily Descartes' complete conception of the act/content ontology of mind. Newton's statement to the effect that we can successively imagine greater and greater extension, and yet can "negatively imagine it [extension] to transcend all limits" (*De Grav.* p. 102) in order to achieve a positive representation of its actual infinity, are clear evidence of this. Moreover, I shall argue that Newton's conception of the extension of space is in no way based on a doctrine of innate ideas, or on a view of the mind's capacity to form *a priori* constructs which are determinative of experience.

But what is it that the understanding grasps when it represents the infinite extension of space? Newton's answer is surprising: the understanding apprehends through its representative content that extension is an "eternal, infinite, [and] uncreated" nature which is revealed by the power of the mind alone (*De Grav.* p. 111). It is the understanding that has an "exceptionally clear idea of extension" (*De Grav.* p. 99) based as it is on a nature which is "real and intelligible" (*De Grav.* p. 110) in and of itself. Newton's conception is

Platonic in spirit. He is claiming that the mind has a non-sensuous representation of an eternal nature, in this case, infinite extension, a nature which in itself is apprehended by the mind prior to any of its specific determinations. As we shall see, however, for Newton this does not mean that sensory experience plays no role in the mind's ability to grasp extension's nature.

Newton's epistemic approach to the infinity of extension, involving as it does the notion of apprehending extension *qua* extension, can usefully be compared with Aristotle's and Locke's. Though the infinite is not among *aistheta* for Aristotle (*Metaphysics*, K. 10. 1066b 25), he nevertheless considers problems of infinity in the context of studying objects of *aísthēsis*. In Aristotle's view, the core notion of the infinite is "that which cannot be gone through (*Physics*, 111. 6. 204a 15, *Met.* K. 10. 1066a 35sqq.), as when it is said paradigmatically of a series of numbers, say the positive intergers, that they are an endless progression. Consequently, he explicates his conception of infinity on the notion of number, or 'the countable', as it pertains to quantity, i.e., to continued quantity construed as the repetition of a determinate unit (*Phys.* 111. 6. 204a 17–19, *Met.* K. 10. 1066b 5–10).

Aristotle is emphatic that the infinite is not a thing which exists sparately; in fact, it is simply a category mistake to think of it as being in any sense 'a thing or object'. It is this conception which motivates his appeal to the 'mode of being' which 'a day' that it is occurring and will occur, and of 'the games' that they are going on and will go on. Of 'Socrates' or 'a horse' this cannot be said; they are individuals which, *qua* individuals, exist as a whole at anytime. But this cannot be said of 'a day' or 'the games', as their 'mode of being' is successive in character. Aristotle is here mobilizing a distinction between completed and uncompleted 'things'. 'A day', 'the games' or 'a walk to somewhere' fall into the second category. Of each it can be said that it is *now* occurring or is *now* going on, as well as that it will occur or will go on. Being successive in character, they possess a characteristic *péras* which indicates that they are processes which can be completed, i.e., we can *at* any time be walking to Thebes, though at a given time the process is incomplete. However, of individuals like 'Socrates' or 'a horse', *qua* individuals, it cannot be said that they are occurring or are going on.

Despite Artistotle's claim that 'the infinite' is unlike a thing, but like the games' is successive in character, we must recognize an important disanalogy between 'the infinite' and 'the games'. As Aristotle understands the *ápeiron*, it of course lacks any characteristic *péras*; as such, it is unlimited and hence *incompletable*. We can say of 'the games' that they are going on *at a time*, as there is a characteristic *péras* applicable to that type of phenomenon. However, it is self-contradictory to say that the infinite is going on *at a time*; that would be to complete the incompletable. So while we can be playing a game at a time, we cannot be said to be infinitely dividing *at* a time, since an infinite operation cannot ever be complete and actual. Furthermore, we can say of 'the games' that they take place in a stretch of time, but temporal notions are inapplicable to the way in which the infinite is successive. It is not a process that takes time; rather it is successive in the way in which a numerical series is successive. In short, for Aristotle, the intelligible root of the *ápeiron* is that it is a well-formed potentiality, the features of which can be understood by means of numerical series and asymptotic procedures.

Though there is no definite text for this interpretation, I think it captures the spirit and thrust of Aristotle's reasoning. His works are everywhere filled with talk of 'number' and 'measure'. For example, when considering infinites in Bk 111 of the *Physics*, Aristotle embarks on an extensive discussion of various rules of division which pertain to a given finite quantity (6. 206b 5–30; 207b 5–25), rules in the formulation of which the notions of number and measure play a prominent part. But also involved in his discussion, at least implicitly, is the notion that a determinate algorithm for generating the positive integers, so we can articulate an algorithm for any determinate rule of division which pertains to quantity. In both cases, in defining the algorithm, appeal is made to an appropriate unit, as well as to a principle of iteration. Thus, to paraphrase Euclid, we can say that given two unequal quantities, from the greater we can subtract more than its half, and from the remainder more than its half, such that a quantity smaller than a given smaller quantity is always reached (*Elements* X. prop I). Here a rule of division *ad infinitum* is stated without appeal to the notion of infinity: all that is mentioned explicitly is an appropriate unit and a rule of iteration. It is procedures of this sort to which Aristotle appeals when he says that a finite quantity is infinitely divisible potentially, a claim which does not entail there being a stage at which the quantity is divided into an infinity of parts.

It would be a mistake to assume, however, that Aristotle's aim is merely to provide techniques which make the potential infinity of quantity conceivable. He must and does go beyond a conceivability claim: he intends that his account of potential infinity pertains to the quantity of things as they exist in nature. For Aristotle there is a significant correlation between objects, considered as quantities which are geometrically conceivable, and the way objects are. Space permits me to give only a brief and programmatic sketch of what I take to be the core of Aristotle's position.

What must be understood are the doctrines of *aphaíresis* and intelligible matter (*noete hýlē Met.* Z. 1036a 10; H. 6. 1045a 33–35) as they relate to Aristotle's claim that the genus is matter (*Met.* Δ. 28. 1024b 36–b4, *Post. Analytics*. A. 10. 76a 31–36, 76b 11–16). When Aristotle says of a genus that it is the matter, he means that the genus term refers to what underlies the differentiae; "for that to which the differentia – belongs is the substratum, which we call matter" (*Met.* Δ. 28. 1024b 9–10). It is important to realize, however, that Aristotle's doctrine of *aphaíresis*, the notion that the mind 'separates off' or 'takes away' certain features of a sensible object, in order to attend to others, involves the procedure of taking a genus together with its differences. Thus, in order to show that a circle is a plane figure we must take the relevant differentiae in the right order. What becomes clear is that a plane – the continuous and quantitative in two dimensions – is the matter and a figure the form, a circle in this case being a determinate sort of plane figure established by taking the right differentiae. Accordingly, when we consider a sensible object under the genus of quantity, what we grasp is not the thing's quantity as such, but a hierarchy of differentiae which move the mind from an initially vague grasp of the object to a progressively more refined determination of it; at all times, however, its intelligible matter, the continuous and quantitative in three dimensions, is presupposed. This is not to say that a thing's intelligible matter can exist devoid of all differentiation; rather, it is to claim that intelligible matter underlies its differentiae. For example, the solid is in the genus of solid figures. And given that each figure is such and such a solid, that which underlies (the reference of the genus term) never exists separately or apart: it is always one or other of some determinate sort of figure. Thus, according to Aristotle, mathematical objects are never separate from their substances in any ontological sense. Moreover, they can only be thought of as 'this – suchs' in the category of quantity. So while we cannot conceive an object as just extended, but only as an extended 'this – such' with a certain shape, it is possible nevertheless to understand its underlying extendedness rationally. I shall return to Aristotle's views in Section 5, where I consider Newton's treatment of infinite extension and its relation to the geometricals.

But how does Aristotle's doctrine of the potential infinite relate to taking matter as genus in the category of quantity? After all, in so far as he construes infinity by an appeal to the way in which a numerical series is generated algorithmically, doesn't Aristotle detach the apparatus for handling infinity from its applications? Here it is important to appreciate the role that measure plays when a genus and difference analysis is given. As an analysis becomes progressively refined, the claim that an object *is* a determinate length, can be substituted for the notion that it merely *has* a length. In other words, together with a consistent rule of operation, we can apply a unit of measure to an object.

But we can go further. A quantity can be marked off in successive intervals which can be halved, doubled, or otherwise considered in accordance with a rule of iteration. Book III, Chapter 6, of the *Physics* provides an important discussion of this. Aristotle says that in a finite quantity "the infinite by addition comes about in a way inverse" to the infinite of division. "For in proportion as we see division going on, in the same propertion, we see addition being made to what is already marked off" (6. 206b 4–10). Aristotle buttresses his claim by drawing a distinction between proportional and aliquot parts in the context in which the divisibility of a finite quantity is bieng considered. An aliquot part is an integral fraction of a whole number which, when multiplied by itself, reconstitutes the number, *viz.* 2 is an aliquot part of 10 but not 3. So an aliquot part of a finite quantity is an equal part of the whole which, when appropriately added to itself, reconstitutes the whole. It is clear that a division by aliquot parts exhausts any finite quantity and so fails to capture the *ápeiron* of divisibility and addition (6. 206b 10–15).

But a rule of division by proportional parts does capture the *ápeiron* of divisibility, as well as its inverse additivity (6. 206b 5–10). Aristotle's reasoning can be illustrated by an example. If a third of a finite quantity is taken, a third of what remains, a third again of the next remainder, and so on, each subtraction proceeds endlessly without exhausting the quantity: so too does the corresponding addition of each part taken. Here is a rule of divisibility which subtracts the same proportional part from what remains of a quantity after each division, as successive remainders converge assymtotically through smaller and smaller intervals. There can be no doubt that Aristotle's approach to infinity is algorithmic in character. Moreover, it involves the claim that the mind exercises its capacity to conceptualize iterative procedures in accordance with rules. Consequently, it need not be asked whether infinity is a genuine possibility, that is, whether or not it can remain unactualized through time: Aristotle's *ápeiron* is not a 'thing', nor is it a process which takes time or exists at a time. Rather, it is a well-formed potentiality, the complete nature of which is captured by the mind's capacity to formulate rules which generate iterative procedures, the individual *lógos* of which is established through a genus and difference analysis in the category of quantity.

Now Locke's account of infinity in the *Essay* has similarities with Aristotle's approach, though ultimately it lacks the latter's sophistication. For Locke, the notions 'finite' and 'infinite' are simply "modes of quantity", that is, characteristics of things which have parts, and are capable of increase and decrease by the "addition or subtraction of any least part" (Bk. II. Chapter 17.1). Moreover, essential to Locke's program is showing how the *idea* of infinity can be constructed on the basis of ideas of sense and reflection, together with the simple and distinct modes of number (*Essay*, Bk III, 16.4). For example, given that we have sensory-based ideas of finite bits of extension, the mind can go on doubling or adding those ideas as often as it pleases, a procedure that in fact applies to anything that has proper parts which can be increased by addition. And since there is no reason to stop these mental progressions, we are led to the *idea* of the *infinity* of space (Bk. II, 17.3).

As for Aristotle, so for Locke, number is basic to the notion of infinity. "But of all other ideas, it is number – which I think furnishes us with the clearest and most distinct idea of infinity we are capable of". For the idea of infinity "is nothing but the infinity of number applied to determine parts of which we have – the distinct ideas" (Bk. II. 17.9 & 10). If the infinite for Aristotle is something which cannot be gone through, for Locke it is a process which requires continued addition, a process "which affords no prospect of stop or boundary" (Bk. II. 17.9).

There are, however, some important differences between the two thinkers. Locke speaks as though the infinite were a process, like walking, which takes time. As we have seen, this is precisely what Aristotle denies: to that extend he has freed himself form metaphors associated with ongoing processes. But metaphors continue to entrap Locke's account of the infinite. To be sure, Aristotle is committed to holding that calculating according to a rule is a mental procedure which takes time: but the rule itself which immediately defines the iterative procedure does not involve time. It is the characteristics of the latter which Aristotle's account of the *ápeiron* relies on.

In the second place, Locke's notion of a positive idea is that which pertains to countable units established through sensory experience; it follows that there can be no positive idea of infinity, whether of number or whatever. "– we have no more a clear idea of infinite parts in matter than we have a clear idea of an infinite number by being able still to add new numbers to any assigned number we have". In neither case can the mind grasp an *idea* of an actual infinity, "they both being only in a power still of increasing the number, be it already as great as it will". However, regarding "what remains to be added (wherein consists the infinity) we have but an obscure, imperfect and confused idea –" (Bk. II. 29.16). Consequently, since number is the foundation of the notion of infinity, and as we cannot have a positive idea of an infinite number, so we cannot have a positive idea of an actual infinity of numberable things. Though he would agree with Locke that it is self-contradictory to suppose that the infinite is *actually* completable, there is nothing in his position which prevents Aristotle from holding that we can have a positive conception of infinity as understood algorithmically. Aristotle's conception of the infinite is fundamentally not imagistic, nor does it depend on a genetic epistemology of ideas such as we find in Locke.

Newton's *questiones* provides evidence that he was concerned to understand infinite processes in terms of the properties of numerical series. Under the influence of John Wallis' *Arithmetica Infinitorum*,[13] Newton explores series in which, for instance, a fraction is said to become arbitrarily large as the denominator x approaches zero (folio $90^r$). In this early manuscript, we witness Newton's attempt to formulate the idea that the 'limit' of an infinite sequence can be evaluated mathematically without the notion of an ultimate or infinite ratio. Though Newton never writes Wallis' symbol for infinity $\infty$, when he says that $a/x$ becomes arbitrarily large as $x$ approaches zero, in effect (put anachronistically) he is saying that as the variable $x$ approaches zero the numerical value of the corresponding variable $a/x$ increases indefinitely, i.e., $a/p = \infty$ (*Questiones*, folio $90^r$). Thus it is clear that Newton is willing to model a notion of 'potential' infinity on the properties of numerical progressions. However, with regard to the quantity of objects that exist extramentally, Newton is an Epicurean actualist. He holds that "nothing can be divided into more parts than it can possibly be constituted of" (*Questiones*. folio 89ʳ). Accordingly, if a finite quantity is said to be infinitely divisible, this states a genuine possibility; so to say that the quantity is infinitely divisible is to say that it *can* be so divided. But this is contradictory, as a finite quantity cannot be constituted of infinite parts (folio 89ʳ). It is also clear that Newton does not confuse infinite extension, regarded as an eternally complete form of being, with the idea of the infinity of quantity construed as "an endless growing idea" (*Essay*, Bk II. 17.7). In Newton's view, "all the extension that is" is actually infinite (folio 90ʳ). After all, the infinity of spatial extension is coeternal with divine existence (*De Grav.* p. 103); as such it is a nature of which the mind can have a positive and not merely a negative conception.

Since Locke strongly opposes a doctrine of infinite space which is reminiscent of Newton's, it will be useful to contrast Locke to Newton. Locke's account differs from Newton's in two important ways. First, Locke holds that we have an idea of the *infinity* of space, but denies that we can have a *positive* idea of an actually infinite space: "– I think it is not an insignificant subtilty if I say that we are carefully to distinguish between the idea of the infinity of space and the idea of space infinite" (Bk. II. 17.7). The former idea results from the mind's power of continuing without end a process of multiplication. The latter idea is impossible. It involves having in mind the idea of a completed *and* endless progression. On Locke's account this is a contradiction in terms (Bk. II, 17.7).

Newton, however, affirms what Locke denies. He claims that we do have a positive idea of an actually infinite extension, and not merely an idea of the *infinity* of extended space (*De Grav.* pp. 101–102). While for Locke the infinity of space is merely an indefinite extension of the finite (the latter the sole basis of whatever positive idea we have of it) for Newton the infinite extension of space is positive and real, the conception of which is presupposed by any notion of finite extension we may have. As Newton puts the point in the context of discussing Descartes' doctrine of the indefinite, "the limit or boundary is the restriction or negation of a greater reality or existence [in respect of] limited beings – " (*De Grav.* p. 102). So rather than the infinite being the negation of the finite, for Newton the finite is itself the negation of the infinite. In this conception, he is probably following Descartes (AT. VII, *Meditation* III. 45), and both betray their Neo-Platonic heritage. Though Newton and Descartes make the limit a negative concept and the infinite a positive one, their position involves more than play with words in a game in which Locke merely reverses the order. As we shall see in Section 3, both Newton and Descartes treat the claim that something is without limits as having its epistemic basis in the understanding's capacity to apprehend that fact non-discursively, without the sort of appeal that Locke makes to endless mental progressions. Furthermore, Locke's account of the boundlessness of extension is concerned to show how that idea is constructed by the iterative powers of the mind; whereas, for Newton, to say that we can understand that extension is boundless, is to say how it is with extension *per se*.

Newton's position differs from Locke's in another important way. If the infinite is a mode of quantity, and if quantity is essentially capable of increase and decrease, there is no actually infinite quantity. This would be a quantity incapable of increase or decrease, and hence contradictory to the notion of quantity itself. On Locke's account of how we get the idea of infinity, there could be no idea of an actually infinite extension: this would be an idea of a quantity than which it is impossible that there be any greater. It is clear, then, that there is an intrinsic connection between Newton's claim that extension is an infinite and eternal nature, of which the understanding can form a positive conception, and his denial that the idea of infinite space is based on the mind's indefinitely repeating sensory based ideas of the finite. In Newton's view, it is the understanding that grasps the unchanging and essential nature of extension through its non-sensuous representations, and not the imagination which merely repeats sensory orginated ideas of the finite.

I conclude that Newton's epistemology, based as it is on the distinction between the understanding and the imagination, is close to Descartes', especially on the question of how the mind grasps the notion of the infinite. After all, it is Descartes who claims, in a statement which expresses a view similar to Newton's, "that the notion I have of the infinite is in me before that of the finite because, by the mere fact that I conceive being or that which is, without thinking whether it is finite or infinite, what I conceive is infinite being; but, in order that I could conceive finite being, it is necessary that I remove something from this general notion of being which consequently must procede" (AT. V. 352). While it is true that Descartes (writing to Clerselier) is commenting on the problem of apprehending being *qua* being, and not talking about extension, as such, Newton will argue that the same epistemic principle applies to extension, i.e., that the finite is the negation of the infinite. So the question that must be asked is this: given that Descartes and Newton have similar epistemological views, why does Descartes hold, but not Newton, that we grasp extension as 'indefinite' in character?

## 3. DESCARTES ON NEGATION AND THE 'INDEFINITE'

I shall begin by considering the principles which lie behind Descartes' claim that extension is to be regarded as 'indefinite'. As regards the limits of extension, the divisibility of a body, and the number of the stars, Descartes tell us in his *Principia Philosophiae* (Part 1, XXVII) that:

– these things we call indefinite rather than infinite in order to reserve to God alone the name of infinite, because in him alone in every way we apprehend [*agnoscimus*] no limits, and because we positively understand [*intelligimus*] that he has none; with regard to other things, we do not in the same way positively understand them to be in every part unlimited, but negatively admit that their limits, if they have them, cannot be discovered by us (At. VIII 15).

Descartes' distinction between the indefinite and the infinite is this: a thing is indefinite just in case there are no reasons by which to show that it has limits, i.e., that it is epistemically possible that it is boundless. On the other hand, a thing that is infinite just in case it can be shown that is impossible that it is bounded in any respect whatsoever. According to Descartes, it is in the nature of God alone that it is possible to ground a demonstration of this sort (*Meditation* IV. At. VII 64–69). This is Descartes' official version of the distinction between the negative and the positive infinite. What can be said of the extension of the Universe, for example, can be expressed negatively, as it involves a successive negation of all limits that can be assigned to it. And though we have "no arguments to prove, and not even being able to conceive, that the world has limits", nevertheless, we are only entitled to call it indefinitely large. (Letter to Chanut, 1647, AT. V. 50). But were it possible to show that the extension of the Universe is in fact unbounded, there would be positive warrant to say that it is infinite.[14]

Descartes' position rests on a particular view regarding the relationship between what is conceivable and what is real. He repeatedly distinguishes between that which the finite mind is capable of grasping – such that what is conceivable is the ground of that which can be asserted or denied – and reality itself as it exists independently of conception. So though the mind is not the measure of reality, it is the measure of what can be consistently asserted and denied of it:

I consider only what I can conceive and what I cannot conceive, and I take great pains that my judgment should accord with my understanding. And so I boldly assert that God can do everything which I conceive to be possible, but I am not so bold as to deny that he cannot do whatsoever conflicts with my understanding – I merely say that it involves a contradiction [in conception]. (To More, 1649, AT. V. 269).

Though Descartes claims that the mind is the arbiter of the conceivable, he is certainly not committed to holding that everything encompassed in our ideas exists in reality, any more than it is true that God "cannot do what we cannot comprehend" (AT. III. 545). So given that there are separate reasons to think that our clear and distinct conceptions are not deceived, "whatever we perceive clearly is true, and so it exists, if we perceive that its existence is possible" (To Mersenne, 1642. AT. III. 545).

So what is conceivable is possible, but what is conceivable is not necessarily impossible: if our beliefs involve clear, distinct and consistent (noncontradictory) ideas, and if there are separate grounds for believing that God is not deceitful, we can truly assert that there is a correspondence between what is said and what is in fact the case. In this sense Descartes holds that there is an intrinsic connection between conceiving and that which is conceived. However, while it is true that in conceiving something clearly and distinctly, what the mind conceives is in the nature of the thing, it does not follow on Descartes' view that its attentive operations necessarily exhaust the real, or that the mind is in a position to know that it possesses complete and adequate knowledge of things.[15]

It is clear that Descartes does not draw unwarranted ontological conclusions about what actually is from his epistemological account of the conceivable. When he says that from the mind's point of view extension is regarded as indefinite, Descartes is saying that that characterization is consistent with extension as we conceive it, but not that the mind necessarily exhausts everything that pertains to the nature of extension itself. In this viewpoint, there is nothing which compells Descartes to conclude that extension is in fact ontologically indeterminate. So though there may be properties of extension of which the mind lacks ideas, there is nothing in extension which is inconsistent with conceiving it as indefinite. In this restricted sense the mind is the measure of reality.

Given these epistemological commitments, it is not surprising that Descartes is noncommittal on whether or not extension is infinite. Moreover, he leaves open the possibility that extension may in fact be infinite. Referring again to the extension of the world, and the number of parts into which matter is divisible, Descartes writes to Henry More in 1649 that, "– I do not know whether or not they are straightforwardly infinite" (AT. V. 274). It is precisely because he thinks that the infinity of extension is *conceivable* that Descartes makes this statement. "It conflicts with my conception to attribute any boundary to the world; and I have no measure of what I should affirm or deny except my own conception". Descartes goes on to point out, however, that this is also consistent with holding that the mind demonstrates whether or not the world's extension is in fact bounded or unbounded (AT. V. 342).

In sum, though it is epistemically possible that extension is infinite, in Descartes' view it is not epistemically possible to show that this is a fact. On the other hand, he sometimes allows that extension could be finite (AT. V. 274). This, however, cannot be a genuine possibility for Descartes, as it violates his fundamental doctrine that wherever there is extension there is material substance. So if Descartes were seriously to posit a finite and bounded world, this doctrine would be contravened. A bounded world can only be demarcated by something other than itself, by something that is not a material substance, i.e., a void or an immaterial space; both sorts of entities, however, are impossible on the principles of Descartes' philosophy (AT. V. 269 & 401). It is true that he tells Chanut in 1647 that the extended world may have limits "which are known to God, though incomprehensible to me: and that is why I do not say decisively that it is infinite" (AT. V. 51–52). Later in 1649, the same possibility is mentioned to Burman, namely, that God may have "a conception and understanding of fixed limits in the world, and, in number, quantity, etc. – and so for him these things may be finite". (At. V. 167). It is difficult, however, not to read these statements as pious gestures to divine power and wisdom, as well as characteristic affirmations that the finite mind is incapable of comprehending God's intentions regarding creation.

Recall, that in the *Questiones* passage (cited in Section 1), Newton seems to accuse Descartes of illicitly moving from a claim about the epistemic indeterminacy of the limits of extension, to the claim that extension is indeterminate in reality. In *De Gravitatione.*, Newton returns to this issue. He tells us that Descartes should be corrected by the grammarians", since "the indefinite is never said of that which actually is, but always pertains to a future possibility –". In Newton's view, the term 'indefinite' is to be reserved for unrealized yet possible forms of being, and is therefore inappropriately said "of that which actually is". Of that which is, it must be possible to say that it "either has limits [*terminos*] or not, and so is either finite or infinite" (*De Grav.* p. 102). As early as the *Questiones*, Newton makes it clear into which category extension falls: "It is indefinite, that is undetermined, how great a sphere may be made, how great a number may be reckoned, how far matter is divisible, how much time or extension we can fancy; but all the extension that is [and] Eternity – are infinite" (folio $90^r$).

It is clear that Newton is challenging Descartes to make a genuine ontological claim. But how are we to interpret his accusation that with respect to extension Descartes applies the principle that "that which actually is, is not to be defined [*non est definiendum*]? (*De Grav.* p. 102). On one interpretation, we might say that Newton is accusing Descartes of holding that extension is indefinable just because he holds that it *is* ontologically indeterminate. On this view, the basis of Newton's criticism is ontological, i.e., extension can't be indeterminate in nature, as that conception is contradictory to what is, whether or not we are able to define it. In order to sustain this view, however, it is necessary to suppose that Newton fails to understand Descartes' epistemological conception of how the conceivable relates to that which is.

On the other hand, as John North observes, Newton may be accusing Descartes of equivocating on the notion of the 'indefinite'.[16] I shall make a related observation. The Greek *horisós* and the Latin *definio*, both convey the notion that to define is to set bounds or limits. Accordingly, in defining something we limit or confine it to a kind or species: we say what it is. But the root meaning of these traditional notions also involves reference to the fact that in reality things are said to have measurable boundaries or limits. So if we say that the extension of the world is defined, this means that we know something about it, i.e., we know that it has limits in reality. On the other hand, if we say that the world is 'indefinite', this can mean *either* that we are not in a position to know whether it is finite or infinite, *or* that it is not defined, i.e., that its limits do not exist. On this reading of Newton's criticism, he is accusing Descartes of equivocating between the notion of epistemological indeterminacy, and the view that extension is 'indefinite' in the sense that it is without limits, i.e., an extended *ápeiron*. On the second view, Descartes can reasonably be taken as saying that extension is infinite. It is clear from my general discussion of Descartes' epistemology that he is able to separate these two sorts of claims. But it also seems likely that he is attempting to run both together under the rubric of the 'indefinite', much to Newton's consternation.

Descartes' ambiguous use of the 'indefinite' allows him to develop another plausible pattern of reasoning. Since we cannot demonstrate that the world is limitless, all that is required is a negative conception of it. To every boundary assigned to the world, "or the totality of corporeal substance", another can be imagined, such that no upper bound can be assumed in the procedure (AT. III. *Principia*. Part II, XXI. 52). But this provides a concept of negative infinity, or, in Descartes language, the 'indefinite'. However, one "cannot infer because limitation involves in itself [*contineat*] the negation of infinity, that the negation of limitation involves knowledge of infinity". On the contrary, that which "makes the infinite different from the finite is something real and positive" (AT. III. 427). But it is only God or "the greatest being" that can be called infinite in this sense. Without doubt Descartes has a two-fold conception of infinity: a negative idea of the indefiniteness of material extension, in virtue of the imagined negation of each boundary assigned to it, and a positive conception of the infinite reality of divine being.

## 4. NATURES, PERFECTION AND INFINITY

Is it the case that Descartes must restrict the notion of positive infinity to the nature of divine essence? Consider first the views of Descartes and Newton on whether infinity can be properly understood when expressed through the vehicle of the grammatically negative. According to Descartes, the mind cannot understand the infinite by "the negation of limitation". His reason is straightforward. If what makes the infinite different from the finite is positively real, "limitation, by which the finite differs from the infinite, is a non-being or a negation of being [*est non ens sive negatio entis*]; and that which is not, cannot lead us to a knowledge of that which is" (AT. III. 427). On the contrary, the act of negating a thing's nature presupposes positive knowledge of the thing itself. So though infinity can be characterized in terms of negating negations, it does not follow that the positive nature of infinity is known through this procedure, but only the nature of what is said to be 'indefinite'.

Newton's comment on this kind of reasoning is as follows:

> Should it be said that we do not understand what an infinite being is (*quid sit ens infinitum*), save by negating the limits of a finite being, and that this is a negative and faulty conception, I deny this. For the limit or boundary is the restriction or negation of a greater reality or existence [in respect to] the limited beings, and the less we conceive any being whatsoever to

be constrained by limits, the more we understand that something applies to it, that is, the more positively we conceive it. And thus by negating all limits, the conception becomes most positive. End [*finis*] is a word negative as to sense, and thus infinity, as it is the negation of a negation (that is, of ends), will be a word most positive according to our sense and conception, though it seems grammatically negative (*De Grav*. pp. 101–102).

There is much here that Descartes can agree with. Both thinkers hold that the procedure whereby the mind negates the negation of a nature in expressing the infinite, depends on a prior and unrestricted conception of that nature. If a restricted conception of a kind defined by a nature presupposes a cognitively prior apprehension of the kind itself, in Newton's view "negating all limits" in our idea of it, is a legitimate way of expressing that prior conception. Thus in conceiving the infinite, "though we negatively imagine it to transcend all limits, yet we positively and most certainly understand [*positive et certissime intelligimus*] that it does so" (*De Grav*. p. 102). Though our language is grammatically negative, it does not follow that what we understand is devoid of positive ground. Descartes agrees on this point, saying that it is "common usage – to conceive of a thing bound by no limits in order to understand the infinite" (At. III. 427). Furthermore, Newton as much as Descartes, is committed to holding that the negation of finite being does not in itself provide a positive conception of infinity. This merely posits a lack of limits in a thing in accordance with our mode of conceiving it, but it cannot change the finite reality from which the mind beings. What is characteristic of Descartes' position, however, is the conviction that the positive infinity of divine being cannot adequately be expressed in the language of the 'not-finite', as divine reality cannot be conceived in terms of what it is not. Descartes admits, as does Newton, that we are nevertheless forced to use the grammatically negative in expressing it.

In the first version of this paper I wrote: "In Descartes' view, extension contradicts the nature of infinity, because it is an imperfection, and infinity cannot be admitted where there is not perfection". As it happens, what I intended to state is elliptically and misleadingly expressed. In the first place, I ought to have claimed explicitly that "extension contradicts the *positive* nature of infinity" where infinity is conceived to be constitutive of divine essence. Unfortunately, the claim is at best contextual implied, and North was understandably led to suppose that I was claiming that the idea of infinite extension is *per se* self-contradictory for Descartes. In the second place, I argued that according to Descartes, extension is an imperfection in comparison with other realities. This, too, was disputed by North in his original reply.

However, nothing in my position then or now commits me to the view that extension is self-contradictory for Descartes. After all, basic to my interpretation is the claim that Descartes is willing to conceive of extension as limitless, and consequently infinite in that sense. Moreover, because he thinks that he lacks positive reasons for showing that it is impossible for extension to be limited, Descartes remains epistemically non-committal regarding its nature. Furthermore, if the argument thus far is correct, Descartes is committed to the principle that grasping the finite of a kind presupposes positive knowledge of the infinite of the kind (AT. VII. 45 *Meditation III*) a principle that applies equally to extension.

There are a number of passages in Descartes' writings in which he suggests that extension is without limits, and where these commitments are apparent. To More he says: "It conflicts with my conception, or, what is the same, I think it involves a contradiction, that the world should be finite or bounded; because I cannot but conceive a space beyond whatever bounds you assign to the world" (AT. V. 341). Similar arguments are put to Chanut and Burman (AT. V. 51 and AT. V. 167), and in each case Descartes seems to admit the possibility that the extended universe is infinite. The structure of his argument is this. For any boundary assigned to the extension of the universe, it is always possible to conceive an area beyond that boundary which much be included in the universe. But for every boundary assigned to accommodate an additional area, the same procedure applies, and so on. It is thus impossible to talk of a bounded universe, or of a universe that has the greatest extension. This is no doubt what Descartes has in mind when he tells Burman that "we can perhaps say that the world is infinite, and the same for number, etc."; for "when the indefinite is multiplied again and again – it is identical with infinity" (AT. V.). This reasoning seems to legitimize the term 'infinity' as an appropriate characterization of extension. Again, it is clear that Descartes sees nothing self-contradictory in the notion that extension is infinite. Moreover, his argument is reminiscent of Newton's claim that though we negatively imagine successive limits to extension, we nevertheless understand *that* there is extension beyond any limit (*De Grav.* p. 101).

Thus far it seems that Descartes ought to accept the view that God and extension, *qua* extension, are both infinite according to their kind. This is precisely the view that Newton argues for. He says:

> But I see what Descartes feared, namely, that if he should consider space infinite, it would perhaps constitute God because of the perfection of infinity. But by no means, for infinity is not perfection except when it is attributed to perfections. Infinity of intellect, power, happiness, and so forth, is the height of imperfection; and infinity of extension is so far perfect as that which is extended (*De Grav.* pp. 102–103).

Here Newton claims that infinity is a character which can be predicated of different things. Moreover, it is a second-level attribute; that is to say, it is a way of characterizing the mode of instantiation of first-level properties in so far so they allow something to be the highest exemplification of its kind. Given these claims, it is clear that Newton construes infinity as an extrinsic attribute which can be said of many things. Thus infinity *in itself* is neither a perfection nor an imperfection, nor is it defining of anything's nature. As Newton says, "infinity is not a perfection except when it is attributed to perfections" (*De Grav.* p. 102).

Newton's reasoning turns on the notion that things exemplify the natures which specify the kind or class to which they belong. Furthermore, he holds that things can be conceived as either perfect or imperfect in accordance with the nature that they exemplify.[17] This allows him to conceive of things as constituting levels of reality in which a difference in level is defined in terms of the specific nature of the kind to which the things belongs. In accordance with this scheme of predication, a thing is said to be infinite just in case it manifests in an exemplary manner the nature of its kind, whether the reality in question is taken to be a perfection or an imperfection.

Thus infinity is not a real or first-level attribute which defines the nature which a thing exemplifies, nor is it identical with either the concept of perfection or imperfection. Newton's formal theory of infinity therefore allows the existence of different kinds of infinity, and is consistent with distinguishing divine infinity from the infinite extension of space. Consequently, Newton argues that God is infinite because it can be said of him alone that he is infinitely wise, infinitely powerful, etc., where to say that God's attributes are infinite is to say that he possesses in an exemplary manner attributes which other agents own deficiently. Accordingly, "not everything eternal and infinite will be God, nor will God be prevented from the eternal and infinite exercise of his omnipotence in the creation and governing of things, by the imperfect nature of created things". In contrast to God, space by "reason of its eternity and infinity" will not be "wise nor powerful nor alive, but will merely be shown to be increased in duration and magnitude". The fact that God and space each possess their attributes infinitely, does not imply that their respective natures are identical, for "no thing is by eternity and infinity made better or of a more perfect nature".[18]

On the face of it, Newton's position on infinity is compatible with Descartes' pronouncements, and indeed may in part have been derived from the latter's philosophy. Why then does Descartes characterize extension as 'indefinite', rather than infinite? Is this merely a pious sop to the theologians? Such an explanation is not sufficient. In the first place, as we have already seen, he holds that extension may be infinite, though we are unable to justify the truth of the statement. And we cannot do so, because we lack positive reasons for showing that extension is necessarily infinite, and are able to claim only that its boundlessness is an epistemic possibility. Accordingly, Descartes' position comes to this: though we can consistently conceive that extension is infinite, we are not in a position to show that it is.

What is at issue can be brought out by considering Descartes' attitude towards the ontological proof for God's existence. On the basis of that proof, Descartes claims that the intellect grasps God's perfections, and sees that he is a necessary being who exists eternally and infinitely (AT. VII. 66–67.

*Metitation V*). Moreover, the act of apprehension is intuitive and non-discursive; in the instant it grasps divine nature, the intellect sees that the mind's objective idea of the determinate, true and immutable nature of divinity is directly grounded in God's formal and persistent reality, the essence and existence of which is necessarily and inseparably connected (AT. VII. 65–66. *Meditation V*). The *cogito* argument has the same formal structure. The intellect moves directly and non-discursively from the objective reality that it is a thinking thing, to the truth that, at the very moment it is thinking, it is impossible that it think without being something that actually exists (AT. VII. 25. *Meditation II*).

It is on this basis, the existence of self and the existence of God, that Descartes grounds the possibility of systematic and discursive knowledge. In his view, the ontological argument shows that divine nature possesses all perfections in a limitless manner, and consequently that God's substantial presence embodies the absoluteness of infinity. He makes this claim emphatically:

> By an infinite substance, I understand substance which has actually infinite and immense and true perfections. This is not an accident superadded to the notion of substance, but the very essence of the substance taken absolutely and bounded by no defects; these defects, in respect of substances, are accidents, but infinity or infinitude is not (AT. V. 252).

Infinity, truly conceived, is intimately connected with possessing perfection; and since all true perfections are possessed by divinity, infinity is constitutive of its essential nature. Moreover, the possession of absolutely all perfection implies supremacy for Descartes. Therefore, only one infinite substance can exist, as it is contradictory to suppose that more than one supreme being exists.

It is clear, then, that Descartes thinks the ontological argument provides positive reasons for holding that God's perfections are absolutely without limit, and therefore that he is truly an infinite being. This does not mean that infinity itself if comprehensible to the human mind, nevertheless "in so far as clearly and distinctly to understand a thing to be such that no limit can be found it in, is to understand clearly that it is infinite" (AT. VII. 112). Notice that Descartes distinguishes comprehending x from understanding or knowing x. To comprehend something is to have adequate knowledge of it, i.e., to know all the properties that belong to it. Moreover, it is possible that we may in fact achieve adequate knowledge of something, while it remains possible that the mind not know that its knowledge is adequate. However, while it is possible to know a finite nature adequately, though impossible to be certain of it, it is impossible that the mind possess adequate knowledge of the infinite itself, since "incomprehensibility itself is contained in the formal concept (*in ratione formali*) of the infinite" (AT. VII. 368). While it is impossible that the finite mind comprehend the infinite, as that entails grasping at once and entirely all that is involved in the notion, it does not follow that we cannot understand the infinite. As we understand the notion of a triangle completely and distinctly, when we conceive of a figure bounded by three lines in a plane, "so it is enough to understand a thing bounded by no limits in order to have a true and complete idea of the whole of infinity (*ut vera et integra idea totius infinite*)" (AT. VII. 368). Again, we see a basic epistemological principle at work: though it is the case that what is conceived clearly and distinctly by the mind is true, the mind's distinct conception does not necessarily exhaust the reality of what is thus understood.

Descartes is clearly distinguishing infinity, as formally understood from things which are infinite (AT. VII. 113). While we can understand infinity *per se* as that which lacks limits, "the thing itself, which is infinite, is indeed positively understood, though not adequately, that is, we do not comprehend the whole of that which is intelligible in it" (AT. VII. 113). So infinity, abstractly and formally considered, must be understood negatively as a conception that pertains to anything without limits. However, when it is said of something that it is infinite, this is positively understood, in so far as it can be shown that a thing's nature cannot possibly have any limitations whatsoever, and even though the reality that is grasped must be expressed in negative language. It is only of divine nature that we can establish positive grounds, by means of the ontological argument, for claiming that its perfections are limitless in all respects. Unfortunately, Descartes' distinction between understanding *that* something is infinite, and the impossibility of comprehending its infinity, raises a difficulty in the context of the ontological argument. When the intellect non-discursively apprehends the nature of divinity, and grasps the fact that it is a true and immutable nature, the locus of all true perfections, and the ground of its own existence, this is surely to comprehend the infinitude of divine nature, and not merely to understand *that* God's nature is infinite. Indeed, the ontological argument demands that the intellect positively grasps God's nature. How else are we to make sense of Descartes' claim that we have positive reasons which show that divine nature is without limits? But how is this to be reconciled with Descartes' claim that the finite mind cannot grasp all that is involved in the nature of infinitude? In other words, how can Descartes reconcile the view that infinity itself is incomprehensible, with the claim that the mind can have "a true and *complete* idea of the *whole* of infinity" (AT. VII. 368, my italics).

But is there anything about the *nature* of extension that leads Descartes to characterize it as 'indefinite'? It is clear that it does not embody infinity *qua* essence, the conception that Descartes thinks is shown of divine nature by the ontological argument. Moreover, it is manifestly not self-contradictory to hold that extension may be infinite in the sense that it lacks an end point. It is well to recognize, however, that (a) extension, *qua* created reality, depends continuously on God's creative power, and (b) that it is the principal attribute of a created nature, i.e., corporeal substance.

Let us consider (b) first. In so far as we think that being extended is both necessary and sufficient for something's being of material kind, we think of a particular this or that which is shaped or bounded. On the other hand, one and the same extension can be considered as space; but this is just a way of regarding the disposition of, or the relation among, a set of shaped, bounded, and extended quantities (AT. VIII. 45–46. *Principia*, II, articles 9–17). In reality, there is no distinction between the three dimensions that comprise corporeal substance and those which comprise the basis of the idea of space; that is, that which is extended and extension are the same. But this is just to deny the existence of an absolute, vacuum, if to be extended is to be material (AT. VIII. 46). So when Descartes speaks of the generic unity of spatial extension, he is not referring to anything that can exist independently of matter, but only to extension, *qua* extension, as that which is abstracted by the power of the understanding. So considered, extension is no longer an object of the imagination, that is of any particular configuration of extension, and can therefore be conceived as lacking limits; i.e., as infinite. But so considered, extension does not exist in *rerum natura*, since this is tantamount to considering the existence does not exist in *rerum natura*, since this is tantamount to considering the existence of an extension not connected inseparably with material reality. Given Descartes' principles, this is contradictory; "for wherever extension is, there, of necessity, is body also" (AT. V. 219). Consequently, there is nothing in the principles of Descartes' philosophy which compells him to posit the infinity of extension, even though the mind can imagine extension beyond any boundary taken.

In his reply to the *Second Objections*, Descartes tells us that "it is self-evident that it is greater perfection to be undivided than to be divided" (AT. VII. 138). While this claim is made in the context of discussing divine attributes, it is clearly a general principle that Descartes adheres to. It must be understood, however, in relation to Descartes' Platonic conception of unity. In *Meditation Five*, Descartes connects the indivisible perfection of the non-extended with unity and simplicity (AT. VII. 66–69). Divine being is an absolute unity because its nature is simple and all-inclusive. It is simple because God is prior to all differentation: in him being is everywhere one and the same. He is an absolute unity because he possess all perfections completely and inclusively: there is nothing *ab alio* on which God depends (AT. *Meditation* V). These characteristics of unity and simplicity are foreign to the nature of extension. Not only can it be differentiated conceptually, but its part can be separated one from another. This means that the unity of any extension can be conceived as contituted by its parts of which the number and disposition can change. Since Descartes holds that any aggregation of extension in divisible, it follows that extension is an aggregate of aggregates of aggregates, without end. Accordingly, its unity is merely notional and relative to the possibility of a changing plurality of parts. It must be admitted that the implications of Descartes' position on extension do not show that it cannot be infinite. After all, granted that the unity of extension is supervenient on its parts, nothing is said as to its infinity or otherwise. I suspect, however, that Descartes, while not equivocating on the concept of unity, choose to restrict talk of positive infinity to that whose being and unity are one, i.e., to the nature and perfection of divinity as shown in the ontological argument.

I turn now to (a), Descartes' claim that extension *qua* created reality, is dependent continuously on divine power. Descartes certainly holds that any form of dependent being cannot be infinite, since "independence, conceived distinctly, involves in itself infinity" (AT. III. 191). It is clear that this statement is prompted by the nature of divine being, and equally clear that Descartes makes a categorial distinction between the independent existence of necessary being and the radical contingency of created being. Anything, he seems to say, that does not embody in its nature the power of existing beyond a given moment is not intrinsically infinite. Clearly, no reality but God can satisfy this conception. If this is a fair reconstruction of Descartes' reasoning, it seems that he has excluded everything from infinity in that sense except God. And we can well imagine, that in this frame of mind Descartes would think it of little consequence to consider whether or not extension is infinite. Suffice it to say that it is indefinite, on the grounds that the mind cannot show that it is bounded, or show that there are limits to its divisibility. Moreover, though he vacillates on whether extension is finite or infinite, Descartes' reasoning surely presupposes the traditional distinctions of being (*divisio entis*) into *necessarium* and *contingens, creans* and *creatum*, and *infinitum* and *finitum*. According to these divisions, infinity is intrinsic to divine perfection, and finiteness is a necessary defect of created entities. So conceived, an infinite but created entity falls under a self-contradictory notion, and is thus an impossible existent which God cannot create. This does not impose any limitation on God's power as the impossible is not an intrinsic limit.

Descartes' position on the ontology of created extension is therefore different from Newton's. For the Frenchman, extension is an eternal archetype which exists eminently in the divine mind, and is that in accordance with which divine power brings created extension into being. For Newton the extension of space is coeval with the omnipresent existence of God. It exists because God exists. It is little wonder, then, that Newton conceives extension as an eternal, uncreated and infinite nature, and as part of the defining essence of space itself.[19] And why he, but not Descartes, is explicit in holding that the parts of infinite extension exist at once and altogether: in the language of the schools Newtonian extension is *simul et semel.*

## 5. GEOMETRY AND INFINITY

Unlike Descartes, Newton holds that we can understand the infinity of extension through geometry. Why does Newton think this, and what relationship does he suppose exists between the geometry of infinity and the nature of space itself? Two important sources provide answers to these questions: the treatise that we have been considering, *De Gravitatione*; and a document of 1692 which I have entitled *Locus et Tempus*.[20]

Immediately after his claims in *De Gravitatione* that space "extends to infinity", and that everywhere it contains inherent, volumetric figures of all sizes and kinds, Newton says that "you may have in truth an example of something infinite [*Verum ut infiniti specimen aliquod habeatis*]": he asks us to consider any triangle under the following conditions:

Imagine (*fingite*) any triangle whose base and one side are at rest and the other side so turns about the contiguous end of its base in the plane of the triangle that the triangle gradually opens at the vertex (*in vertice gradatim apperiatur*): and meanwhile animadvert to the point where the two sides meet, if they are produced that far: it is manifest that all these points are found on the straight line along which the fixed side lies, and that they become perpetually more distant as the moving side turns further until the two sides become parallel and can no longer meet anywhere. Now I ask: What was the distance of the last point where the sides meet? It was certainly greater than any assignable distance, or rather none of the points was the last (*vel potius nullum e punctis fuit ultimum*), and so the straight line in which all those meeting points lie is in fact greater than finite (*est actu plusquam finita*) (*De Grav.* p. 101).

Newton's argument involves some interesting intuitions which can be illustrated by considering the following triangle:

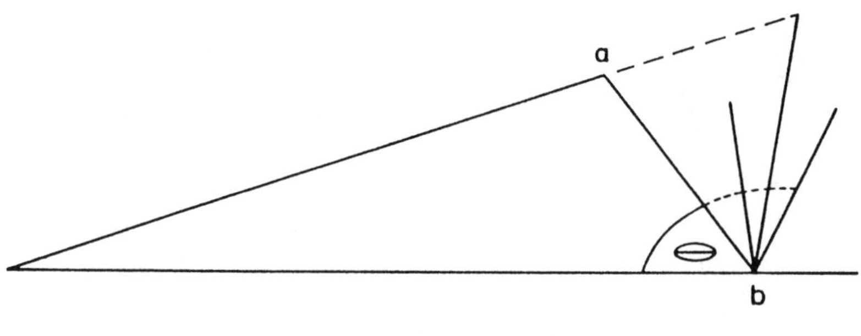


Point $a$ is the intersection of sides $ac$ and $ab$, and sides $ca$ and $cb$ are fixed. Now allow angle $\ominus$ to increase, so as to open at the vertex $b$. Also allow sides $ac$ and $ab$ to adjust so that $ab$ and $ac$ intersect as $ac$ is produced. As the angle $\ominus$ opens out, sides $ac$ and $ab$ approach parallel, so that the array of intersection points transversed before parallelism is infinite. Thus in the geometry we can *understand* that a distance which has no upper bound is infinite.

What are we to make of this argument? Notice first what Newton is not maintaining. He is not saying that the two parallel lines meet *at* infinity; rather, he holds that as one line is rotated towards the position of parallelism with the other, there is *no* last point where they meet. What is infinite is not the distance of the last point, but the array of intersection points traversed before parallelism is reached: it is the line *ac* itself "in which *all* those meeting points lie" that is infinite.

Newton takes himself to have established the following: (a) that the line exists as a clear and distinct object of the understanding; (b) that all the intersection points traversed, lie along it; and (c) that the totality of intersection points is infinite. Can he maintain these claims? I think that the geometers of his period would give Newton the existence of the line as an object of understanding. After all, the property that his argument attributes to it, is consistent with its nature, namely, that since there is no last intersection point on the line traversed, it is infinite. From a modern set-theoretical perspective, however, we would demand a separate proof, one which shows that there is a line which satisfies the claim that an infinite totality of intersection points lie along it. In other words, we would want to know whether it is possible for a line satisfying this condition to exist. In Newton's case, this would mean a proof that any denial that the totality of intersection points is infinite, is open to a *reductio ad absurdum* argument.

But what response would Descartes make to Newton's argument? He would probably construe it from a constructivist perspective. For example, he could say that though Newton's argument shows that we cannot determine that there is a last point on the line traversed, this in itself does not warrant the claim that the line is infinite, but rather the claim that it is greater than any assigned number of intersection points traversed. Now a construal of this sort is certainly consistent with Descartes' conception of the nature of quantity. It does not, however, capture the intuition on which Newton's argument rests, which is simply that there is no last point on the line where the sides meet. And if it is impossible that there be a last point, this is tantamount to the claim that the line is infinite. For the length of the line is greater than any finite length, and so we can understand an infinite length or distance. In fact, Descartes' claim that we can speak of an intersection point beyond any assigned, rests on the Newtonian intuition that the line is infinite.

But Newton has another triangle argument. This argument claims that the infinite line is not only understandable, but that it actually exists:

Nor can anyone say that this is infinite only in imagination, and not in fact (*non actu infinitam esse*); for if a triangle is actually drawn, its sides are always in fact directed towards some common point, where both would meet if produced, (*ubi concurrerent ambo si modo producerentur*) and therefore there is always a point where the produced sides would actually meet, although it may be imagined (*fingatur*) to be beyond the limits of the corporeal universe. And so the line traced by all these points will be real (*erit actualis*), though it extends beyond all distance (*De Grav.* p. 101).

This is a strange argument, for it rests on a mis-stated conterfactual. To say that if the sides of the triangle are in fact drawn, they are always directed to some common point where they *would* meet if produced, makes no sense. If the sides make a closed figure – in this case a triangle – they must *actually* meet. Newton must mean that 'if two non-parallel line segments are drawn in a plane, they are always in fact directed towards some common point, where both would meet if produced'. The argument now makes sense, but we are given no reason for believing that the common point exists. To say "it may be imagined to be beyond the limits of the corporeal universe" does not help. Of course we can *imagine* the point to be wherever we please, but this is no argument to show that it actually *is* anywhere.

Some twenty-four years after writing *De Gravitatione*, Newton returned to his claim that infinity is understandable through geometry in the document entitled *Locus et Tempus*. As I have argued, this document is in all likelihood part of a revision of the *Principia* which engaged Newton during the early 1690's. In it he presents arguments that are related to the Scholium on space and time of the *Principia*, and others that anticipate the theological views put forward first in the General Scholium to the second edition of *Principia* (1713).

Once again Newton is concerned to show that infinities are understandable, especially the notion of an infinite distance. While admitting that infinities are difficult to conceive of, Newton nevertheless maintains that "there are many things concerning numbers and magnitudes which to men not learned in mathematics will appear paradoxical, and yet are entirely true" (*Locus et Tempus*, p. 119). He then proceeds to give examples of infinities, which fall into four categories: (a) the properties of infinity as determined by the geometry of conic sections; (b) the generation of infinite loci by moving points and lines; (c) the divisibility of lines to infinity; and lastly, (d) the mutual approach of things converging to limit at infinity.

It is an example of the kind of infinity mentioned in (a) which concerns us here. Newton characterizes it in general terms as follows: "that a point by moving according to a given law in a finite time, can go off to infinity, and conversely can return to us from infinity in the same time". The example he takes is this: "that if any straight line is revolved round a fixed point with uniform motion, and meanwhile cuts another straight line fixed in position, the point of intersection in each revolution will twice go to infinity and twice return from infinity" (*Locus et Tempus*, p. 119).

This example is geometrically isomorphic with the opening vertex example of *De Gravitatione*, and the logical structure of the argument is the same. As the moving line (fixed at one end) sweeps along the side of the given line, there is a moment such that the point of intersection of the two lines is infinitely far out. This can be understood as before. The moving line generates a sequence of points until it becomes parallel to the given line, the points being the points of intersection of the two lines determined by this motion. The length of the given line is greater than any finite length; hence we can understand an infinite length or distance. Once again the geometry enables us to express the intuition that a distance which has no upper bound is infinite. In this example, however, Newton conceives that the rotating line completes a total revolution around its fixed end. In this case there is an asymmetry in the motion of the rotating line at opposite ends of the line given in position. Thus at one end of the given line, the array of intersection points traversed *before* the rotating line changes direction is infinite; at the opposite end, the array of intersection points traversed *after* the rotating line changes direction is infinite. So given that the points of intersection generated by the rotating line "in each revolution will twice go to infinity and twice return from infinity", the length of the given line in either direction is greater than any finite length; once again we can understand an infinite distance. In this example, Newton has dispensed with the triangle as a device for generating a succession of intersection points, and speaks directly of proceeding to greater and greater distances along the line. There can be no question here that he is trying to make the notion 'actually more than finite' understandable by seeing it as a function of more readily intuited concepts exemplified by the triangle, i.e., other lines, angles, intersection, parallelism, etc. As a matter of fact, many of the same intuitive notions are here appealed to, only now we simply have a line sweeping out 360° as it intersects a fixed line!

It is clear from these arguments that Newton holds not only that we understand an infinite length or distance, but also (as against Descartes) that what we understand is positively conceived. This immediately prompts the following question: what is the ground of conceivability for the geometrically expressed and positive idea of infinite distance which we have? In other words, what is the relationship in Newton's view between conceiving an infinite distance and the basis in reality for that conception? This question is especially pressing, since Newton's attempt to show that the 'infinite' triangle exists not only in the imagination, but also in actuality, does not succeed. The question at hand comes to this: is there a correspondence between the geometry of infinite distance and Newton's conception of the nature of infinitely extended space?

One response open to Newton is one evident in both *De Gravitatione* and the document *Locus et Tempus*. Mathematics provides us with a sure foundation in terms of which to conceive the properties of the infinite. On the traditional assumption that geometry is the science of continuous magnitude, anything that can be said about magnitude consistently in geometry is at least consistent with the nature of spatial extension. So if we can speak of infinite distance in geometry, equally we can speak of infinite distance in space itself. This, of course, does not show that space is in fact infinitely extended, but it shows that we can coherently conceive of space as infinite.

There is no doubt that at a certain level this is Newton's view. From *De Gravitatione*, however, a more metaphysical approach to the question of how geometry and space correspond can be reconstructed. It involves three commitments: (a) the claim that space actually exists; (b) the claim that space possesses an infinitude of inherent, actual, and geometrical figures which are everywhere present in its infinite expanse; and (c) the implicit thesis that the objects of geometry exist in the reality of space itself.

I shall not examine the first commitment. Suffice it to say that Newton advances theological arguments by which he purports to show that space is actually infinite just because it is a consequence of God's real and omnipresent existence.[21] Given that this claim can be made out, namely, that space is actually infinite, what is the import of commitment (b)? After arguing that space contains a three-dimensional network of figures of all kinds which are everywhere fused together at their common boundaries Newton states:

> And hence there are everywhere all kinds of figures, everywhere spheres, cubes, triangles, straight lines, everywhere circular, elliptical, parabolical and all other kinds of figures, and those of all shapes and sizes, even though they are not delineated by sight (*non ad visum delineatae*). For the material delineation of any figure (*Nam materialis delineatio figurae alicujus*) is not a new production of that figure as regards space, but only a corporeal representation it of, so that what was first insensible in space now appears to the senses to exist (*ut jam sensibus appareat esse quae prius fuit insensibilis in spatio*). For thus we believe all those spaces to be spherical through which any sphere ever passes, being progressively moved from moment to moment, even though a sensible trace of the sphere no longer remains there. We firmly believe that the space was spherical before the sphere occupied it. (*imo spatium credimus prius fuisse sphaericum quam sphaera occupabat*), so that it could contain the sphere; and hence as there are everywhere spaces that can adequately contain any material sphere, it is clear that space is everywhere spherical. And so of other figures (*De Grav.* p. 100).

A number of claims are made in this passage. In the first place, Newton outlines the ontological conditions that obtain in space so that God can be said to bring matter into being there. Briefly put, space is so structured that God, in virtue of his creative power, can 'pick out' and condition various of space's inherent figures so that they in turn can cause the mind, via sensory experience, to form representations of material phenomena. In other words, Newton is putting forward a theory to show how God can be said to create material phenomena, given that space's inherent figures are omnipresent 'place-holders' in which active powers of the sort requisite to cause the mind's inner representations of material phenomena can come to be .

This is not the place to examine the details of Netwon's theory regarding the creation of matter, nor to consider whether the ontology of matter of *De Gravitatione* is ultimately coherent. But what is important here is that we get a firm grasp on his views concerning the structure of space. The figures in space actually inhere *in* space everywhere and eternally, and are immutably present in the sense that they cannot be separated from space itself. Thus Newton is not claiming that space is the potentiality of figures in the sense that bodies of different sizes and shapes can occupy it either successively or at the same time, but is claiming that figures of all sorts and sizes are *actually* omnipresent in its infinite expanse. Accordingly, we can conceive of space as having structure because it in fact has an immutable structure.

Again we see that Newton's conception of extension is very different from Descartes'. In Descartes' view extension is inextricably connected with material reality. Cartesian *Res extensa* does not contain actual parts, but it can be conceived as having parts because we can actually divide it, or lay down an algorithm for dividing it in theory. But in Newton's view extension is not inextricably connected with material reality; on the contrary, it exists eternally. Moreover, it is *eo ipsa* a condition presupposed by the very existence of matter itself. While it is not necessary to consider here whether Newton begs the question against Descartes in denying the latter's claim that extension is necessarily connected with the corporeal (Newton in fact considers the problem in *De Gravitatione*), it is abundantly clear that his conception of how spatial extension is structured differs fundamentally from Descartes' theory of material extension with respect to which space is defined derivatively.

In Newton's view the presence of matter presupposes ontologically the infinitude of spatial extension; whereas, in Descartes' case, space is merely a mode of material extension. But if space is a form of real being, as Newton claims, it must possess a differentiated structure which the mind can grasp. Otherwise it is in danger of slipping into the realm of indeterminacy and non-being, of which everything or nothing can be said. Moreover, in order to be a knowable reality space's structure must be intelligible. In this regard, there is no other candidate for Newton but the framework of Eucidean geometry.

As we might expect, however, Newton does not hold that Euclidean geometry merely affords a set of descriptions consistent with the nature of space. The constraints which delimit the scope of conceptual well-foundedness are to be drawn more determinately than that. Accordingly, in Newton's view, for anything that can be said consistently of space, there must exist a real and corresponding ground in the nature of space itself. In other words, in conceiving the quantitative nature of anything, Newton accepts a form of the real distinction, which holds that the mind can differentiate something quantitative just in case there are direct counterparts in the thing itself which allow it to be conceived thus and so (*Questiones*, folio $88^v$). The contrast with Aristotle's view of how the mind differentiates quantity could not be more marked. In answer to question (c) Newton is forthright. Where do the objects of geometry exist? They exist in the omnipresent expanse of infinite space. Accordingly, they are not *entia rationis*, or ideal entities whose *esse* is simply to exist in the mind alone. Rather, they exist, extra-mentally and independently of human conception, in the reality of space itself; whereas they exist in the mind derivatively in so far as it has the capacity to apprehend them.

But what connection is there between these sorts of ontological claims and Newton's argument that an infinite distance is a distinct object of the understanding? Do Newton's geometrical arguments presuppose this ontology? If so, is he not presupposing what he takes the geometry to demonstrate, i.e., that we can understand an infinite distance. Furthermore, what role does experience play in relating the geometrically established conception of infinite distance and Newton's claim that "there are everywhere – straight lines" in the expanse of space itself.

Newton's geometrical arguments for establishing an understanding of infinite length stand alone, and, in that context presuppose nothing beyond the geometrical framework which he employs. However, if we ask on what basis it is that the mind forms geometrical concepts, we are asking a different question. The empiricist would answer that geometrical objects are idealizations of shapes and lengths abstracted from sensory particulars by the power of the mind; or, put more crudely, that they are simply ideas arising from observations of recurrent features of experience.

This, of course, is not Newton's answer. Given that he accepts the epistemic principle that knowing that something is imperfect of a kind presupposes a conception of the kind's perfection, Newton claims that the understanding grasps the nature of extension *through*, but not in abstraction *from*, the deliverances of sensory experience. Put more precisely, the understanding, not the senses or the imagination, apprehends by means of its representative content the general nature of extension as it is variously and imperfectly exemplified by 'materially delineated' bits of extension, i.e., it apprehends that the extension of space is in itself infinite. Thus the relationship between the understanding's distinct idea of infinite distance and the infinity of distance in space itself is this. When the mind attends to its idea of infinite distance in the context of a geometrical argument, that idea of itself is not *what* we know; rather, it is the vehicle by *which* the thought of infinite distance takes place. In other words, the content, as such, of the mind's conception of infinite distance is not what the thought is *about*; it is about the infinity of distance in space itself, though this, of course, involves the mind mobilizing or exercising an ability to think a thought with that content. Furtherfore, the likeness between our geometrically expressed idea *of* infinite distance and infinite distance itself is not one of portrait to original, but rather that of the resemblance between effect and cause as mediated via active dispositions present in the objects of sensory experience. It cannot be otherwise, if for Newton our understanding of the infinity of distance is an affair of the intellect, and not of pictures or images in the imagination as it is for Locke. In this sense, then, the understanding possess a non-sensuous representation of infinite distance which has its ultimate ground in the real but uncreated nature of extension itself. Epistemological alternatives are thus rejected by Newton, i.e., that the mind possesses a special mode of intuition capable of recognizing directly the objective essence of space, or, alternatively, that it has an innate idea of space itself.

As I have reconstructed Newton's reasoning there appears to be a kinship with Descartes' arguments concerning the nature of the wax. Someone might object: Descartes' arguments are about the essence of a particular piece of wax, which in its entirely cannot be an object of experience. Moreover, Descartes throughout insists that extension is that which is imaginable and "has parts outside other parts of a determinate size and shape –" (AT. V. 268). So although the understanding alone grasps that the essential nature of wax is extension, at any given time the wax has some determinate shape or other which falls under the scope of the imagination, while in the case of the chiligon the imagination, though it cannot conceive it all at once, can at least attend to its sides successively. In no sense, however, can the infinite extension of space, as such, fall under the scope of the imagination.

These objections are well made, and they indicate a fundamental diffence in how Descartes and Newton conceive of the relationship between knowing the nature of extension and that nature itself. The difference can be put precisely: Newton, but not Descartes, is committed to the principle that the nature of extension, considered in itself, is ontologically prior to all finite determinations consistent with its nature. Given that Newtonian space is in fact an infinitely extended expanse, at once containing an infinitude of inherent and extended figures, in perceiving these figures through their 'material delineation' the mind comes to grasp the ontological priority of extension's infinity to that of its inherent determinations.

Reconstructed, Newton's position is this. By means of geometrical arguments the understanding can distinctly and positively conceive that space is infinitely extended; but since conceiving *that* involves conceiving *of* (and conversely), when the mind conceives *that* extension is infinite it conceives *of* extension. And consistent with the Euclidian conception *of* infinitely extended space is Newton's claim *that* it is differentiated by means of geometrical figures of various sorts and sizes. But the unity of infinite extension, which can be grasped distinctly by the understanding, is prior to the various figures which inhere in it. Thus, according to Newton there is a parallel between the ontological priority of infinite extension itself and the fact that understanding grasps it distinctly, a view consistent with claiming that confused sensory ideas of finite extensions are prior in the order of conceiving to those which are more distinct.

Newton's position can be compared with Aristotle's as there are similarities and differences. Moreover, Newton himself thinks that his position has affinities with the Aristotelians (*De Grav.* pp. 106–107). For Aristotle, intelligible matter, the continuous and quantitative in three dimensions, is that which underlies all differentiation in terms of the genera of quantity and their specific differences. Furthermore, intelligible matter, the subject of all determinate figures, is always figured in some determinate way, and can never be said to exist apart from quantified individuals. Newton expressly states that there is an analogy between Aristotle's notion and his own doctrine of infinite extension, on the ground that the latter is the intelligible ground and subject of an infinitude of determinate figures (*De Grav.* pp. 106–107). But here the similarity ends. According to Newton's conception, extension and its determinate figures can and do exist separately from the active powers of matter with which God endows certain figures in space. Moreover, though infinite extension underlies the 'material delineation' of space's inherent figures, Newton nowhere provides a detailed account of how the mind moves, in genus and species terms, from the perceptible material delineation of space's figures to a grasp of the intelligible reality (extension) which underlies them. As I noted earlier in Section 2, Newton holds that natures are prior to natured things, and thus that extension is prior to the extended figures that inhere in it. But he gives no clues (as does Aristotle), as to how, for example, extension itself is prior to a particular triangle and how they are related by means of a genera/species patterning. Accordingly, I have tried to reconstruct it in a manner consistent with Newton's text.

What Descartes cannot accept is Newton's claim that infinite extension itself (conceived as existing all together and at once) is prior to the actual determinations of its inherent figures in the sense that their being depends on its existence. For Descartes, the metaphysical conclusion to be derived from his epistemic 'appraisal' of the wax is that extension in itself is a changeable quantity which is capable of increase or decrease in either thought or reality. So considered, it is an imperfect, dependent and created nature which is divisible potentially to infinity. And since Descartes conceives of material extension (generically) as a homogeneous, undifferentiated plenum, which is everywhere indistinguishably the same, its unity is that of a whole to its potential parts, a conception very different from Newton's view that the unity and infinity of extended space is prior to its inherent figures, a claim compatible with the conception that finite cases of extension are cognitively prior in the order of knowing.

There would appear, however, to be a difficult in Newton's position. If extension is an uncreated and infinite being which exists necessarily (*De Grav.* p. 105) is its infinite not qualified by the finiteness of its inherent figures? In response, Newton can mobilize his conception of infinity as a second-level characterization of existing natures. Since extension and its actually indivisible figures are mutually compatible natures, one can be characterized as infinite in magnitude, the other as infinite in number. In other words, given Newton's account of how infinity is attributable, the infinity of space's extension is compossible with the infinitude of its inherent 'parts'. Although there are deep problems concerning the unity of extended space (i.e., Newton denies that it is a substance), which go beyond the scope of this paper, in terms of its gross structure space is a compositional unity in which infinity and the immutability and non-separability of its inherent figures are linked.

## CONCLUSION

To say the least, the issues which I have been considering are complexly interrelated. I have attempted to show that they can be dealt with, effectively, only if we willingly engage in conceptual analysis which is sensitive to the intellectual and metaphysical framework of the thinkers under consideration. More is involved, of course, then merely exposing the shortcomings in the arguments of a Newton or a Descartes. To be sure, this must be done. But it cannot be done without displaying the coherence and implications of their reasoning in relation to the assumptions and principles which each thinker accepts or denies. And this means getting hold of the traditions that they assume, desperate though it seems.

It is probably the case that today we will reject any representative theory of knowledge which is based on an unanalyzed notion of mental stuff; probable, too, that we will deny that the mind 'intuits' objective natures; and it is more than likely that we will discard as misguided the metaphysical views of Newton and Descartes on infinities and geometry. However, if we take seriously the possibility of doing disciplined conceptual history we have got to enter the 'dynamics' that drive the thinking of men such as Newton and Descartes. This will enable us to see why they argue as they do, allow us to isolate the motivations for their arguments, and provide an understanding of the tacit and explicit implications which they draw from their intellectual commitments.

Newton and Descartes alike view knowledge as the mind's confrontation with an outer reality via its inner representations. For both, the 'grounding' of the known is not in terms of what is abstracted from the given of experience, but in terms of the understanding's power to apprehend the outer world *through* the medium of sensory experience.

## NOTES

1 A. Rupert Hall and Maria Boas Hall, *Unpublished Scientific Paper of Issac Newton*, Cambridge University Press, 1962, *De Gravitatione*, pp. 99, 100–101, and 110. Subsequent page references to *De Grav.* will be in the text.

2 U. L. C. MS. Add. 3996. folio $83^{r-v}$. For this manuscript see J. E. McGuire and Martin Tamny, *Certain Philosophical Questions: Newton's Trinity Notebook*, Cambridge University Press, 1983.

3 John Locke, *An Essay Concerning Human Understanding*, London, Everyman's Library, 1972, Vol. I, Bk. II, Chapter XVI and XVII. Subsequent references are to this edition will be in the text.

4 See Chapter 1.

5 London, 1687, Part XIII, Chapter 11, p. 858.

6 Diogenes Laertius, *De Vitis Dogmatis et Apophthegmatis – Libri X*, London, 1664. Epicurus

Lib. X, p. 376. This page is dog-eared by Newton, as indeed are many pages in the edition having to do with the infinity of the world.

7 London, 1654, Bk. I, Chapter 11, Section 11, 9–15. There are many late 16th century and early 17th century copies of Lucretius' poem in the Trinity College Library. There is no evidence, however, that Newton used them. Nevertheless, it seems improbable that he did not know of Lucretius, at an early stage. He did, however, own the 1686 London edition.

8 Nicolas de Cusa, *Opera Omnia*. Heidelberg Academy Edition, ed. Ernest Hoffmann & Raymund Klibansky, Lipsig, 1932, Vol. I, *De Docta Ignorantia*, Chapter XVII, p. 33. Since nothing determinate can be infinite, Cusanus never affirms the infinity of the world, as Descartes mistakenly claims he does (AT. V. 50, To Chanut, 6, June 1647). D. W. Singer, *Giordano Bruno, His Life and Thought*, New York: Abelard-Schuman, 1950. *On the Infinite Universe and the Worlds*. First Dialogue, p. 257. Gottfried Liebniz, *New Essays Concerning Human Understanding*, the Open Court Company, Chicago, 1916, the critique of Bk. II of Locke's *Essay*, Chapter XVII, pp. 161–164.

9 *Ibid.*, Chapter XV, pp. 158–159 and Chapter XVII, pp. 169–174. Leibniz's approach to infinity must be understood in the context of his claim that ostensibly extended phenomena are not true unities in the manner of indivisibles whose unity and being are one.

10 U. L. C. MS. Add. 3995. folios $9^{r-v}$–$18^{r}$.

11 Folio $83^{r-v}$. Both folios are copiously filled with citations from, and page references to, Descartes' *Meditations* and the *Objections and Replies*. In every case Newton's citations and page references are to the *Opera Philosophica*, 3rd edition, Amsterdam, 1656, which includes Descartes' complete works. The exact citations and references to this edition leave no doubt that Newton had read Descartes' philosophical writings carefully. See McGuire and Tamny, *op. cit.*, note 2.

12 *Oeuvres de Descartes*, Charles Adam and Paul Tannery (eds.), Libraire Philosophique J. Urin, Paris, 1973. *Meditation II*, pp. 28–29, Vol. VII. In the future, all references will be in the text as (AT. VII. 28–29).

13 John Wallis, *Operum Mathematicorum Pars Altera*, Oxford, 1656, *Arithmetica Infinitorum in pars altera*, propositions 34 and 40.

14 For a useful discussion of Descartes' distinction between the 'indefinite' and the infinite, see John Cottingham (ed. and transl., *Descartes' Conversations with Burman*. Clarendon Press, Oxford, 1976), Commentary, p. 101.

15 *Ibid.*, See Cottingham's discussion of the notion of adequate knowledge in Descartes, p. 65.

16 In his reply to my paper at the Montreal meeting of the Union of the History and Philosophy of Science, see *Nature Mathematized*, ed., William Shea, Reidel, Dordrecht, Holland, 1983, pp. 18–19.

17 J. E. McGuire, 'Newton on Place, Time, and God: An Unpublished Source', *British Journal for the History of Science* **II** (1978), 114–129, p. 119.

18 *Ibid.*, pp. 119 and 121.

19 *Op. Cit.*, note 4. See Section Three entitled 'Existence and natures', pp. 474–487.

20 *Op. Cit.*, note 17.

21 *Ibid.*

CHAPTER 5

# FORCE, ACTIVE PRINCIPLES, AND NEWTON'S INVISIBLE REALM

In recent years, scholarship has made it possible to achieve a more extended interpretation of the development of Newton's thought in the years following the publication of the *Principia*. It is now time to attempt the task of offering a framework within which the results of certain of these investigations can be linked together in a synoptic way. I shall concentrate on what I take to be one of the central problems which Newton attempted to solve in this later period, namely, the ontological problem of the causation of force.

This included, for Newton, speculation on different sorts of aethereal media and on various kinds of immaterial agents: and both types of speculation had firm roots in his thought of the sixties and seventies. The key passages on which my analysis will primarily base itself are to be found in the 1706 Latin edition of the *Opticks* and in the second English edition of 1717–18. These passages are concerned with the ontological connexions among motion, force, and active principles, and are reproduced in Section II of this paper. There are a considerable number of manuscript variants of these, which clearly show the important role that chemical and biochemical phenomena played in Newton's attempts to clarify the properties of the "smaller" forces of nature. These and other related documents will be analysed in conjunction with the printed sources.

Lastly, I shall attempt to argue that the basic concepts of Newton's natural philosophy can be ultimately clarified only in terms of the theological framework which guided so much of his thought. This thesis is not original.[1] But it will be empty of content unless these concepts can be tightly linked with his theology. This I believe can be done, by showing the fundamental role that the traditional Judeo-Christian doctrines of creation and providence played in Newton's thought.

## I

From the time of the publication of the *Principia* in 1687 to about 1707, Newton seems largely to have abandoned attempts at a quasi-material explanation of force. Many of the arguments in the 1687 edition of the *Principia* turn on the acceptance of a matterless vacuum in nature,[2] and the closing section of Book II is a sustained criticism of Descartes's theory of vortical motion in a fluid aether.[3] In the emendations for an abortive edition of the *Principia* undertaken in the early 1690's , the doctrine of void space is explicitly considered by Newton.[4] Also in the 1706 *Optice*, the doctrine is well represented, being used in explaining a host of phenomena by means of an interstitial vacuum.[5] About 1707, however, Newton's thought seems to have returned to his earlier speculations on the aether as a means of explaining certain phenomena.[6] By 1713, in the *General Scholium*, there is an explicit mention of a "certain spirit" as a causal agent.[7] And finally in the 1717–18 *Opticks* there is a return to a quasi-mechanical aether, diffused throughout space, as an explanation of optical effects, and as a conjectural cause of gravity.[8]

This story is now well known and is generally accepted by Newtonian scholars. Still largely shrouded in mystery are the various reasons for these changes in Newton's views. Added to the usual difficulties of establishing sources which will serve to clarify this sort of problem,[9] a student of Newton's thought is faced with a special difficulty. Many of his characteristic ideas were adumbrated early in his career, and then progressively refined into their mature form. This is so with the nature of force and its possible causes. The main approaches to the ontological problem of causation, developed at various times in later writings, are found in an embryonic stage in an early treatise. This is *De Gravitatione et Aequipondio Fluidorum*,[10] written no later than 1670, and the first sustained expression of Newton's philosophy of nature. Are forces to be explained by some intermediary agent such as a world soul or a quasi-mechanical aether? Or are they a direct and continuous expression of Divine Providence? After considering the hypothesis of a world soul as a plausible but improbable source of activity in the world, Newton inclines to the view that God acts in nature without intermediaries: "But I do not see why God himself does not directly inform space with bodies; so long as we distinguish between the formal reason of bodies and the act of divine will.[11] Nevertheless, a subtle aether is not dismissed as a possible means of transferring the "mutual actions"[12] of bodies. In the later 1670's various aether hypotheses embodying elements of alchemical and platonic ideas were formulated as possible explanations for a variety of phenomena, and as a source of activity in nature.[13] Earlier, in *De Aere et Aethere*, written about 1673, Newton attempted to explain the nature of forces by means of an intrinsic repulsive power attributed to particles of air.[14] By 1679 in his letter to Boyle,[15] he had transferred repulsion to aether particles and, as the Halls suggest, probably conceived the reduction of all force for a time to this primordial agent.[16] In this period the possible existence of void spaces and the nature of God's intervention in the world tended to move to the periphery of Newton's thought. As in the mid-seventies, aethereal media also probably played an important role in his analysis of optical phenomena.[17]

As already indicated, at least by the time of publishing the first edition of the *Principia*, Newton seems to have rejected the aether speculations of the 1670's. This rejection no doubt came in the wake of an exacting analysis, developed in that treatise, of motion in resisting and non-resisting media.

The quantitative expression of the law of gravitation and of the laws of Kepler were thought by Newton to be incompatible with the existence of a vortically structured fluid aether.[18] Thus the dynamics of the heavens seemed to pre-suppose the near vacuity of planetary spaces. And since he considered these interplanetary spaces to be relatively empty, Newton no longer sought the source of motion and activity in the diffused alchemical aether of 1675, but in a very tenuous spirit or vapour emanating from the tails of comets which move throughout the planetary system.[19]

It is possible to conclude that one of the important stages in Newton's thought – an apparent rejection of the aether and a clear commitment to the void – began in the early nineties with revisions intended for the third book of the unimplemented edition of the *principia* and later in draft materials relating to the new Queries of the 1706 *Optice*. These ideas first found public expression in the new Queries to the *Optice* and in the corrigenda at the front of the book.[20]

The significance of Newton's commitment to the void, from 1687 to 1706, has not been sufficiently appreciated. In the first place intimately related to this doctrine, in Newton's mind, was the view that there exists in nature a paucity of matter. Again there is evidence from early post-1700 to support this interpretation. In a draft Query, probably written about 1705 and related to many similar drafts connected with the *Principia*,[21] Newton gives an elegant expression of his belief in the extreme porosity of matter. The first part of the unpublished Query describes a plausible progression through continuing combinations of parts with vacuities, to show that a body could "have above a thousand thousand thousand times more pores than parts & so on perpetually".[22] These ideas are probably the basis of David Gregory's observation in 1705, "that water is not one part of twenty full matter, & yet by the strongest machine incompressible. Sir Isaac Newton proposed this as one theory of making bodies in any degree porous & yet solid".[23] Newton concludes the Query with a reference to the contrivance of matter being a token of the wisdom of God:

And it [porosity] may be further argued from the great capacity of (one & the same matter) by corruption & generation to put off & put on all manner of forms, & be changed into all sorts of shapes. The contrivance of the bodies of living creatures is admirable. Not a member but has its use & is mighty well fitted for that use. And (since the forms & uses of matter are infinite) can we believe that he who contrived ye bodies of animals with so much artifice was not as skilful & curious in (the contrivance of matter) contriving the texture (of matter so as to fit it, make it capable by corruption of all manner of forms & uses applicable) of matter for those ends for which it is fitted. For the forms & uses of matter are innumerable & therefore we are not to consider it as composed of irregular particles casually laid together like stones in a heap, but as formed wisely for all those uses. One use of matter is to admit menstruums easily into its pores in order to new mixtures & actions by fermentation, putrefaction, corruption & generation. Another to conceive heat easily & keep it long for promoting changes by (fermentation) putrefaction & generation. Another to transmit, refract & reflect light for producing all the (phaenomena) of colours of bodies (without which) wherein the beauty of nature (mightily) chiefly consists. Without these uses the earth would have been a dead lump void of heat & motion & alternation & variety of colours: & therefore it is reasonable to allow that he who contrived all things with wisdom, framed matter in such a manner as to fit it best for these uses, & by consequence made it very porous.[24]

It would be difficult so find another passage in Newton's writings which conveys these ideas with the same intensity of commitment. Not only is the porosity of matter essential grounds for explaining a variety of phenomena, many of them chemical in nature, but being so structured it manifests God's wisdom. Accordingly, this doctrine, as many others, is subordinate to Divine purpose and foresight. And it is therefore possible that Newton's acceptance of atomism was from the beginning conditioned by the Christian doctrine of Providence.[25] For as we will see later, he never considered that any part of nature was either purposeless or bereft of the constant supervision of God.

In the second place, the paucity of matter in relation to the void, is connected with another significant characteristic of Newton's thought – his belief in the existence of a variety of forces in nature. It is within the internal vacuities of matter that these forces operate. And "almost all the phenomena of nature . . . depend on the forces of particles".[26] Newton was led to this belief because "if Nature be simple and pretty conformable to herself, causes will operate in the same kind of way in all phenomena, so that the motions of smaller bodies depend upon certain smaller forces just as the motions of larger bodies are ruled by the greater force of gravity".[27] In this passage written in 1687 the "analogy of nature" prompted Newton to investigate the properties of these "invisible" forces interacting with particles in the interstitial vacuities of matter. A clear reference to these speculations on force and the paucity of matter is found in the 1704 edition of the *Opticks*. In the closing paragraph of Proposition VIII, Book II, Part III Newton observes that "we may understand that Bodies are much more rare and porous than is commonly believed". He concludes:

And he that shall find out an Hypothesis, by which water may be so rare, and yet capable of compression by force, may doubtless by the same Hypothesis make Gold and Water, and all other Bodies as much rarer as he pleases, so that Light may find a ready passage through transparent substances.[28]

In a typically cautious mood, Newton did not mention his own hypothesis, but there is no doubt that he believed in the rarity of matter. By 1713 in Book III of the *Principia*, Corollary III, Proposition VI, he became more emphatic about his doctrine: "And if the quantity of matter in a given space can . . . be diminished, what should hinder a diminution to infinity?"[29] This did not, of course, include the primordial particles: they are immutable. The Newtonian explanation of phenomena demanded that these particles be affected by forces which cause them to approach or recede from one another: hence increasing or diminishing the number filling a place occupied by any given body. The paucity of matter, therefore, seems to have been clearly connected in Newton's mind with an increasing tendency to stress the way in which forces arrange the constituents of matter in the explanation of phenomena. thus in such explanations little emphasis was placed on the geometrical qualities of the constituents.

The primacy of force in Newton's thought during this period was a consequence of another characteristic doctrine. This was the view that the smaller a particle of matter the stronger the forces associated with it. This doctrine is intriguingly employed in an hypothesis to the effect that once light "enters the composition of all bodies, why may it not be the chief principle of activity in them?"[30] In draft materials relating to Query 22 of the 1706 *Optice* there are three substantially similar versions of the hypothesis:

> Now in bodies of the same kind & virtue attraction is stronger in the smallest bodies in proportion to their bulk. It is found stronger in small magnets for their weight than in great ones. For the parts of small ones being closer together unite their forces more easily. And therefore since the rays of light are the smallest bodies known to us we may expect to find their attractions very strong. And how strong they are may be gathered by this rule [after briefly analysing the strength of a ray of light from its interaction at a reflecting surface in analogy to the bulk, velocity and curve of a projectile with respect to the horizon Newton concludes] and so great a force in the rays cannot but have a very great effect upon the particles of matter with which they are compounded for causing the particles to attract one another & for putting them in motion amongst themselves: for the better understanding of which I will propose the following questions.[31]

Unfortunately this Query, if written, has not survived. The force of the particles in rays as a cause of microscopic attractions is, as will be seen below, another speculative attempt to solve the problem of "smaller" forces, and the internal composition of matter. What is of interest here, however, is the connection posited between the size of a particle and the forces surrounding it. For this leads to the possibility that in the limit the dimensions of particles could be vanishingly small, leaving space containing forces clustering around a changing network of foci. Although this possibility almost certainly occurred to Newton, he could never have taken it seriously because the transmission of force pre-supposes inertia:[32] a vanishingly small particle could not be endowed with such a property.

In any event, as the properties and size of particles became less important, so the reduction of matter in any given place became more plausible. Thus, not surprisingly, forces became for Newton agents of change in phenomena, and the cause of activity in nature. Therefore his commitment to the void helped to shape the related doctrines of the paucity of matter and the primacy of force. Consequently as the totality of matter in the world seemed to fade in significance, the operation of various forces not reducible to it increased in importance. Not only did Newton enrich his ontology by including forces, but in the nineties they became essential to his natural philosophy. By 1706 he seemed to consider them, rather than matter, to be the primordials of nature.

We can now return to his views on the causative nature of force in this period. Was gravity, for example, a direct fiat of God's will or was it explicable within the natural order through the Divine use of second causes? What little evidence we possess indicates that from 1687 to about 1705 the first alternative seems to have been Newton's opinion. His tendency to vacillation at the beginning of this period is clear. Fatio de Duillier observed in 1689–90 that Newton "did not scruple to say that *there is but one possible mechanical cause of gravity to wit that which I had found out*: tho he would often seem to incline to think that Gravity had its foundation only in the arbitrary will of God . . .".[33] In a letter to Leibniz on 16 October, 1693, Newton may have had in mind Fatio's hypothesis which is based on the vacuity of matter. Having stated that all the phenomena of the heavens and sea follow solely from the laws of gravity Newton declares:

> . . . and since nature is very simple, I have myself concluded that all other causes are to be rejected and that the heavens are to be stripped as far as may be of all matter, lest the motions of planets and comets be hindered or rendered irregular. But, if meanwhile, someone explains gravity along with all its laws by the action of some subtle matter, I shall be far from objecting.[34]

A year later on 30 March, 1694, Fatio wrote to Leibniz via De Beyrie: "Monsr. Newton est encoure indeterminé entre ces deux sentimens. Le premier que la cause de la Pesanteur soit inherente dans la matière par une Loi immédiate du Createur de l'Univers; et l'autre que la Pesanteur soit produite par la cause Mechanique que j'en ai trouvée".[35]

In characteristic fashion Newton is keeping possible explanatory options open. By 1716 his thought prompted by new evidence would return to a more serious consideration of a quasi-aether. But in the nineties, with the strong commitment to the void, he did not take the "action of some subtle matter" seriously. The earliest direct rejection of it is found in a memorandum written by David Gregory in December, 1691. Referring to Fatio's "mechanical" explanation of gravity read to the Royal Society on 26 February, 1690, he observes that "Mr. Newton and Mr. Halley laugh at Mr. Fatio's manner of explaining gravity".[36] Two manuscript passages from the early nineties support Gregory's observation. The first is an unpublished Hypothesis inserted as part of a conclusion to an unfinished Fourth Book of the *Opticks*. There Newton unambiguously declares, "All solid bodies are Atoms. Prove this first adhominem by assuming ye vulgar supposition that gravity is mechanical, & then by the difficulty of making the parts of solid bodies cohere by any other principle".[37] At the about the same time, Newton explicitly dismissed the fluid aether in unmistakable terms. The following fragment is part of an emendation associated with the revision of the *Principia* in the early 1690's.

> The heavens and universal space some persons fill with a most subtle fluid matter, but the existence of this is neither evident to the senses nor is proved by any arguments, but is precariously assumed for the sake of an hypothesis. Moreover, if we may trust reason and the senses, that matter will be in exile from the nature of things. For one cannot understand how movement can be accomplished in a plenum.[38]

In both these passages it is evident that Newton is against any form of mechanical or fluid aether on grounds of reason and evidence. Thus, though he remained interested in the possibility of a mechanical explanation of gravity, he seemed more inclined to think that it was a manifestation of the Divine presence in nature.

By the end of the 1690's, his ideas began to crystallize in favour of the latter opinion. On 20 February, 1697–98, David Gregory wrote in a memorandum that Christopher Wren had a mechanical method of explaining gravity: "He smiles at Mr. Newton's belief that it does not occur by mechanical means but was introduced originally by the Creator".[39] Lastly, on 21 December, 1705, Gregory wrote that the question, "*what the space that is empty of body is filled with*", could only answered for Newton in theological terms:

The plain truth is, that he believes God to be omnipresent in the literal sense . . . for he supposes that as God is present in space when there is no body, he is present in space when a body is also present. But if this way of proposing this his notion be too bold, he thinks of doing it thus: *What cause did the ancients assign to gravity?* He [Newton] believes that they reckoned God the cause of it, nothing else, that is no body being cause; since every body is heavy.[40]

It seems, then, that a firm commitment to a non-mechanical mode of explaining gravity had occurred by 1705. And as we shall see, an important problem became the ontological status of the various agencies operating on matter.

Before 1694, however, there is abundant manuscript evidence to support the view that Newton was interested in a non-mechanical mode of explaining gravity. In draft scholia intended for Proposition IV to IX of the Third Book of the *Principia*, Newton hoped to show that the ancients had anticipated the doctrines contained in these Propositions. As Dr. P. M. Rattansi and I have shown elsewhere,[41] these scholia were used to establish four basic theses which Newton attributed to antiquity: that matter is atomic in structure and moves by gravity through an infinite void; that gravitational force acts universally: that its action diminishes in the inverse square of the distance between bodies: and that God's direct action is the true cause of gravity.

This material certainly confirms my interpretation of Newton's opinion regarding the causative nature of gravity in the nineties, namely, God himself who is ubiquitously present in space. The "classical" material also throws light on Gregory's memorandum of 1705. Passive, inert matter could never originate activity in nature: thus matter itself presupposes an immaterial principle "since every body is heavy": therefore God is the direct cause of activity. In commenting on Newton's belief that nothing material could be the cause of gravity, Gregory was probably also referring to another addition that was intended for the unfinished revision of the *Principia* undertaken in the nineties. It was meant to strengthen the doctrine of void space discussed in all editions of the *Principia* in Corollary III of Proposition VI, Book III.[42] Newton argues that the only means of explaining gravity mechanically is to postulate the existence of a less dense set of particles, but the gravity of these in turn can only be explained by a third set of particles less dense again: and so a regress begins. On the other hand, if an explanation by postulating the existence of non-gravitational matter is attempted, not only is the notion without physical significance, but it violates the conception of universal gravitation.[43] Thus at this stage in Newton's thought the only viable explanation of gravity seemed to involve an immaterial agent operating in a nearly vacuous universe. From the mid-nineties to 1706, this was the Divine Will. The harmonious order of nature immediately revealed the continuous influence of the Divine power actually present in nature. For the exquisite structure of nature seemed antithetical to the presence of intermediaries such as a world soul or a quasi-mechanical aether.

## II

With this background established, it is possible to turn to the central concern of the present study. This arises from a problem which, from 1687, exercised Newton's mind in a systematic way. How can we obtain experimental knowledge of hidden forces in the invisible realm of nature? With almost a cry from the heart he exclaimed in his preface to the *Principia*:

> I wish we could derive the rest of the phenomena of Nature by the same kind of reasoning from mechanical principles as in the case of universal gravitation, for I am induced by many reasons to suspect that they may all depend upon certain forces by which the particles of bodies, by some causes hitherto unknown are either mutually impelled towards one another and cohere in regular figures, or are repelled and recede from each other; which forces being unknown, Philosophers have hitherto attempted the search of Nature in vain. But I hope the principles here laid down will afford some light either to that or some truer method of Philosophy.[44]

In a draft *Preface* and *Conclusion* intended for the *Principia*, Newton elaborated his philosophy of force.[45] Each of these documents argues that chemical evidence will most probably provide the key to the invisible realm. another key which he discussed extensively in the *Opticks* was light. In the Second Book,[46] Newton reports his patient experiments on thin films which produce different spectral colours. He correlated the coloured rings with the thickness of the film, and then argued by analogy from the colours of natural bodies to the relative dimensions of their particles: these he held to be responsible for the reflection and absorption of light rays. This method offered a possibility of indirectly ascertaining the properties of the associated forces. But chemical evidence offered a much broader and richer foundation of possibilities and seemed to Newton, as we shall see, a more feasible way of attempting to determine the "properties, quantities and effects"[47] of invisible forces. The clearest expression of this programme as Newton conceived it in the early 1690's, is to be found in another hypothesis intended to form part of a projected Fourth Book of the *Opticks*. It reads:

Hypoth. 2. As all the great motions in the world depend upon a certain kind of force (wch in this earth we call gravity) whereby great bodies attract one another at great distances: so all the little motions in y$^{e}$ world depend upon certain kinds of forces whereby minute bodies attract or dispell one another at little distances.

Here is Newton's characteristic appeal to the "analogy of nature" which abounds in his writings. Nature "observes the same method in regulating the motions of smaller bodies", he continues in his remarks on the Hypothesis, "wch she doth in regulating those of the greater":

This principle of nature being very remote from the conceptions of Philosophers I forbore to describe it in that Book [*Principia*] least I should be accounted an extravagant freak & so prejudice my Readers against all those things wch were y$^{e}$ main designe of the Book: but & yet I hinted at it both in the Preface & in y$^{e}$ book itself where I speak of the inflection of light & of y$^{e}$ elastick power of y$^{e}$ air but the design of y$^{y}$ book being secured by the approbation of Mathematicians, I had not [doubted *crossed out*] scrupled to propose this Principle in plane words. The truth of this Hypothesis I assert not, because I cannot prove it, but I think it very probable because a great part of the phaenomena of nature do easily flow from it wch seem otherways inexplicable: such as are chymical solutions, precipitations, philtrations, . . . volatizations, fixations, rarefacations, condensations, unions, separations, fermentations, the cohesion, texture, fluidity and porosity of bodies, the rarity & elasticity of air, the reflexions & refraction of light, the rarity of air in glass pipes, & ascention of water therein, the permiscibility of some bodies & impermiscibility of others, the conception & lastingnesse of heat, the emission & extinction of light, the generation & destruction of air, the nature of fire & flame, ye springinesse or elasticity of hard bodies.[48]

It is difficult to imagine a more frank and fascinating passage from Newton's pen. It shows how "unphilosophical" he considered his use of the grand analogy of nature. Yet he could not forbear appealing to it, though he clearly knew it could never be justified according to his standards of rigour and certainty.[49] It was, nevertheless, an assumption that made the world intelligible. The passage shows, furthermore, the beguiling attractiveness that the invisible realm had for Newton. He felt his work would not be complete unless he could indicate a plausible path into the secret bowers of nature. Lastly, it is important to notice the considerable list of chemical phenomena probably caused by hidden forces. These are for the most part the same as those discussed in the draft *Preface* and *Conclusio*, the *De Natura Acidorum* of 1692,[50] and alter in Query 31 of the *Opticks*. Earlier in the 1670's, Newton had devoted considerable time to chemistry and alchemy. Again many of the same phenomena are explained in terms of a "certain secret principle in nature, by wch liquors are sociable to some things, & unsociable to others".[51] By 1686 this "secret principle" had become the attractive and repulsive forces of Newton's mature philosophy.[52] There can be little doubt that chemistry played a central role in Newton's attempts to extend his methodology beyond observable phenomena.

Between 1704 and 1706, Newton began to employ a wider framework of ideas in his treatment of the problem.

The elaboration of these ideas, regarding chemical evidence as a key to the nature of hidden forces and the connexion of the latter with active principles, occurs in a series of manuscript drafts of some important passages which appear for the first time in Query 23 of the Latin *Optice* of 1706. Since there are some important differences in these passages as they appear in the 1717–18 edition, I shall reproduce them in pairs of parallel columns. Hereafter I shall cite the number of the pair when it is necessary to make a reference:

I

(*a*)
*Query 23*
*Optice, 1706, p. 341*

*Vis inertiae*, est Principium passivum quo Corpora in Motu suo vel Quiete persistant, recipiunt Motum Vi moventi semper proportione respondentem, & resistunt tantum quantum sibi resistitur. Ab hoc solo Principio nullus unquam in rerum Unviesitate oriri potuisset Motus. Alio aliquo Principio omnino opus erat ad movenda Corpora; & jam, cum moventur, alio itidem Principio opus est, ad Motum ipsorum conservandum.

(*b*)
*Query 31*
*Opticks, 1717–18, pp. 372–373*

The *Vis inertiae* is a passive Principle by which Bodies persist in their Motion or Rest, receive Motion in proportion to the Force impressing it, and resist as much as they are resisted. By this Principle alone there never could have been any motion in the world. Some other Principle was necessary for putting Bodies into Motion; and now they are in Motion, some other Principle is necessary for conserving the Motion.

II

(*a*)
*p. 343*

Quoniam igitur varii illi Motus, qui in Mundo conspiciuntur, perpetuo descrescunt universi; necesse est prorsus, quo ii conservari & recrescere possint, ut ad actuosa aliqua principia recurramus: Qualia utiq; sunt *Gravitatis causa*, qua Planetae & Cometae Motus suos in perpetuis Orbibus conservant, Corporaq; omnia Motum magnum sibi acquirunt, cadendo; *Fermentationis* Causa, quo Cor & Sanguis Animalium & Calore perpetuo consoventur, partes interiores Terrae perpetuo calesiunt, corpora permulta ardent & lucent, Montes ignem concipiunt, cavernae Telluris ictibus subitis disjiciuntur, & Sol ipse perpetuum vehementer candet & lucet & Luce sua omnia calefacit ac fovet. Nam admodum paullum Motus in mundo invenimus praeterquam quod vel ex his Principiis actuosis, vel ex imperio *Voluntatis*, manifesto oritur.

(*b*)
*p. 375*

Seeing therefore the variety of Motion which we find in the World is always decreasing, there is a necessity of conserving and recruiting it by active Principles, such as are the cause of Gravity, by which Planets and Comets keep their Motions in their Orbs, and Bodies acquire great Motion in falling; and the cause of Fermentation by which the Heat and Blood of Animals are kept in perpetual Motion and Heat; the inward Parts of the Earth are constantly warm'd, and in some places grow very hot; Bodies burn and shine, Mountains take Fire, the Caverns of the Earth are blown up, and the Sun continues violently hot and lucid, and warms all things by his Light. For we meet with very little Motion in the World, besides what is owing to these active principles.

III

(*a*)

*p. 344*

Porro, videntur mihi hae particulae primigeniae, non modo in se *vim inertiae* habere, Motusq; *Leges passivas* illas, quae ex Vi ista necessario oriuntur; verum etiam *Motum* perpetuo accipere a certis *Principiis actuosis*; & qualia nimirum sunt Gravitas, & Causa Fermentationis & cohaerentiae corporum. Atq; haec quidem Principia considero, non et *occultas Qualitates*, quae ex *Specificis* rerum *Formis* oriri Singantur; sed ut *universales* Naturae *Leges*, quibus res ipsae sunt Formatae. Nam Principia quidem talia revera existere, ostendunt Phaenomena Naturae; licet ipsorum Causae quae sint, nondum fuerit explicatum.

(*b*)

*p. 376*

It seems to further, that these Particles have not only *Vis inertiae*, accompanied with such passive Laws of Motion as naturally result from that Force, but also that they are moved by certain active Principles, such as is that of Gravity, and that which causes Fermentation, and the Cohesion of Bodies. These Principles, I consider not as occult Qualities, supposed to result from the specifick Form of Things, but as general Laws of Nature, by which the Things themselves are form'd: Their Truth appearing to us from Phaenomena, though their Causes be not yet discover'd.

IV

(*a*)

*p. 326*

. . . utiq; intelligere poterimus, omnino ita comparatum esse Terram, ut in visceribus ejus abundent Vapores sulphurosi, qui cum mineralibus Fermentescere debeant, & interdum ignem concipere, cum subita coruscatione & displosu; &, si Forte in cavernis subterraneis arcte inclusi contineantur, vehementer conquassare Terram, & cavaernas ejus disrumpere, quemadmodum cuniculi pulvere tormentario repleti terram suffossam mira cum violentia disjiciunt; Quod cum accidit tum vapores explosione ista generatos per occultos Terrae meatus expirare, aestuososq; sentiri & suffocantes; procellasque, turbines & tempestates ciere, efficereq; nonnunquam ut terrae Tractus de locis suis transportentur, ebulliatq; Mare, & guttatim subvehantur in caelum Aquae, quae deinceps acernvatim & vorticoso pondere corruant, quasi e nubibus effusae Praeterea, Exhalationes quaedam sulphurosae omni tempore, quando Terra sit siccior, in Aerem ascendentes, Fementescunt ibi cum Acidis nitrosis; & nonunquam ignem concipientes, Fulmina generant, & Tonitrua, aliaq; Meteora ignea: . . . Jam vero hi, quos diximus, Motus, tanti sunt tamq violenti, ut ex eis satis appareat, utiq; in Fermentationibus particulas corporum, quae Ferme quieverant, novis motibus cieri a Principio aliqu praepotente. . . .

(*b*)

*p. 354*

. . . we may learn that sulphureous Steams abound in the Bowels of the Earth and ferment with Minerals, and sometimes take Fire with a sudden Coruscation and Explosion; and if pent up in subterraneous Caverns, burst the Caverns with a great shaking of Earth, as in springing of a Mine. And then the Vapour generated by the Explosion, expiring through the Pores of the Earth, feels hot and suffocates, and makes Tempests and Hurricanes, and sometimes causes the Land to slide, or the sea to boil, and carries up the Water thereof in Drops, which by their weight fall down again in Spouts. Also some sulphureous Steams, at all times when the Earth is dry, ascending into the Air, Ferment there with nitrous Acids, and sometimes taking Fire cause Lightning and Thunder and Fiery Meteors. . . . Now the above mention'd Motions are so great and violent as to shew that in Fermentations, the Particles of Bodies which almost rest are put into new Motions by a very potent Principle. . . .

There are a number of problems which arise from these passages. Foremost among them, of course, is the doctrine of active and passive principles. Since it is clearly related to Newton's conception of motion and force, the latter is better considered first. Preliminary to this, however, it is necessary to consider some draft passages which bear on II and IV. All were most probably written in 1705, when Newton was preparing the new Queries for the 1706 *Optice*, and are thus related in the first instance to Query 23 of that edition; moreover, they throw considerable light on the connexions Newton made between chemical phenomena, motion, force, and active principles. The following passage is an expanded version of II, but it contains ideas which are found in IV and afterwards cancelled:

Seeing therefore the variety of motion (wch we see) in the world is always decreasing, there is a necessity of conserving & recruiting it by active principles; such as are (the power of life & Will by which animals move their bodies with great & lasting force;) [this is bracketed and cancelled] the cause of gravity by which Planets & Comets keep their motions in their Orbs & bodies acquire great motion in falling; & the cause of fermentation by wch the heart and blood of animals are kept in perpetual motion & heat, the inward parts of the earth are constantly warmed, bodies burn & shine, mountains take fire, the caverns of the earth are blown up, & the Sun continues violently hot & lucid & warms all things by his light (For we meet with very little motion on the world besides what is owing to these active principles & therefore we ought to enquire diligently into the general Rules or Laws observed by nature in the preservation or production of motion by these principles as the Laws of motion on wch the frame of Nature depends & the genuine Principles of the Mechanical Philosophy & the inward parts of the earth are constantly warmed & generate hot sulphureous unhealthful exhalations wch breaking forth with violence cause earthquakes, tempests, & hurricanes, raise or subvert Islands & Mountains, sink Lakes & carry up the sea (partly) in columns, (partly) in drops & thick mists wch convening above fall down in spouts, & sulphureous steams, set mountains on fire & the inward parts of the earth are constantly warmed) [all the material in round brackets is *cancelled*.] For we meet with very little motion in the world besides what is (visibly) owing to these active principles, & the power of the will.[53]

As in the case of II and IV, many of the phenomena cited in the passage are chemical in character. In the second place, Newton speaks of laws of motion directly produced by active principles. And lastly he refers to these laws as the "genuine Principle of the Mechanical Philosophy". But what are these laws and to what sort of motion do they refer? Considering the type of phenomena discussed in the printed passage and in the draft, it would seem that Newton does not mean local motion – movements of terrestrial and planetary bodies as treated in the *Principia*. In IV he refers to such phenomena as "Motions . . . so great and violent". Does Newton mean that these phenomena give rise to a type of motion that cannot be reduced to the doctrine of matter in motion characteristic of the classical mechanical philosophies? The evidence in the following passage suggests an affirmative answer. In two draft variants related to I, Newton is explicit:

I have hitherto been arguing from the effects to their causes & carried the argument (as high as) up to certain forces (the powers) by wch little bodies act on one another at small distances.

These Forces may be reckoned among the laws of motion (& referred to an active principle) but whether they depend on bodies alone may be a question. For Bodies (alone considered as long, broad & thick . . .) are passive. By their vis inertiae they continue in their state of moving or resting & receive motion proportional to ye force impressing it & resist as much as they are resisted; but they cannot move themselves; & without some other principle than the vis inertiae there could be no motion in the world. (And what that Principle is & by (means of) laws it acts on matter is a mystery or how it stands related to matter is difficult to explain). And if there be another Principle of motion there must be other laws of motion depending on that Principle. And the first thing to be done in Philosophy is to find out all the general laws of motion (so far as they can be discovered) on wch the frame of nature depends. (For the powers of nature are not in vain [two words are illegible]. And in this search metaphysical arguments are very slippery. A man must argue from phenomena). We find in o$^{r}$selves a power of moving our bodies by o$^{r}$ thoughts (but the laws of this power we do not know) & see y$^{e}$ same power in other living creatures but how this is done & by what laws we do not know. And by this instance & that of gravity it appears that there are other laws of motion (unknown to us) than those wch arise from Vis inertiae (unknown to us) wch is enough to justify & encourage our search after them. We cannot say that all nature is not alive.[54]

The second passage occurs in a draft of Query 23 which argues that the ancient philosophers believed in the doctrine of atoms and the void, and that God was the source of gravitation:

Whence it seems to have been an ancient opinion that matter depends upon a Deity for its (laws of) motion as well as for its existence. The Cartesians make God the author of all motion & its as reasonable to make him the author of the laws of motion. Matter is a passive principle & cannot move itself. It continues in its state of moving or resting unless disturbed. It receives motion proportional to the force impressing it, and resists as much as it is resited. These are passive laws & to affirm that there are no other is to speak against experience. For we find in o$^{r}$selves a power of moving our bodies by o$^{r}$ thought. Life & Will (thinking) are active Principles by wch we move our bodies, & thence arise other laws of motion unknown to us.[55]

And lastly, in passage related to IV, Newton again makes the point with clear emphasis:

Now the above mentioned motions are so great & violent as to shew that (in Fermentations) there is new motion in the world generated from other Principles than the usual laws of motion [Newton bracketed this sentence, then repeated it again verbatim as far as the term 'Fermentation' after which he wrote] bodies (w$^{ch}$ almost rest) are put into (new) motions by a (very potent active Principle which acts upon them only when they approach one another & wch is much more potent than are the passive laws of motion arising from the Vis inertiae of the matter).[56]

Since Newton's appeal to "a power of moving our bodies" is connected with active principles, and the *sensorium* analogy, I shall comment on it later. From the above evidence it is clear that by 1706 he had come to the conclusion that there are active and passive principles in nature which produce two different sorts of motion, falling under two different categories of law. Thus Newton's ontology contains four fundamental types of entities existing in void space; *matter*, a passive principle, which is movable; *active principles* which "conserve" and "recruit" *motion*: and a variety of *forces*.

What is of great interest, and adds to the unexpected complexity of Newton's ontology, is the contention that there are two sorts of motion in nature. In the Preface to the First edition of the *Principia*, we are informed that the central problem of philosophy is to move "from the phenomena of motions to investigate the forces of Nature, and then from these forces to demonstrate the other phenomena".[57] Motion is important, therefore, in determining the properties of forces as considered in the *Principia*. "*Vis inertiae*", we are told in Query 31, is "accompanied with such passive Laws of Motion as naturally result from that Force".[58] These are the laws of the *Principia*, which only apply to the *atomic* part of reality – matter in motion. Newton certainly hoped, nevertheless, that this system of quantitative laws could be transferred to the motions of the invisible particles of matter. Perhaps this was all that the "analogy of nature" would bear. The draft passages, however, contrast the "passive laws of motion arising from the *vis inertiae* of the matter" with "new motion", or as Newton had called it in Query 8 "vital Motion",[59] arising from a "very potent active principle". Moreover, unlike the published passages, they suggest a possibility of knowing the laws of such motion, and thereby something of the micro-forces which give rise to it; perhaps in the manner that the laws and properties of gravity were "deduced" from planetary and terrestrial motion by the techniques of the *Principia*.

The connextion between force, active principles and the new category of motion is implicit in the published Queries 23 and 31. In all cases, fermentation, generation, explosions and exhalations are described as "great and violent" motions. Such phenomenon, Newton seems to hold, are not explained by mere matter in motion. Unlike the mechanical philosophers, he did not think that all phenomena could be reduced to this simple basis. In so far as phenomena of this sort occur as a result of the changing configurations of particles, passive matter is important. But their sheer activity suggested to Newton the existence of powerful "agents in Nature", which it was the "business of Experimental Philosophy" to find out.

Not only do these drafts provide evidence for Newton's belief in an ontology richer and more complex than any version of the mechanical philosophy, but they also suggest that he considered his methodology could be extended to cover the establishment of these new laws of motion. Thus Newton conceived the possibility that, by reasoning and experimentation based on non-mechanical phenomena, these laws could be characterized. A systematic investigation of nature was not in principle restricted to phenomena which fell under the categories of the *Principia*. For, "if there be another Principle of motion there must be other laws of motion depending on that Principle".

A fascinating interpretation of Newton's intentions can now be considered. The evidence strongly suggests that by 1706 he believed that a natural philosophy based more directly on the phenomena of force – a natural agency not reducible to matter – was a distinct possibility. As the Queries to the

*Opticks* alone justify, he did not think that the philosophy of the *Principia*, with its rigourous treatment of the motions of gross bodies, was the only type of physical knowledge open to the inquiring mind. The ontology of 1706 was far richer than that. Nor did he, as it now apparent, think that future advances in natural philosophy would necessarily have to be made within the mathematical framework of his classic treatise.[60] The possibility was now open, not only of applying the laws of the *Principia* to the motions of invisible particles, but of establishing another framework of laws characterizing motions caused by agents different from gravitation. In this way the range of phenomena on which a sound philosophy could be based, might be considerably broadened by future generations.

A problem appears to arise concerning the measuring and quantifying of these agents in terms of the motions they cause. The present interpretation claims that Newton conceived the possibility of determining the mode of action of laws not based directly on matter: laws of forces causing such motions as fermentation and putrescence, phenomena not reducible to the mere translation of matter or to mechanical impulse. Since forces causing these motions exist separately from matter, these laws should also be independent of material phenomena. But such agents are invisible and intangible: though causing uniform effects, they are not in themselves phenomena. How, then, could the properties of such forces be measured except through those of matter?

There is nothing strange in speaking of laws of invisible agents, or even of measuring them. After all, the law of gravitation is itself a clear example; the force whose action it characterizes is measured only through its effects. And in his drafts Newton is proposing that the laws of hidden forces be based on "new motions" different form the "phenomena of motions" upon which the laws of the *Principia* were established. As these laws determine the effects of large-scale forces such as gravity, so the laws of the other sort of motion would, when established, partially characterize the hidden forces of nature. Here in 1706, the matter rested. The philosophy of the invisible realm was to be squarely based on chemical motions, and a new physics was conceived which would attempt to infer the laws of these hidden forces. There are sufficient outcrops from these drafts, evident in the passages from Query 31, indicating that for Newton, this approach to the problem remained a possibility as late as 1717–18.

## III

His commitment to a philosophy of force and the void from 1686 to 1706 was certainly a major cause of Newton's apparent abandonment of earlier attempts at a quasi-material explanation of forces. In character incorporeal, these forces were never such that could be explained away by an aethereal medium operating by mechanical impulses. Even when such an aethereal medium was introduced in Query 31 in 1717–18, as a tentative means of explaining gravity, immaterial active principles still sustained the motions to which it gave rise.[61] With force so conceived, it is hardly surprising that Newton should be concerned to devise a means of bringing it within the scope of experimentation.

But Newton's restless and fertile mind did not remain content with the solution of 1706. By early 1707 he began to consider the possibility that the hidden forces of nature might be electrical in character.[62] These speculations mark an interesting shift in his thought. His general view from 1686 to 1706 was that a host of short-range forces existed in nature – as many as phenomena to be explained. However, on the occasion of the experiments of Francis Hauksbee in 1706–7 on electro-luminescence and on electrical attraction and repulsion, Newton began to consider again[63] not only that short-range attractions and repulsions might be electrical in nature, but perhaps other hidden forces as well. The difference between this approach to the physics of the invisible realm and that based on chemical phenomena is important. The latter approach attempted to develop a procedure which would establish the laws of motion of non-mechanical phenomena, and thereby reveal some of the properties of hidden forces. Newton remained agnostic, however, abut the nature of these forces. But between 1706 and 1713, he seriously considered the possibility that the forces producing insensible motions among particles were electrical, as large-range motions were produced by gravity. In 1713, this conception was announced in the concluding paragraph of the *General Scholium* in terms of a description of "a certain most subtle Spirit", electric and elastic in nature, which "lies hid in all gross bodies".[64] Short-range attractions causing cohesion, electrical bodies attracting at greater distances, the emission, reflection, refraction, diffraction of light, the heating of bodies, the causing of sensations and the power of moving muscles were all attributed to "the force and action" of this Spirit. In characteristic fashion, Newton claimed that he was not yet able to determine the laws of action of this agency. There can be little doubt, however, that he considered its existence as highly probable.

In a fascinating draft which is probably an early version of new Queries finally to appear in 1717–18 (nos. 17–24), Newton indicates the nature of this force:

Quaest 24. May not the forces by w$^{ch}$ the small particles of bodies cohere & act upon one another at small distances for producing the above mentioned phaenomena of nature [the same as those mentioned in Hypothesis II of the intended book IV of the *Opticks*] be electric? For altho electric bodies do not act at a sensible distance unless their virtue be excited by friction, yet that virtue may not be generated by friction but only expanded. For the particles of all bodies may abound with an electric spirit w$^{ch}$ reaches not to any sensible distance from the particles unless agitated by friction or by some other cause & rarified by the agitation. And the friction may rarify the spirit not of all the particles in the electric body but of those only w$^{ch}$ are on the outside of it: so that the action of the particles of the body upon one another for cohering

& producing the above-mentioned phaenomena may be vastly greater than that of the whole electric body to attract at a sensible distance by friction. And if there be such an universal electric spirit in bodies, certainly it must very much influence the motions & actions of the particles of the bodies amongst one another, so that without considering it, philosophers will never be able to give an account of the Phenomena arising from those motions & actions. And so far as these phaenomena may be performed by the spirit w^ch causes electric attraction it is unphilosophical to look for any other cause.[65]

Without doubt, then, some time between 1707 and 1713, electricity became the chief contender for the role of a primordial microscopic force. The importance which Newton placed on this possibility is evident by his saying that "it is unphilosophical to look for any other cause" of electrical attraction. The only other place in which he uses this form of words is in Query 31 when he states that God created nature as it is: "if he did so, it is unphilosophical to seek for any other Origin of this world . . .".[66] In any event, by agitation or some other means this latent spirit may be expanded from bodies over a sensible distance, and by such means it could influence the interaction among particles of matter.

The universality and flexibility of this spirit is illustrated by the beginning of a draft Query which, at this stage, Newton probably planned to follow after "Quest. 24":

Quest 25. Do not all bodies therefore abound with a very subtile, but active, potent, electric spirit by w^ch light is emitted, refracted, & reflected, electric attractions & fugations are performed, & the small particles of bodies cohere when continguous, agitate one another at small distances & regulate almost all their motions amongst themselves. For electric . . . uniting the thinking soul & unthinking body. This spirit may be also of great use in vegetation, wherein three things are to be considered, generation, nutrition & preparation of nourishment.[67]

In this truly extraordinary conjecture we witness the power of Newton's creativity, controlled by disciplined scientific imagination. The breadth of the conception is remarkable: he has in mind not only the chemical and optical phenomena mentioned in Query 31 and the draft *Preface* and *Conclusion* to the *Principia*, but also principles of life and a substitute for animals spirits – an electrical *arche* connecting mind with matter. Two and a half sheets are then devoted to a more detailed development of these ideas.[68] By means of this primordial agent or spirit, Newton was speculating on the possibility of uniting under one principle, life and nature, vitality and matter. Thus this unifying principle, electrical in nature and latent in all matter, might form a bridge between the animate and inanimate. We could have no better evidence for the importance of force in Newton's thought, especially with respect to the problems of the invisible realm of nature.

In a Latin manuscript of six Folio pages, Newton developed these ideas in even greater detail. The following passage illustrates the extent to which he considered, at this time, referring chemical phenomena to the operation of this electrical spirit:

Hence also, in proportion as the particles of bodies the more easily or with more difficulty glide among themselves, the bodies by the agitation of this spirit either become soft and ductile or else become fluid and are converted into liquids. And some bodies become fluid and put on liquid form by very small agitations of this spirit, as do water, oil, spirit of wine, quicksilver; others by greater agitations, such as wax, cebum, resin, bismuth, pewter: others by very great agitation, as clay, stones, copper, silver, gold. And in proportion as the particles of the bodies the more firmly cohere or are the more easily separated from each other, some of them by the agitation of this spirit are quickly converted into vapour and smoke, as are water, spirit of wine, spirit of turpentine, spirit of urine: others with more difficulty, as oil, spirit of vitriol, sulphur, quicksilver; others with very great difficulty, as lead, copper, iron, stones; while others even with the strongest fire remain stable, as gold and diamond.

By the action of the same spirit some particles of bodies can the more strongly attract one another, others less strongly, and thence can arise varying congregations and separations of particles in fermentations and digestions, especially if the particles are agitated by slow heat. Heat in fact, brings together homogeneous particles on account of greater attraction, and separates heterogeneous particles on account of lesser attraction. And by these operations living bodies attract parts of aliment which are like themselves, and get nourishment. Between acid particles and fixed particles deprived of acid by fire, there is very great attraction. Out of these are compounded particles less attractive, and out of these by slow fermentation and continuous digestion are compounded the bodies of animals and vegetables and minerals.[69]

It is again clear that chemical, and other phenomena which in earlier writings were simply attributed to attractive and repulsive forces are referred to this subtle electrical spirit. Newton has simplified his problem considerably. Not only, by 1707–13, has he attempted to reduce the cause of microscopic forces to one primordial – something not yet conceived in the 1690's – but he has thereby narrowed the problem of determining the proportion and rule of action of these hidden forces. Though, as Newton says, it was the "experiments shown by Mr. Hauksby before ye R. Society"[70] that recalled his mind to earlier speculations on electrical phenomena, it is clear that he thought this electrical primordium could be extended to cover the explanation of phenomena other that those on which Hauksbeen experimented.

But Newton's thought did not rest content with this possibility. As Professor Guerlac has shown in an important paper,[71] Newton became more convinced by 1716 of the possible existence of an all-pervading and continuous medium which he called an aether. Guerlac also urges persuasively, following the evidence of a letter from Desaguliers to Hans Sloane in March, 1730–31, that Newton took his aethereal medium more seriously for explaining optical phenomena than for explaining gravity. The evidence for this change in his thought is six folio sheets relating to a revision of the 1717–18 edition of the *Opticks*, in which Newton develops some "Observations" on the experiments of Hauksbee and gives an account of the origin of the "two-thermometer" experiment of Query 18.[72]

Not only did Newton distinguish an optical from a gravitational aether, and the aethereal medium from electrical spirits, but he made some other important distinctions which help to clarify his conception of force in 1717–18. He tells us in *Observation* I that, since this medium "readily pervades the solid body of Glass, it is reasonable to believe that it is not bounded by any solid body but readily pervades them all & is diffused through all the heavens".[73] Thus it is different from the earlier electrical spirit, since the latter was continuous only in the vicinity of bodies. After describing the operations of this "very subtile active substance or medium"[74] which are those formerly attributed to the "elastic spirit" Newton defines it: "To distinguish this Medium from the bodies which flote in it, & from their effluvia & emanations & from the Air, I will hence forward call it Aether & by the word bodies I will understand the bodies which flote in it, taking this name not in the sense of the modern metaphysicians, but in the sense of the common people & leaving it to the Metaphysicians to dispute whether the Aether & bodies can be changed into one another".[75]

The aethereal medium, therefore, is to be distinguished from other media such as air, and from bodies, effluvia and emanations. Newton's firm reference to what he understands by the term "body" is almost certainly related to definitions of body and void which were intended to preface Book III in the third edition of the *Principia*.[76] These were probably written late in 1716, at the same time as the *Observations*. Like the *Observations* these definitions are anti-Cartesian and anti-Leibnizian in spirit. Body is defined as "everything that can be moved and touched, in which there is resistance to tangible things . . . it is indeed in this sense that the common people always accept the word . . .".[77] And anything that is "hypothetically termed body is more properly treated of in metaphysics and hypothetical philosophy".[78] The definitions also indicate that Newton must have thought his aethereal medium, as defined in *Observation* I, was quite different from other conceptions of the aether since he says: "This subtle matter in which the planets float, and in which bodies move without resistance is not a phenomenon. And what are not phenomena, and subject to none of the senses, have no place in experimental philosophy".[79] But this is the point. As Professor Guerlac has argued,[80] the aether, because of the experimental work of Hauksbee and Desaguliers, became for Newton a phenomenon, something perceptible and tangible. Therefore, though he speaks of bodies floating in an aethereal medium in *Observation* I, this is not intended literally. Newton is attempting, by contrast, to state the difference between the two entities, for at this stage the aether, as he conceived it, was a possible phenomenon. If the Cartesian fluid aether, however, which he has in mind in the passage cited from the definition did exist, bodies could literally float in it. But as the last section of Book II in the *Principia* amply demonstrates such an aether being a physical impossibility, could not be a phenomenon.[81]

Newton has not, however, rejected his earlier speculation on the cause of interstitial attractions and repulsions. In his customary and cautious fashion, he again keeps his alternatives open. This is evident in an interesting passage from the *Observations*:

There are therefore Agents in Nature able to make the particles of bodies attract one another very strongly & to stick together strongly by those attractions. One of these Agents may be the Aether above mentioned whereby light is refracted. Another may be the Agent or Spirit which causes electrical attraction. For tho this Agent acts at great distances (only) except when it is excited by the friction of electrick bodies: yet it may act perpetually at very small distances without friction & that not only in bodies accounted electric, but also in some others. And as there are still other mediums which may cause attractions, such as are the Magnetick effluvia; it is the business of experimental Philosophy to find out all these Mediums with their properties.[82]

Thus, by 1717, Newton had once again changed his attitude towards force. In the *Observations* he argues for an optical aether diffused throughout space. He is, however, less certain that a continuous medium of this sort can explain gravity. With respect to electrical and chemical phenomena, however, Newton still thinks that they might be explained by an electrical spirit seated in bodies. But as the use of the term medium indicates more assuredly than with either agent or spirit, Newton's thought becomes focussed at this stage more on the interplay of force in spaces surrounding matter. Lastly, the doctrine of the paucity of matter developed from the 1690's, in large measure provided the general background in terms of which he came to concentrate on the problems of force, and its possible explanations.

## IV

The last two sections have concentrated on Newton's concern with activity in nature, in terms of the "downgrading" of matter and his changing attitudes towards hidden forces. In order to analyse his conception of active principles, it is first necessary to consider in some detail what he meant by force in contrast to the possible causes of force. Was force material or immaterial, or something else besides? A. R. Hall and M. Boas Hall have called our attention to the fact that Newton has two ways of speaking about attractions between bodies: in terms either of a language of force or that of an aether.[83] Since he never doubted the sheer existence of a hierarchy of forces, they rightly argue that the former mode of speaking was more fundamental. Aether hypotheses, on the other hand, where postulated as possible causal mechanisms of force, were always cautiously qualified. The Halls, however, tend to agree with Leibniz's reaction to Newtonian forces. He attempted, in the *Correspondence*, to persuade Clarke to accept the following dichotomy: either forces are material and mechanical, or else they are truly a direct manifestation of God's will and thus miraculous. If the validity of the dichotomy is denied by attempting to introduce a *tertium quid*, for Leibniz this constituted a reversion to occult qualities.[84] So the Newtonians were charged with making force either "a perpetual Miracle" or an inexplicable quality in nature. Of Clarke's view that "the means by which two different bodies attract each other, may be invisible and intangible and of a different nature from mechanism, and yet, acting regularly and constantly, may well be called natural".[85] Leibniz observes cynically: "He might as well have added, inexplicable, unintelligible, precarious, groundless and unexampled".[86]

Thus Leibniz rejected the Newtonian criterion of natural action. To be construed as an explanation of physical phenomenon a force must be intelligible; for Leibniz this meant that it must be an efficient and mechanical cause. The non-mechanical transcends the "powers of creatures" and therefore according to Leibniz's criterion is miraculous: it does not matter that it manifests uniform effects. But it is precisely this issue that divides leibniz and the Newtonians. For the latter the regular operations of gravity and other forces can be characterized by law, and are therefore in some sense real and natural. Furthermore, since many forces cannot be reduced to mechanical impact, the Newtonians argued that they must be construed as non-mechanical. "If the word *natural forces*, means here mechanical; then all animals, and even men, are as mere machines as a clock".[87]

Is there any reason for supposing that Newton must accept one or other alternative of Leibniz's dichotomy? This would be so if he could only say that forces exist and that they are neither mechanical nor a "perpetual miracle". Negatively defined in terms of such general characteristics, Newton's conception of force would not be an entirely acceptable *tertium quid*. But such forces can be determined by law, because of the regularity of their effects operating uniformly through long periods of time. Thus the Newtonians could point to three criteria to indicate the existence of an invisible agency in nature different from matter. Leibniz's denial that a non-mechanical agent is natural, though it can be shown to act on matter, is a prejudice: anything which is subject to law and which acts regularly, can in a perfectly meaningful sense be called natural. Newton did not, after all, deny the existence of mechanical forces. He was, however, concerned to maintain the existence of other sorts of forces.

Nor did he accept the idea of action at a distance. Change in nature must be the result of action and this means contact. Newton, however, was resisting a tendency to conceive all change in terms of the interaction of matter. For him the existence of invisible agents was evident from phenomena: the problem was the experimental determination of their properties. Certainly, measuring and quantifying forces could only be done mathematically by determining their magnitude and direction in terms of interactions among gross bodies. From this it does not follow, as Newton perceived, that force simply means impact or change of momentum. Or that the laws of motion determined by a force are the laws of impact: rather they are the laws of bodies in *motion* with impact as a special case. This is evident from Section II, where Newton's conception of different sorts of laws based on the motion of invisible particles was discussed.

Though he was not compelled to accept Leibniz's dichotomy, were there positive characteristics, independent of its effects on matter, that his conception of force possessed? In common with contemporaries, Newton used the terms "power", "virtue", or "agent" synonymously with force. Often, however, the mechanical philosophers used these terms merely as a convenient shorthand in referring to the changing configuration of matter. Newton, we have seen, held that there was activity in nature which could not be explained solely in terms of the ontologies of the mechanical philosophy. Gravity is clearly a phenomenon of this sort. It proceeded from a cause "that penetrates to the very centers of the Sun and Planets" and which "operates not according to the quantity of the surfaces of the particles upon which it acts (as mechanical causes are accustomed), but according to the quantity of the solid matter which they contain . . .".[88] Unlike matter, then, gravity is something that can spread or diffuse itself through the solid particles of bodies. And since it can affect them, it is in some sense physical. For given that the laws of action of *some* forces are known from their effects on matter, the problem of their operations on matter is different from that of how the soul influences the body. Form this point of view mechanical causes, such as impulse, are really the passive manifestation of activity in nature. Again, in the *Opticks* Newton holds that light and matter do not mutually interact at a "single point of the reflecting Body, but by some power of the Body which is evenly diffused all over its Surface, and by which it acts upon the Ray without immediate Contact".[89] In Query 29, he observes that "Magnetism may be intended and remitted"[90] and the same phenomenon is described in the *Principia* as the "intensione virtutis".[91] Lastly, in Query 31, he speaks of attractive forces in analogy to negative and positive quantities in algebra: where one vanishes the other begins, "so in Mechanicks, where Attraction ceases, there a repulsive Virtue ought to succeed" as "Rays are repelled by Bodies . . . without the immediate Contact of the reflecting or inflecting Body".[92]

What can be concluded from these descriptions of force taken at random from Newton's published work? In general, the Newtonian conception of force, at least from 1687, is that of a subtle and diffuse agent, able to penetrate through solid bodies, which can manifest varying degrees of intensity, and whose mode of operation is attraction or repulsion between the particles of matter. Like matter, force so conceived is a physical fact. Though like matter, dependent on the will of God, it can originate motion and activity. Not surprisingly, his conception of non-material agents affecting matter was not yet emancipated from notions such as alchemical powers. We have sufficient evidence to show that in the 1670's he was considerably influenced in his view of activity by alchemical and chemical literature. Since this view had not changed substantially, Newton's conception of force remained a meaningful but not yet a distinct physical concept. His physics was not yet that of the late eighteenth century.

He was also influenced by a pervasive thrust in English thought of the latter part of the seventeenth century – its belief in the realm of the spiritual.[93] This is especially evident in a group of thinkers who influenced Newton, the Cambridge Platonists. For example, in many of his works, Henry More devoted considerable effort to clarifying the notion of "spirit". For More, it had positive characteristics: it was penetrable, indiscerpible (no part of it removable), self-active, indivisible, and it had "essential spissitude" – a disposition to contract and dilate.[94] More was deeply concerned to emphasize the reality of the spiritual, stressing that there were many phenomena in nature "as cannot be resolved into mere mechanical powers".[95] Moreover, the world was populated with different forms of the spiritual: God, created spirits, plastic spirits and the spirit of nature. Only God, the angels and the soul of man possessed the higher attributes of spirituality, thought and will. Like More, Newton stressed the spiritual aspects of nature. At the level of phenomena this is shown by his emphasis on the activity of forces, and the consequent minimizing of the role of passive matter. Newton disclaimed, however, that his conception of force had anything in common with More's spirit of nature or Hylarchic Principle.[96] The notion of an unreflecting intermediary which blindly performed Divine edicts, struck Newton as not only an ontological extravagance, but also as scientifically impotent not being based on any experimental evidence.

Admittedly, Newton's general conception of force shared many characteristics in common with More's concept of spirit. Both concepts were forged in the belief that not otherwise could such non-mechanical phenomena as cohesion, electricity and gravity be explained. Spirit and force both had a disposition to penetrate matter, were indivisible, and admitted of degrees of intensity. Neither shared the higher attributes of spirituality, and unlike "purely" spiritual substances, they overtly interacted with phenomena, as exemplified by light and matter. Still, Newton's doctrine of force, though probably influenced by More, was based on a careful analysis of a wide range of phenomena, many of which were chemical in nature. Thus, it is clear that no straightforward answer can be given as to whether Newton's forces are material or immaterial. They are manifestly different from both matter and spiritual substance, and thus they seem to occupy a twilight zone between the corporeal and the incorporeal. Relative to ponderable bodies they are incorporeal; but relative to the higher realms of the spiritual they are corporeal. From the time of the first edition of the *Principia* to 1706 in the *Optice* Newton's conception of force rested here.

Before 1717–18, however, his ideas began to change. And this was largely brought about, as we have seen, by his attempting to explain invisible motion in terms of the interplay of electrical forces. In this period, Newton spoke of spirits and of aethereal media, and very little of virtues, powers and forces. The electric spirit was "so subtle as to pervade the solid body"; it was "capable of contraction & dilation expanding itself to great distances"; capable of being "agitated like a turbulent wind"; also "elastic and susceptible of vibratory motions", and "exceptionally active"; sometimes said to be continuous, sometimes likened to effluvia.[97] Again, however, Newton did not accept the strict dualism which stated that a thing must be either material or immaterial. In a draft for the "Account" of the *Commercium* this is clearly stated. Referring to the "subtile spirit pervading the pores of bodies" of the 1713 edition of the *Principia*, Newton observes that he has "nowhere denyed that the cause of gravity is Mechanical nor affirmed whether that subtile spirit be material or immaterial".[98]

The term "spirit", as used in the seventeenth and early eighteenth century, is notoriously slippery and ambiguous. Denying, in the company of most English thinkers, the implication of strict Cartesian dualism, Newton was employing the full range of connotations associated with the term. Thus he could move from spirit of vitriol to electric spirit, through animal, aerial, and aethereal spirits, to angelic spirits, and God. If Newton were not fully conscious of the subtle shifts in meaning it mattered little; for his abiding concern was to assert the existence of the non-material in the world. The vivifying effects of chemical spirits merged imperceptibly in Newton's world, through a chain of gradations into the spiritual realm: he and the seventeenth century made no clear distinction between the "spirituous" and the "spiritual".

Nevertheless, Newton was concerned with an electric spirit which he took to exist on the evidence of Hauksbee's experiments, and on his own observations made in the 1670's. Though he claimed to make no affirmation about the nature of the spirit, it was nevertheless allied more to matter than to God and the higher spiritual realities. Thus after 1706 Newtonian forces became more material-like in character, and were more directly based on the perception of such phenomena as electro-luminescence and electrical attraction, rather than on inferences from non-mechanical phenomena such as fermentation.

Not only, however, did Newton believe in the existence of forces which brought about two different sorts of motions, he also consistently held that forces themselves were caused by agents which were characterized by the general rubric of active principles. As mentioned above, relative to matter, a force was non-material, but relative to its cause it was material. Again, this conception is not surprising when it is realized that Newton implicitly accepted in general outline the ancient doctrine regarding a chain of being stretching from the throne of God into nature.[99] Though he rejected overt intermediaries such as plastic natures, Newton tacitly assumed that existing entities move imperceptibly up the chain from the material through the immaterial to the greatest reaches of spirituality.

Newton's conception of the cause of force changed again, we have seen, just prior to the second English edition of the *Opticks* in 1717–18. In the *Observations* he spoke of a tenuous medium filling all space. This aethereal medium was put forth as a plausible explanation of long-range forces such as gravitation and, in particular, optical phenomena. As described in the new Queries of the *Opticks* (nos. 17–24) it became at once quasi-mechanical and particulate in structure, and is characterized by the following terms: "rare", "subtle", "elastick", and that which "dilates and contracts", "condenses", has graduated "density" and "vibrates".[100] In all cases the basic operation of this aether is in terms of the gradations of a medium with the dispositional property of differential density, a property not mentioned in the *Observations*.[101] It is different from but modelled on, some of the characteristics of continuous media such as air and to a lesser extent water. Thus it has no imaginary properties and in good Newtonian fashion it is based on sensory experience.

Though this aethereal medium, because its particles mutually repelled one another, could be analysed mathematically, it is difficult to suppose that Newton took it seriously. It was subject to obvious conceptual inconsistences; it was a flagrant example of the sort of intermediary entity which Newton had always tended to reject; and more significantly it repudiated his basic metaphysics of God in an empty universe. To whatever extent it may have exonerated him from the continental charge of employing gravity as an occult quality, the 1717–18 aether remains the most puzzling of Newton's attempts at the problem of the cause of force.

It is important to notice, however, that Newton did not explicitly apply this aethereal notion to the hidden forces of nature. Apart from its tentative use in explaining neurological processes such as the operation of nerves and muscles,[102] he maintained various other media and spirits, of an electrical and magnetical sort, as possible explanations of the origin of invisible motion. But these speculations do not appear as such in the Queries.[103] Rather, in Query 31 of 1717–18, Newton again discussed the role of chemical phenomena as a basis from which to investigate the properties and mode of action of micro-forces. We can only conclude that "vital" motions still provided the basic experimental route to the invisible realm.

## V

In order to appreciate fully the significance of the changes in Newton's conception of matter, force, active principles and the aether, it is necessary to reconsider the salient characteristics of the theological framework within which his physical thought was developed. This framework remained relatively constant throughout his intellectual development, and influenced the formation and transformation of many of the basic concepts of his natural philosophy.

English theological and natural philosophical thought in the post-restoration period continued to be concerned with the relation between God's Providence and the working of secondary causes in nature. Boyle, the apologists for the Royal Society Sprat and Glanvill, the Cambridge Platonists, and the physicotheologians[104] were all cautious in their use of second causes, for fear of unduly diminishing God's Power over nature. Omnipotence was "the cardinal Praerogative of Divinity"; and the "chaining up his arms in the adamantine fetters of Destiny"[105] was blasphemous. The position of the English thinkers is admirably expressed by Thomas Burnet in his *Sacred Theory*:

> . . . we are far from excluding Divine Providence, either ordinary or extraordinary from the Causes and conduct of the Deluge. I know a Sparrow doth not fall to the ground without the Will of our Heavenly Father, much less doth the great World fall into Pieces without his good Pleasure and Superintendency. In him all things live, move and have their Being. Things that have Life and Thought have it from him, he is the Fountain of Both. Things that have Motion only, without Thought, have it also from him. And what only naked Being without Thought or Motion, owe still that Being to Him. And these are not only derived from God at first, but every Moment continued and conserv'd by him. So intimate and universal is the dependence of all Things upon the Divine Will and Power.
>
> In the second place, they are guilty . . . of a great Error or indiscretion, that oppose the course of Nature to Providence . . . nor are things less Providential because constant and regular: on the contrary such a disposition or establishment of second Causes as will in the best Order, and for a long succession, produce the most regular effects, assisted only with the ordinary concourse of the First Cause, is a greater Argument of Wisdom and Contrivance, than such a Disposition of Causes as will not in so good an Order, or for so long a time, produce regular Effects without an extraordinary concourse and interposition of the First Cause. . . .
>
> Thirdly, besides the ordinary Providence of God in the ordinary Course of Nature, there is doubtless an extraordinary Providence that doth attend the greater Scenes and the greater Revolutions of Nature. This methinks besides all other Proof from the Effects, is very rational and necessary in itself; for it would be a limitation of the Divine Power and Will so as to be bound up to Second Causes, as never to use, upon occasion an extraordinary influence or direction.[106]

Apart from a strong assertion of God's immanence in nature, Burnet is concerned to distinguish between two modes of Divine Providence; his ordinary and extraordinary concourse. In mediaeval theology this distinction was made by the terms *potentia ordinata* and *potentia absoluta*: God's ordained and absolute power.[107] In the theology, the natural philosophy, the legal and natural law theories of the seventeenth century, the distinction was held to be relevant to all operations of the Divine Will.[108] According to the tradition of voluntarist theology God, by his absolute power, established the natural and moral order. Within and with respect to that established framework, second causes and moral laws are absolute and immutable. It is to the greater glory of Divine activity that it operates through the constant and regular laws of that order. For as Burnet tells us, God, has created by his wisdom and foresight, an order able to produce regular effects for a "long succession". Though all laws and causes have their continual source in the Divine Will, that Will itself is not completely arbitrary and capricious. If it were, there could be no law in the natural order. So in terms of natural regularities – his "ordinary concourse" – God's omnipotence normally expresses itself; for we are assured that he will work through the scriptural dispensation. God is not, however, bound by the natural order. He could, by his absolute power, abrogate the present economy of the world: there is nothing ultimate about the order of nature. The absolute mode of Divine activity is seen in the case of miracles, whereby God imposes his sovereign will not through physical causes, but by suspending them or using them in an extraordinary way.

Though he nowhere uses the terms "ordinary" and "extraordinary" concourse, or *potential ordinata* and *potentia absoluta*, the distinction and the theological tradition behind it was accepted by Newton. His clearest statement of it is found in a draft of some ideas written in the early nineties, which in another form are found in the *General Scholium* of 1713:

> That God is an entity in the highest degree perfect, all agree. But the highest idea of the perfection of an entity is that it should be one substance, simple, indivisible, living and life-giving, always everywhere of necessity existing, in the highest degree understanding all things, freely willing good things; by his will effecting things possible; communicating as far as is possible his own similitude to the more noble effects; containing all things in himself as their principle and location; decreeing and ruling all things by means of his substantial presence (as the thinking part of a man perceives the appearances of things brought into the brain and thence rules its own body); and constantly co-operating with all things according to accurate laws, as being the foundation and cause of the whole of nature, except where it is good to act otherwise.[109]

In language more sublime than that of the *General Scholium* itself, the central idea is God's absolute Dominion over nature. All things have their source and continuing existence in his Will; for He is "living and life-giving". But again, once the natural order is established, God operates "according to accurate laws". These, of course, could be abrogated by Divine discretion: his Will is omnipotent. For Newton will, not reason, lies at the heart of God's decrees upon nature which are manifested by physical and moral laws. Only "an intelligent entity, acting by the force of its own will according to the intellectual ideas of objects in terms of final causes could have introduced the variety of objects".[110] Such variety could never have arisen from "blind necessity".

Natural objects are present to God in a way that can only be hinted at in analogy to images of things in the human mind. As we rule our bodies in terms of our ideas and intentions, God in an absolute fashion rules nature, being present to the world as the mind is present to the body. By thus evoking a macrocosm-microcosm analogy between God and man, Newton was drawing upon an unstated assumption still current amongst educated classes of his time. Later in using this analogy, he spoke of space as the *sensorium* of God. But in the above passage, however, and also in *De Gravitatione* of 1670, the term is not used.

An aspect of the analogy is of interest here, as it is concerned directly with the modes of God's omnipotence. This is the relation between the Divine Ideas and their copies in nature hinted at in the manuscript. A fuller indication of this relation between archetype and ectype is found in the 1675 letter to Oldenburg in which Newton reports that the world was:

. . . after condensation wrought into various formes, at first by the immediate hand of the Creator, and ever since by the power of Nature, wch by virtues of the command Increase & Multiply, became a complete Imitator of the copies sett her by the Protoplast.[111]

The main contention is that after the initial creation *ex nihilo* the whole of nature originated by Divine fiat from a primitive and spiritous aether[112] which can transmute by alchemical means into multitudinous forms. The statement itself is platonic in character abounding as it does with such notions as imitation, copy, protoplast, and creator, and it is made explicitly Christian by reference to the biblical command "increase and multiply". Thus by a sovereign decree God brought the world into existence. Once established, however, that order maintains itself by "the power of nature". Which in turn exacts the Divine decrees, by generating phenomena according to the protoplast or archetype. These copies are therefore physical imitations, manifesting physical laws, of the Divine Ideas.

An interesting accounting of the creation of matter by God's absolute power is found in *De Gravitatione*. Newton advances an hypothesis involving a notion of "immaterial" matter, to show how God could have created something which, without being material, could manifest all the effects produced by physical phenomena:

Since each man is conscious that he can move his body at will, and believes further that all men enjoy the same power of similarly moving their bodies by thought alone; the free power of moving bodies at will can by no means be denied to God, whose facility of thought is infinitely greater and more swift. And by like argument it must be agreed that God, by the sole action of thinking and willing, can prevent a body from penetrating any space defined by certain limits.[113]

Thus God is able to endow certain determinate parts of space with such properties as impenetrability and mobility, so that if "this world were constituted of this kind of being, it would seem hardly any different"[114] from matter. As we can move created bodies by an act of the will, God by his omnipotence can directly inform space with bodies. Moreover, their motion is regulated "hither and thither according to certain laws" which "God has imposed"[115] upon the established order. Once again, we see the distinction reflected in Newton's thought.

This hypothesis concerning the creation of "immaterial" matter throws light on the doctrine of matter's paucity and on the related doctrine concerning its imperfection as an existing thing. In contrast to space, force and God, as I have discussed elsewhere,[116] matter according to this conception is an entirely impoverished being embodying the lowest degree of reality and perfection. Newton was certainly in agreement with More's contention that the higher operations of reality could not have sprung form "so contemptible a Principle as *bare Body* or *Matter*".[117] It is evident that he wished to advance a conception of matter which, for theological reasons, emphasized its complete dependence on the will of God. In company with many English contemporaries,[118] he held that the Cartesian, the Hobbesian, and later the Leibnizian doctrines of matter offered a sure path to atheism, since they all could be interpreted as supporting its independence of God, and even its eternity.[119] Newton never abandoned either his hypothesis regarding the creation of "immaterial" matter, or his view of its low degree of reality and perfection.[120] With an adherence to these theological doctrines, it is hardly surprising that he should come to regard space as containing little solid matter. By 1690, as we have seen, he was convinced that there was sufficient evidence to support the hypothesis and its implications. Among the latter, was a firm commitment to the existence of force and a belief prior to 1706 that God was the direct cause of their operations in a matterless void. There can be little doubt, then, that the entrenched theological commitments of Newton's thought had a deep influence on his conception of space, force and matter.

By way of summing up it may be said in general that for Newton as for many of his contemporaries there was nothing necessary or ultimate about the present state of nature. Though there was no reason to suppose that God would abrogate nature's laws, it was in his power to do so. This is implied in Query 31 of the *Opticks*. Everything in the world is "subordinate to him, and subservient to his will". Since "Matter is not necessarily in all place, it may be also allow'd that God is able to create Particles of Matter of several Sizes and Figures, and in several Proportions to Space . . . and thereby to vary the Laws of Nature, and make Worlds of several sorts in several parts of the Universe. At least, I see nothing of Contradiction in all this".[121] For Newton, therefore, the world comprehended by the *Principia* could have been different than it is.

## VI

The passages, and their related drafts from the *Opticks* quoted in Section II, show that matter and force equally depend upon the will of God. In them, Newton also speaks of active principles as "causes" of such forces as gravitation and fermentation. This doctrine raises one of the most difficult issues in Newton's philosophy of nature, and the following questions must be considered. Are active principles natural agents which act within the physical order? That is, are they a sort of physical cause? Or are they an intermittent manifestation of God's power in nature, in the character of special divine interventions? But before considering this, we must ask: what did Newton mean in the context of these passages by the term "principle"?

In the late seventeenth century, three traditional distinctions were still current: a principle could mean something formulated or asserted: the primitive *arche* out of which all things were thought to originate: or the primary cause of existing phenomena as, for example, the "principels of matter and motion" celebrated in the mechanical philosophies. The second sense of the term embodies the notion of being prior in time. As Newton himself said in the letter of 1675 to Oldenbury, "Thus perhaps may all things be originated from aether". The first sense ranged from primary truths and mathematical propositions, through normative expressions, to laws of nature.

Both this sense and the third are found in the passages partially considered in Section II. *Vis inertiae*, the passive principle of matter, is contrasted with some "other Principle" which changes the state of matter, and which is responsible for the "conserving and recruiting" of the "variety of Motion"[122] in nature. In all places where Newton uses the term "principle" in conjunction with the adjective "active", he has in mind various causal agencies which bring about the manifest effects of phenomena, and which maintain by intermittent action a conservation of motion. He also refers to active principles when speaking of the genesis of the world. "Now by the help of these Principles, all material Things seems to have been composed of the hard and solid Particles above mentioned, variously associated in the First Creation by the Counsel of an intelligent Agent". He ends by maintaining that it is vain "to pretend that it might arise out of a Chaos by the mere Laws of Nature; though once form'd, it continue by those laws for many Ages".[123] Thus active principles are meant to depict some evident characteristic of reality. In no way are they to be confused with occult qualities which only impede the progress of knowledge. For Newton, therefore, they are not mere verbal formulations, but aspects of reality, partially categorized by the mind. Consequently, these principles are laws in the sense of being features of reality rather than confirmed assertions about it. For unlike the laws of the *Principia* they were never fully formulated. Furthermore, they can neither be reduced to changes in matter nor be originated from such changes. Lastly, in their most general character, active principles are a clear manifestation of God's activity in nature conceived, ontologically, within a classical neoplatonic framework.[124]

As causal agents, these principles are basic to Newton's immaterial ontology. This point can best be developed by considering two passages which occur only in Query 23 of the *Optice*.[125] Firstly, II(a) of 1706 in Section II is notably different from II(b) of 1717–18. In the latter, Newton simply says that motion arises from active principles; but in II(a) he says: "For we find very little motion in the world, except what evidently arises from these active Principles or from the power of a will". Instead of the customary opposition of God's will with a mechanical aether as means of explaining gravity and other phenomena, the Divine will is opposed to active principles. In III(a) he is more forthright about active principles than in III(b), saying that "natural phenomena show that such principles really exist: thought it has not yet been explained what their causes are".

Both passages strongly suggest that active principles are within the order of nature rather than existing solely as divine interventions. In the first place, Newton used *vel* – the exclusive "or" – to contrast active principles with the divine will. Characteristically, he is being cautious. Motion may arise from active principles or it may be the *direct* result of Providence, but not both together. Though when God does work through the mediation of these principles, his concourse with nature is not abrogated. Secondly, Newton strongly links active principles with physical and chemical phenomena, which again suggests that they are natural agents the characteristics of which are not yet entirely known.

Before 1706, however, the situation was somewhat different. In a manuscript probably related to the "classical" scholia of the abortive 1690's edition of the *Principia*, Newton was quite unambiguous about the status of active principles:

> The Epicureans making a distinction of the whole of nature into body and void, denied the existence of God, but very absurdly. For two planets distant from one another by a long and empty interval will not approach one another by any force of gravity, nor will they act upon one another, except by the mediation of some *active principle* which comes between them both, and by means of which force is propagated from each into the other . . . consequently those ancients who more rightly held unimpaired the mystical Philosophy as Thales and the Stoics taught that a certain infinite spirit pervades all space *into infinity*, and contains and vivifies the entire world: and this spirit was their supreme divinity, according to the Poet cited by the Apostle, in him we live and move and have our being.[126]

In this passage, which strongly emphasises the existence of divinity in nature, Newton unmistakenly refers to God, as the cause of the "force of gravity", by means of the term "active principle". We could scarcely wish for a clearer expression of Newton's view that active principles are the cause lying behind the variety of forces in nature, and especially of those operating in the invisible realm. In 1705 in a draft for Query 23 of the *Optice*, he still exclusively used the term in referring to God: "Life and will are active principles" and "if there be an universal life and all space be the sensorium of a thinking being . . . then laws of motion arising from life or will may be of universal extent".[127] Without doubt Newton's voluntarist theology is to the fore.

The Query as published in 1706 states, as we have seen, an exclusive disjunction between active principles and the will of God. This is not surprising. In the first place, at no stage in the development of his thought did Newton achieve a full solution, within the natural order, to the ontological problem of the causation of force. Thus he always maintained a cautious posture, and in this instance made a direct reference to the power of Providence in nature. Secondly, in keeping with this attitude, he continued to entertain alternative possibilities for a solution to the problem of the causation of force and motion. So at this stage, active principle became a general term for any sort of fundamental causal gent, and became less a theological term signifying God's direct action in the world.

By 1717–18 in Query 31, Newton once more strongly implied that active principles were part of the natural order since, as we have seen, the reference to "the power of a Will" was suppressed. The reason for this is probably two-fold. Leibniz in the *Correspondence* had severely criticized the orientation of the Newtonian theology around the sovereign will of God. He especially objected to the notion that God was compelled to act from time to time to reform the system of the world. For Leibniz, this impugned Divine foresight, and constituted "a very mean notion of the wisdom and power of God"[128] in creating the world. In reaction to such sustained criticism, Newton probably decided to delete any direct reference to God's will. Thereby he began consciously to relate the intermittent reformation of motion and force to active principles themselves, now characterized as operating entirely within the established order at God's behest. Secondly, at Newton's own direction, as we have seen, the work of Hauksbee and Desaguliers on electro-luminescence and related phenomena occasioned the serious return to earlier speculations on the causes of force. Having increasingly considered from 1687 to 1717 various sorts of spirits, different types of media, and vapours streaming from the tails of comets, as sources of activity, he was now more inclined to talk in terms of natural agents as causes of heat, fermentation, gravity, cohesion, light, and life. Thus in this way Newton hoped to indicate a possibility of extending knowledge within the established order. All these various speculations, therefore, on the causes of such phenomena were subsumed in Query 31 under the term "active principle".

I have discussed the meaning, status and changing emphasis which Newton placed on active principles, but their ontology will only be fully understood in connexion with the doctrine of the *potentia Die absoluta et ordinata*. Newton's position can best be clarified by contrasting it with the theology of Descartes. Descartes' system of theology and natural philosophy are both intimately connected with his affirmation of God's sovereign and transcendent will. In the timelessness of divine unity, God's understanding and will are inseparably blended. Thus, nature, as well as nature's laws are completely dependent on divine decree for their existence and continuance in existence; laws of nature are established, "just as a King establishes the laws in his Kingdom".[129] Once established, these laws according to finite comprehension are "immutable and eternal",[130] and the established order thus maintains itself by God's *potentia ordinata*. Accordingly, human understanding can comprehend nature as a self-regulating system operating by changeless laws in terms of an imperishable fund of motion. To the transcendent mind, however, the essence of dependent existence is otherwise. Not only could the established order have been decreed to exist in a different manner, but all created things are so dependent on God that they must be re-created from moment to moment:

For the whole duration of life is divisible into countless parts, all mutually independent; so from my having existed a little while ago it does not follow that I need exist now, unless some cause creates me anew at this very moment, in other words preserves me. For it is clear, when one considers the nature of time, that just the same power and agency is needed to preserve any object at the various moments of its duration, as would be needed to create it anew if it did not yet exist.[131]

Apart from this rather curious conception of time, Descartes is simply asserting the traditional distinction between *causa secundum fieri* and *secundum esse*.[132] In the reply to the *Fifth Objections*, he puts the point emphatically:

God is the cause of created things, not only in respect of their coming to be but also in respect of their being, and so he must always expend his activity on the effect in the same way, to make it remain the same thing.[133]

Thus from a logical point of view the initial act of creation is of the same type as the continuance of something in existence. One could not expect less from a thinker who adopted so uncompromisingly the theory of unconditioned creation. Creation out of nothing in the Christian sense meant for Descartes, as for many of the schoolmen, that no finite thing once created was self-enduring, or self-subsisting in space. Moreover, should a body be allowed even a momentary self-continuing mode of existence, there would be no reason for denying it power to exist longer. Only God and the activity of his omnipresent agency can be continuous and enduring. Thus within a framework which strongly contrasts eternal and independent existence with the finite and dependent, Descartes collapses the distinction between the two modes of God's activity:[134] the divine will continually exercises its *potentia absoluta*, since everything is actively recreated from moment to moment.

There are some broad affinities between the theological views of Descartes and those of Newton. Both thinkers stress the absolute and sovereign will of God which involves a strict adherence to the doctrine of *creatio ex nihilo*. By means of the traditional distinction in the modes of God's *potentia*, both make the doctrine of the creative will fundamental to their theology and natural philosophy. But with respect to their intentions there are important differences which divide them. Newton's position regarding the general relation obtaining between God and nature is more difficult to interpret than that of Descartes. Certainly he did not wish to make God transcend the world as an *intelligentia supramundana*, or to make Him identical in any sense with nature. Following the main stream of Christian thought, Newton's God was immanent in creation: and he would have followed St. Augustine in regarding the natural and spiritual orders as "interpenetrating one another in all the phenomena of the world".[135]

While insisting on an intimate connexion between God and nature, Newton repeatedly attempted to characterize a mode of Divine existence which did not make his omnipresence one with creation. Thus space and time in Newton's philosophy are physical manifestations of God's existence; neither, however, are properties or qualities of God. In an *Advertissement au Lecteur* which he sent to Des Maizeaux, Newton was emphatic on this point:

> . . . wherever in the following papers, through unavoidable narrowness of language, infinite space or immensity & endless duration or eternity are spoken of as *Qualities* or *Properties* of the substance which is immense or eternal [God]; the terms *Quality & Property* are not to be taken in that sense wherein they are vulgarly, by the writers of Logick & Metaphysicks applied to finite & created beings. . . . They have a nearer relation to the Predicaments of *Ubi* and *Quando* when applied to uniquity & Eternity.[136]

Thus God is everywhere and always: his "where' and "when" are categorically different from finite things, and his relation to nature is not akin to that obtaining between dependent things and their attributes. Language, however, is too impoverished to allow the finite mind to express this difference fully: it can only be hinted at.

Newton, however, was never easy about his conception of the relation between God and the world, and he often used biblical references, similes, and figures to characterize it. Thus unlike the Cartesian conception, God does not dwell in a timeless and transcendent realm of eternity, but exists semi-eternally in unending time; the one "who is and who was and who is to come".[137] But it was the Newtonian conception of space that drew the most vehement criticism. Newton's use of the *sensorium*, and the macrocosm-microcosm analogies was certainly misleading. His real position was perhaps best stated by two figures which never appeared in any of his published work. The first is his use of the Hebrew conception of *MAKOM* which occurs in two places. In the *Advertissement*, in order to clarify God's substantial presence, Newton says that "when the Hebrews called God *MAKOM* place, the place in w$^{ch}$ we live & move and have our being 'they' did not mean that space is God in a literal sense".[138] Thus the point is made sufficiently clear that space is the *Makom* of God wherein he substantially dwells, not a property that characterizes his nature. In a passage from one of the classical scholia intended for the abortive 1690's edition of the *Principia* Newton elaborated this figure: "This one God they [the ancients] would have it dwelt in all bodies whatsoever as in his own temple, and thence they shaped ancient temples in the manner of the heavens".[139] God was therefore in the world but not of it, for "The entire universe was rightly designated a Temple of God".[140]

The most important question, however, is Newton's conception of God's causal connexion with nature. Does he hold that the world once created is thereby actively sustained by God from moment to moment? Is his position akin to that of Descartes; or is the world sufficiently maintained by laws except when the total order needs a "reformation"? Newton's thought is not explicit on this point. It is clear, however, that he argued for God's substantial presence since "The human race is prone towards mysteries, and considers nothing holy and perfect which can be easily understood". Hence "the conception of God from the disputation of the scholastics emerges sterner as when it is said that God is everywhere according to his power and yet nowhere according to substance . . .".[141] This conception Newton could not abide. God's essence and substance are inseparable: since nothing acts where it is not, his power and presence are everywhere. For God has "a propensity to action, that he should never and nowhere be idle especially concerns his glory and majesty; though what God did before the creation of this visible work and outside its limits, we cannot think".[142] God is thus necessarily and substantially omnipresent and "thereby able to act in all times & places for creating & governing the present Universe".[143]

The "Lord of the Universe or Pantokrator"[144] had, therefore, in terms of Newton's interpretation of divine immanence, absolute dominion over all. Though he could never have accepted Descartes' atomic view of time involving as it did a rigorous recreationalist hypothesis, Newton was committed to a doctrine of *concursus* as a mode of Providence. This could not be otherwise since he accepted creation *ex nihilo*, which involves the preservation of dependent creation in existence. The position is clearly put by Newton's contemporary the Cambridge Platonist Peter Sterry:

Learned men and Divines teach us, that the Preservation of the world is *continuata Creatio*, a continued Creation in every moment of time from the Beginning of the world to the end, the Divine Act of Preserving and Governing the world according to the Present form proper to it for that Season is entirely the same with the Act of Creation.[145]

Thus Providence acts to conserve nature which, relative to the initial creation by decree, is a sort of *continuata creatio*, though the essential identity of the established order is not abolished: for Newton, too, the same essence remains and is preserved by God. Thus unlike Descartes' doctrine the world is not strictly recreated from moment to moment nor is it maintained, once created, by physical laws alone.

For God does not disdain to work through second causes. This is accomplished by another mode of Providence the divine concurrence with the established order – God's *potential ordinata*. Thus from the design of nature Divine power and wisdom are knowable. Therefore to know the laws of nature is, for Newton, to know the decrees of God's will, since he concurs with second causes by continually implanting efficacy in them. They are thus a manifestation of Divine volition put into execution by the power of God: there are ". . . no powers of nature independent upon God, as the powers of weights and springs are independent upon men".[146]

Clarke repeatedly points out in the *Correspondence*: "*natural* and *supernatural* are nothing at all different with regard to God, but distinctions merely in our conceptions of things".[147] There can be no doubt that this was Newton's view. In a letter of May, 1712, intended for Leibniz, he forcibly describes this position. He is attempting to counter Leibniz's opinion, as expressed in a letter to Hartsoeker, that if gravity is not explained mechanically the only alternative is the unintelligibility of miracles or occult qualities:

So then gravity & hardness go for unreasonable occult qualities unless they can be explained mechanically. And why may not the same be said of the *vis inertiae* & the extension, the duration & mobility of bodies & yet no man ever attempted to explain these qualities mechanically, or took them for miracles or supernatural things or fictions or occult qualities. They are the natural, real, reasonable manifest qualities of all bodies seated in them by the will of God from the beginning of the creation & perfectly incapable of being explained mechanically, & so may the hardness of primitive particles of bodies. And therefore if any man should say that bodies attract one another by a power whose cause is unknown to us or by a power seated in the frame of nature by the will of God, or by a power seated in (an immaterial) a substance in w$^{ch}$ bodies move & flote without resistance & which has therefore no *vis inertiae*, but acts by other laws than those that are mechanical, I know not why he should be said to introduce miracles & occult qualities & fictions into ye world. For Mr. Leibnitz himself will scarce say that thinking is mechanical as it must be if to explain it otherwise be to make it a miracle, an occult quality & a fiction.[148]

In this fascinating statement of his philosophical position, Newton plainly holds that all natural phenomena – force, matter, and their qualities – are equally manifestations of God's *potentia absoluta*. As Clarke puts it: "With regard to God, no one possible thing is more miraculous than another; and that therefore a miracle does not consist in any difficulty in the nature of the thing done, but merely in the unusualness of God's doing it".[149] The view of creation and its preservation developed in the passage is similar in spirit to that expressed much earlier by Newton in *De Gravitatione*. Again, the various possible solutions to an ontological explanation of force are open – God, non-mechanical causes, or the aether. Moreover, Newton's outright denial that basic causal explanations are mechanical, unmistakably places him outside the traditional "new" philosophies: no more than thought itself are the basic features of nature accounted for by systems such as those of Descartes and Hobbes. Nevertheless, the present natural order having been established by divine fiat, the enquiring mind can attempt to explain it by means of second causes.

With this background considered, we can now investigate further the ontology of active principles. Typical of the expressions in the draft passages of Section II is the statement that "We cannot say that all nature is not alive". These expressions are significant. Not only do they affirm the ontological priority of spirituality in Newton's philosophy, but they provide a further link with the Cambridge Platonists, especially with Cudworth. The significance of this link will be placed in perspective if the connexions between "life", "will", "thought", and active principles are explored. Newton's use of the term "life" in the draft passages of Section II surely reflects the influence of Cudworth. For example, in summing up the true philosophy of the ancients Cudworth tells us that they divided the world into passive matter and active power. "To the latter of which belongs both cogitation, and the power of moving matter, whether by express consciousness or no. Both of which together may be called by one general name of life; so that they made these two general heads of being or entity, passive matter or bulk, and self-activity or life. The former of these was commonly called by the ancients the Tὸ πᾱσχον 'that which suffers and receives' and the latter the Tὸ ποιοῦν 'the active principle'."[150]

Newton also divides the world into two principles, and like Cudworth uses "life" to cover the volitions and intentions of God and man, as well as the various causal agencies which move matter and replenish activity in nature. Thus the term refers to all manifestations of the non-material from forces to the higher attributes of the spiritual. In another important respect the influence of Cudworth is evident. In the Queries, Newton uses active and passive principles in contexts where he wishes to show that atoms and the void are insufficient principles with which to explain nature: material principles of this sort in themselves presuppose an immaterial realm. In the writings of Walter Charleton and Henry More active and passive principles are used in this context,[151] but Cudworth's use of them is characteristically closer to that of Newton. In the context of tracing the atomic philosophy of Moses, Cudworth concludes that the ancient atomists "joined both active and passive principles together, the corporeal and incorporeal nature, mechanisms and life, atomology and pneumatology; and form both of these untied, they made an entire system of philosophy, correspond with, and agreeable to, the true and real world without them".[152] As causal agents, however, active principles are more closely linked with will, thought, and final causes, which in turn are characterized by the *sensorium* and volition analogies which occur in many passages considered above. In yet another draft passage related to Query 23 of the 1706 *Opticks*, Newton plainly states these relations:

> Life and will are active principles by which we move bodies & thence arise other laws of motion not yet known to us. And since all matter duly formed by generation & nutrition is attended with signs of life, & all things are framed with perfect art and wisdom, & Nature does nothing in vain; if there be an universal life, & all space be the sensorium of a immaterial living, thinking, being, who by immediate presence perceives things in it as that wch thinks in us perceives their pictures in the brain and whose Ideas work more powerfully upon matter than the Imagination of a mother works upon an embrio, or that of a man upon his body for promoting health or sickness, the laws of motion arising from life or will may be of universal extent.[153]

Again, there is a clear reference to the laws of another sort of motion discussed in Section II. And as the other draft passages to Query 23 indicate, Newton was attempting in principle to show how these laws could be discovered, so that something might be known about the physical operations of active principles within the established order. For these "other laws of motion" were conceived by Newton as the laws of active principles, working through such forces as generation, fermentation and gravity.

But active principles are also periodic revelations of God's agency in the world. Here the analogies at once reveal the difficulty in clarifying how active principles affect natural phenomena, and the extent to which Newton connected causation with the volition of an active agent. True causation must originate something, make something happen, and the paradigm of this is the exercise of the will.[154] Thus "mechanical causation" in the sense of Hobbes or Locke did not constitute, for Newton, causation at all; it did not involve agency, but was merely a description of events, like the "passive laws of motion". God's immanence and power are reflected by the analogies, and since Newton held that God actively governed the world, "universal life" was omnipresent. Within this framework Divine agency is strikingly characterized by the Platonic image of mind working upon matter as nature upon her offspring. Thus active principles express the "intending" causality[155] of God which always acts to bring about its ends "with perfect art and wisdom": the mind of God has worked and is working upon creation.

We can at last draw an important conclusion. Active principles are intimately connected with the causation of Divine agency. By virtue of God's concurrence with second causes the laws of nature sustain the physical order: "Where natural causes are at hand God uses them as instruments of his work . . .".[156] But nature, according to Newton's interpretation of Providence, has been so conceived by God, that from time to time he will more directly intervene to reform it and replenish motion and activity.[157] Thus Providence planned the world so that divine power should not be limited, and so that it would be manifest to those who studied nature with piety. From this point of view active principles are the expression of God's *potentia absoluta* in nature, such that laws and forces in the present natural order are used intermittently to actualize the pre-conceived ends of Providence.

To Newton's contemporaries, therefore, there would be nothing surprising in this interpretation of Providence; they, too, shared many of Newton's unstated theological assumptions. In fact many contemporaries were gratified by Newton's achievement in harmonizing the traditional doctrine of Providence with "scientific" evidence from nature. Thus God's *potentia absoluta* was given a new meaning since motion "is always upon the decay" and planetary irregularities will increase "till the system wants a reformation".[158] Once Newton's thought is seen within the framework of this traditional theology, it is readily evident that there is no conflict between Providence and natural philosophy.[159] Just as God pre-ordained that nature could be partly explained by second causes, so he pre-ordained that he would intervene in creation to manifest his power. As Clarke put it "this amendment is only relative, with regard to our conceptions. In reality, and with regard to God; the present frame, and the consequent disorder, and the following renovation, are all equally parts of the design framed in God's original perfect idea".[160] For too long, Newton's conception of Providence in relation to nature has been judged from the point of view of Leibniz, and the eighteenth-century deists. Thus the natural philosophy of Newton was compatible with the traditional Christian interpretation of creation and Providence: indeed, as I have tried to show, his interpretation of nature was deeply influenced by this tradition.

## CONCLUSION

I have endeavoured in this study to offer a framework in which to interpret Newton's conception of the natural world as it developed in the post-*Principia* period. Beginning then, he became more consciously involved with the basic principles, stated or assumed, of his physical thought. It must be again emphasized, however, that his characteristic ideas were forged in embryonic form in the 1660's, and then developed and brought to fruition through decades of reflection. Thus, for example, his consideration of electrical spirits and aethers in the second decade of the eighteenth century was occasioned rather than caused by the work of Hauksbee and Desaguliers: these ideas had been form the seventies a possible answer to the ontological problem of causation.

In the tradition of Renaissance thought embodies in England by the Cambridge Platonists, one of Newton's central concerns was the existence of the spiritual in nature. Apart form the existence of gravitation, Newton saw in the wide range of chemical and biochemical phenomena evidence of the spiritual. As we saw, these "vital" motions inclined his mind to the possibility of a system based on laws categorically different from the *axiomata* of the *Principia*: a natural philosophy of the hidden realm based on microforces. Though Newton for a time considered these forces to be electrical or magnetical, or to be generated by the activation of light particles, chemical activity remained the best evidence on which to ascertain their characteristics.

In the 1690's the void was linked with the operations of Providence and Divine omnipresence in nature. In order that Divine purpose might be displayed in the physical realm, Newton conceived that God had made matter extremely porous. Thus contrived, it best exemplified the Divine idea of how its multitudinous functions would be performed. As its actual properties became less important for explaining phenomena, so forces operated more and more in the vacuum left by the diminished amount of matter occupying space. And Newton probably conceived particles in the limit as insubstantial foci about which clustered attractive and repulsive forces. Intimately related to this, was the theological doctrine of the creation of "immaterial" matter, and its consequent ontological imperfection.

Newton's doctrine of creation and the connected doctrine of Providence almost certainly influenced these developments. All created things, matter and force alike, continually depended for their existence on the will of God; thus at each moment they bespoke His power and Providential foresight. Thus, though God will not abrogate his scriptural dispensation to work through the created order, he had foreordained that form time to time he will intervene in nature to manifest his *potentia absoluta*: it is in this context that we must see the ontology of active principles and the various instruments of

Providence for replenishing nature, such as comets. To Leibniz's general criticism that his God was a cosmic tinker continually repairing a faulty artifact, Newton replied:

He insinuates that it is the fault of the workman & not of the materials that a Watch will at length cease to go & in like manner that it would be God's fault if the world should ever decay & want an amendment. And by the same way of arguing a man may say that it would be God's fault if matter do not think.[161]

In a sense this statement represents the high water mark of a long theological tradition centering on the Sovereign Will.

## NOTES

[1] See the fundamental studies of H. Metzger, *Attraction universelle et religion naturelle chez quelques commentateurs anglais de Newton*, Paris, 1938, and A. Koyré, *From the Closed World to the Infinite Universe*, Harper edition, 1958. Also see H. Guerlac, "Newton et Epicure", *Conferences du palais de la découverte*, no. 91, Paris, 1963; an excellent study by David Kubrin, "Newton and the Cyclical Cosmos: Providence and the Mechanical Philosophy", *J.H.I.*, **XXVIII** (1967), 325–346; J. E. McGuire and P. M. Rattansi, "Newton and the Pipes of Pan", *Notes and Records of the Royal Society of London* **21** (1966), 108–143; J. E. MaGuire, "Body and Void and Newton's *De Mundi Systemate*: some new sources", *Archive for History of Exact Sciences* **3** (1966), 206–248; an important lengthy study by A. Koyré and I. B. cohen, "Newton and the Leibnitz-Clarke correspondence", *Archive Internationales d'histoire des Sciences* **15** (1962), 63–126; and A. R. and M. B. Hall, *Unpublished Scientific Papers of Isaac Newton*, Cambridge, 1962, part 3.

[2] See Chapter 3.

[3] *Ibid.*

[4] *Ibid.*

[5] Arnold Thackray, "'Matter in a Nut-Shell': Newton's *Opticks* and Eighteenth-Century Chemistry", *Ambix* **XV** (1968), 29–53; and Chapter 3.

[6] Henry Guerlac, "Francis Hauksbee – expérimentateur au profit de Newton", *Archive Internationales d'histoire des Sciences* **16** (1963), 113–128.

[7] Isaac Newton, *Philosophiae Naturalis Principia Mathematics*, trans. A. Motte, London, vol. II, 1729, p. 393.

[8] Henry Guerlac, "Newton's Optical Aether", *Notes and Records* **22** (1967), 45–57. I cannot in this study undertake the major task of analysing Newton's ideas regarding the aether from the beginning of his career: thus I have restricted myself to the post-*Principia* period. See B. J. T. Dobbs, *The Janus Faces of Genius: The Role of Alchemy in Newton's Thought*, Cambridge University Press, 1991, for a complete analysis of Newton's alchemy.

[9] Of primary importance are the internal circumstances of Newton's thought: that he, for example, continually reconsidered his ideas in the light of his increasing knowledge of phenomena. The 1670's and 1680's generated various social and religious tensions: in response to atheism and enthusiasm, Anglican divines such as Glanvill, Casaubon, Stilling fleet, More, Rust, Hallywell and Tillotson developed apologetics based primarily on Descartes and Platonism. The stress on the immaterial and the spiritual may well have influenced Newton's view of the relation between philosophy and theology. In the 1690's the nature of apologetic controversies changed. The empiricism of Locke's *Essay* replaced the crude sensationalism of Hobbes; and a more effective type of polemic against the social and religious consequences of freethinking was based on the order of nature rather than on Cartesian conceptualism. Newton certainly

approved of the "Newtonian" apologetics of the Boyle lectures based as they were on his experimental empiricism.

[10] Hall and Hall, *op. cit.*, note 1a, pp. 91–156.

[11] *Ibid.*, p. 142.

[12] *Ibid.*, p. 146.

[13] Kubrin, *op. cit.*, and Chapter 7.

[14] Hall and Hall, *op. cit.*, 187–189.

[15] H. W. Turnbull (eds.), *The Correspondence of Isaac Newton*, 1960, vol. II, p. 288.

[16] Hall and Hall, *op. cit.*, p. 189.

[17] J. Lohne, "Newton's 'Proof' of the Sine Law and Mathematical Principles of Colours", *Arch. for the History of Exact Sc.* **I** (1961), 401.

[18] See Chapter 3 and Lohne, *ibid.*

[19] Kubrin, *op. cit.*, note 1. This study is fundamental for an understanding of the development of Newton's ideas on the sources of activity in nature. See also McGuire, *op. cit.*, note 12.

[20] See Thackray, *op. cit.*, note 5, p. 34, and Guerlac, note 1, for a discussion of the void in the new Queries.

[21] See Chapter 3.

[22] The full account of this progression as a possible way of showing how matter might be extremely porous, did not appear in the body of the *Opticks* until 1717–18, pp. 242–244, see Thackray, *op. cit.*, note 4.

[23] W. G. Hiscock (eds.), *David Gregory, Isaac Newton and Their Circle*, Oxford, 1937, p. 29. Gregory concludes his observation with a full description of the progression of void to matter.

[24] U.L.C. Add. 3970, fols. 234$^{r-v}$. In another intended Query, written about the same time, Newton expresses these ideas in a form very similar to the manner they were to be expressed in the *Optice* in 1706. It is found on fol. 296$^{r}$, Add. 3970.

[25] There are entries in one of Newton's early commonplace books to support this conjecture. It is the *Quaestiones* found in U.L.C. Add. 3996. Not sufficient attention has been paid to the variety of atomic hypotheses in the seventeenth century, or to the variety of reasons which led philosophers to adopt atomism.

[26] Hall and Hall, *op. cit.*, p. 345.

[27] *Ibid.*, p. 307. The passage is from partial draft preface intended for the first edition of the *Principia*.

[28] Isaac Newton, *OPTICE sive de reflexionibus, Refractionibus Inflexionibus Coloribus LUCIS*, London, 1706, see Thackray, *op. cit.*, note 5, and Chapter 3.

[29] Isaac Newton, *Mathematical Principles of Natural Philosophy*, trans. A. Motte, London, vol. I, 1728, p. 224.

[30] U.L.C. Add. 3970, fol. 292$^{r}$. The thesis regarding "the smallness of the Rays of Light" is mentioned in the new Query 21 of the 1717–18 edition of the *opticks*. Newton also mentions that attraction is stronger in the smallest magnets, and "in the surfaces of small planets than in those of great ones in proportion to bulk". He concludes, not surprisingly, that the particles of the aether being the smallest of all, have the strongest force.

[31] U.L.C. Add. 3970, fols. 259$^{r}$, 260$^{r}$. These could have been written as late as 1708. The third reference to the light hypothesis is on fol. 291$^{v}$.

[32] Kant was the first thinker to give a fully coherent alternative to the transmission of force from one place to another without presupposing solid, inert matter. Indeed, he conceived "matter" as arising from the combination of attractive and repulsive forces, the latter manifesting degrees of intensity. See *Metaphysical Foundations of Natural Science*, trans. E. B. Bax, London, 1883.

[33] Bernard Gagnebin, "De la Cause de la Pesanteur. Memoire de Nicolas Fatio de Duillier Presenté à la Royal Society le 26 Fevrier 1690", *Notes and Records of the Royal Society* **6** (1949), 117.

34 *Correspondence*, vol. III, p. 287. Professor I. B. Cohen has pointed out to me that in all three editions of the *Principia* Newton speaks of a "Medii interea, is quod fuerit, intestitia particum libere pervadeantis . . ." in Definition I, which defines the Quantity of Matter. This would *seem* to indicate that some sort of aethereal medium remained a possibility from 1687 to 1727, though as such it became less significant from 1687 to 1704.

35 *Correspondence*, vol. III, p. 309.

36 *Correspondence*, vol. III, p. 191.

37 U.C.L. Add. 3970, fol. 338$^{r-v}$. The full set of five hypotheses are published in I. B. Cohen's "Hypotheses in Newton's Philosophy", *Physis* **8** (1966), 163–184.

38 Gregory MS. 247, fol. 14a, Library of the Royal Society. the "classical" scholia, part of which were published in "Newton and the Pipes of Pan", [note 1] are in the same manuscript book. there is little doubt that the fragment is connected with the rejection of the aether explicit in these scholia. The Latin was published in James Crauford Gregory's "Notice concerning an autography manuscript of Isaac Newton . . .", *Trans. R. Soc. Edinb.* **12** (1834), 64–76.

39 *Correspondence*, IV, p. 267.

40 Hiscock, *op. cit.*, p. 29.

41 "Newton and the 'Pipes of Pan'", *op. cit.*, note 1.

42 Is. Newton, *Philosophiae Naturalis Principia Mathematica*, London, 1687, p. 411.

43 This intended addition to Corollary III is discussed in Chapter 7.

44 *Op. cit.*, note 7, vol. I, p. A2. I shall not deal with the so-called problem of "transdiction", as it is independent of the evidence on which such inferences are based.

45 Hall and Hall, *op. cit.*, pp. 302–308, 321–327. Also, of course, it was developed later in Query 31 of the *Opticks*.

46 Isaac Newton, *Opticks*, New York, 1952, Book II, Section III.

47 Hall and Hall, *op. cit.*, p. 307.

48 U.L.C. Add. 3970, fol. 338$^{r-v}$. This is also published in I. B. Cohen's paper cited in note 37 above.

49 There are at least twenty-five separate references to the "analogy of nature" in Newton's published and unpublished writings. For its role in Rule III, Book III of the *Principia* see Chapter 6.

50 I. B. Cohen (ed.), *Isaac Newton's Papers and Letters on Natural Philosophy*, Cambridge, 1958, pp. 255–259.

51 *Correspondence*, vol. II, p. 292. Earlier in his letter to Oldenburg of 1675 Newton used the "Sociablenes" to cover the action of the nerves and animal juices, *Correspondence*, vol. I, pp. 368–369.

52 See M. Boas-Hall's introduction to "Newton's Chemical papers" in Cohen, *op. cit.*, note 50, p. 246.

53 U.L.C. Add. 3970, fols. 255$^{r}$–256$^{r}$. These drafts are all written in English.

54 U.L.C. Add. 3970, fol. 620$^{r}$.

55 U.L.C. Add. 3970, fol. 619$^{r}$.

56 U.L.C. Add. 3970, fol. 244$^{r}$.

57 *Principia*, *op. cit.*, note 7, p. A2, vol. I.

58 *Opticks*, p. 376.

59 *Opticks*, p. 340.

60 For an opposing view see Henry Guerlac, "Where the Statue Stood: Divergent Loyalties to Newton in the Eighteenth Century", *Aspects of the Eighteenth Century*, Johns Hopkins Press, 1965.

61 In the 1717–18 edition of the *Opticks*, eight new Queries were added (17–24) and 21 is concerned with the aether as a possible cause of gravity. Query 31 of that edition is concerned, in part, with active principles as the cause of a wide range of phenomena. Thus the latter term is much wider in connotation than the term "aether". I cannot, therefore, entirely agree with David Kubrin's opinion that by 1717–18 Newton "tended to use the concept of the aether and that of

active principles interchangeably": see his study cited in note 1. My disagreement with Kubrin's interpretation is found in Section VI below.

62 Guerlac, *op. cit.*, note 6.

63 A similar approach to the problem is to be found in the letter to Oldenburg in 1675, *Correspondence*, vol. I, pp. 362–386. In a series of *Observations* which are to be discussed in this section, Newton said of the attractive and repulsive virtue of a rubbed piece of glass, "Above forty years ago I sent the following Observation to Mr. Oldenburg and have now copied it from one of the Books of the Royal Society". U.L.C. Add. 3970, fol. $626^{v}$.

64 *Principia, op. cit.*, note 7, vol. II, p. 393. See M. B. Hall and A. R. Hall, "Newton's Electric Spirit – Four Oddities", *Isis* **50** (1959), 473–476; also A. Koyré and I. B. Cohen, "Newton's 'Electric and Elastic Spirit'", *Isis* **51** (1960), 337.

65 U.L.C. Add. 3970, fol. $235^{v}$.

66 *Opticks*, New York, p. 402.

67 U.L.C. Add. 3970, fol. $235^{v}$. Similar ideas are to be found in fols. $241^{v}$, and $295^{r}$.

68 *Ibid.*, fols. $235^{v}$, $241^{r-v}$. Newton is primarily concerned with the growth and generation of vegetable and animal life in terms of his electrical *arche*. Similar ideas are treated in greater detail in a series of Latin manuscripts: Add. 3970, fols. $236^{r}$, $237^{r}$, $238^{v}$, $240^{r-v}$. These are entitled, "De vita and morte vegetabili". None of this material found its way into the Queries. The electric spirit as a vehicle for uniting the soul with the body is also considered by Newton on fol. $241^{r}$. "This spirit therefore may be the medium of sense & animal motion and by consequence of uniting the thinking soul & unthinking body". It is interesting to note the residual influence of Descartes on this formulation.

69 U.L.C. Add. 3970, fol. $604^{r}$. These and related ideas in various stages of preparation are found in fols. $240^{r}$, $241^{v}$, $597^{r-v}$, $601^{r}$, and $603^{r}$. The Latin of fol. $604^{r}$ reads:

Hinc etiam corpora perinde ut eorum particulae facilius vel difficiluis interlabuntur, per agitationes hujus spiritus vel mollescunt et ductiles evadunt vel fluunt et in liquores vertuntur. Et aligua quidem per spiritus hujus agitationes quam minimas ut aqua, oleum, spiritus vini, argentum vivum, alia per agitationes majores ut cera, cebum, resina, bismutum, stannum, alia per maximas ut argilla, lapides, cuprum, argentum, aurum, fluunt et forman liquorum induunt. Et perinde ut particulae corporum vel fortius cohaerent vel facilius ab invicem separantur, eorum aligna per agitationem hujus spiritus cito vertuntur in vapores et fumos ut aqua, spiritus vini, spiritus terebinthi, spiritus urinae, alia difficilius et oleum, spiritus vitrioli, sulphur, argentum vivum, alia difficillime ut plumbum, cuprum, ferrum lapides, alia igne fortissimo fixa manent,ut aurum et adamas.

Per ejusdem spiritus actionem, aliquae corporum particulae a invicem fortius attrahere possunt, aliae minus fortiter, et inde variae particularum congregationes et separationes in fermentationibus ac digestionibus oriri, praesertim si particulae per colorem lentum agitentur. Calor utique particulas homogeneas ob attractionem majorem congregat, et heteogeneas ob attractionem minorem segregat. Et per has operationes corpora viva attrahent partes alimenti sibi similes et nutruiuntur. Inter particulas acidas & particulas fixes acidis per ignem privatas maxima est attractio. Ex his componuntur particulae minus attractivae, & ex his per fermentationem lentam & digestionem continuam componuntur animalium vegetabilium & mineralium corpora.

70 U.L.C. Add. 3970, fol. $241^{v}$.

71 "Newton's Optical Aether", *op. cit.*, note 8.

72 U.L.C. Add. 3970.9, fols. 623–629. Desaguliers letter is in Sloane MS. 4051, fol. 200, British Museum.

73 *Ibid.*, fol. $623^{r}$.

74 *Ibid.*

75 *Ibid.*

76 See Chapter 3. These definitions were never entered into any personal copy of Newton's *Principia*.

77 *Ibid.*
78 *Ibid.*
79 *Ibid.*
80 Guerlac, *op. cit.*, note 8.
81 See Chapter 3.
82 *Op. cit.*, note 72, fol. 622r. In fol. 240r Newton combines the force of both electrical and magnetical particles: All dense bodies in fact seem to consist of electric particles and also to have a certain number of magnetic particles. And just as the attraction of gravity suffices to explain the greater movements of Planets and Comets and of our sea, even so do electric and magnetic forces seem to suffice to explain the interactions and movements of the particles of each several body." We could want no clearer indication as to how far Newton's thought had temporarily moved away from chemical affinities as the binding forces of particles. I do not wish to suggest that Newton only began to concentrate on media in the second decade of the eighteenth century. For example, in a draft variant for Prop 3, Book II, part III of the *Opticks*, probably written before 1695 he describes " a certain medium by the force of wch surfaces of bodies act upon light". Add. 3970, fol. 374r.
83 Hall and Hall, *op. cit.*, note 1, p. 190.
84 H. G. Alexander (ed.), *The Leibnitz-Clarke Correspondence*, Manchester, 1956, p. 94.
85 *Ibid.*, p. 14.
86 *Ibid.*, p. 84.
87 *Ibid.*, p. 14.
88 Motte, *op. cit.*, note 7, p. 392.
89 *Opticks*, *op. cit.*, note 46, Book II, part III, p. 266.
90 *Ibid.*, p. 373.
91 *Principia*, *op. cit.*, note 42, p. 3.
92 *Opticks*, *op. cit.*, p. 395.
93 McGuire and Rattansi, *op. cit.*, note 1. For an interesting discussion of the role of the spiritual int he literature and Church of England polemics of the period see P. Harth: *Swift and Anglican Rationalism: The Religious Background of a Tale of a Tub*, Chicago, 1961.
94 Henry More, *A collection of several philosophical writings of Dr. Henry More*, London, 1712. *The Immortality of the Soul*, chap. II, p. 6; chap. III, pp. 8–9. Newton owned the 1662 edition of this work.
95 *Ibid.*, The preface, p. xiii.
96 *Philosophical Transactions of the Royal Society*, London, 1714–19, p. 223.
97 These phrases abound in the manuscripts cited in notes 65 and 69 above.
98 U.L.C. Add. 3968, fol. 125r. For a similar analysis of Henry Power's conception of *Spiritus* see my colleague's study: C. Webster, "Henry Power's Experimental Philosophy" *Ambix*, 1967, *XIV*, 150–178.
99 See Chapter 2. For a general discussion of the chain of being in seventeenth-century thought see: M. Macklen, *The Anatomy of the World: Relations between Natural and Moral Law from Donne to Pope*, Minnesota, 1958. The doctrine is also clearly presented in the work of Locke and Berkeley.
100 In terms of the graduated density of the aether, Newton suggested in Query 21 that the aether was rarer in the pores of bodies than in the surrounding space. Therefore the differential between the two places might cause bodies to move towards one another. This idea is only fully developed in the 1717–8 *Opticks*.
101 Guerlac, *op. cit.*, note 8.
102 *Opticks*, *op. cit.*, New York, Queries 23 and 24, pp. 353–344.
103 There are faint outcrops in Queries 8, 18 and 22.
104 T. Sprat, *History of the Royal Society*, London, 1667; R. Boyle, *Notion of Nature, Works*, IV, pp. 362, 370, 371, 367, 373; *Christian Virtuoso*, V, pp. 45–46, and *Excellency of Theology*,

III (mostly written in 1665; published in 1673), p. 411; Joseph Glanvill's *Essays on several important subjects in Philosophy and Religion*, London, 1676; Henry More, *A Collection of Several Philosophical Writings*, London, 1712, fourth edition; The *General Preface*, and also *An Antidote against Atheism* and the *Immortality of the Soul*; Ralph Cudworth, *The True Intellectual System of the Universe*, Cambridge, 1678; Samuel Clarke, *A Demonstration of the Being and Attributes of God: more particularly in answer to Mr. Hobbes, Spinoza and their Followers*, London, 1705. William Whiston, *A New Theory of the Earth*, London (fifth edition), 1734, pp. 293–295; William Whiston, *Astronomical Principles of Religion natural and revealed*, London, 1717; William Derham, *Physico-Theology of a Demonstration of the being and attributes of God from his work of Creation*, London, 1714; Edward Stillingfleet, *Origines Sacrae, or a Rational account of the Grounds of Christian Faith*, London, 1662. Both Hobbes and Spinoza were outside the tradition of voluntarist theology evident in the work of these writers.

105 Walter Charleton, *The Darkness of Atheism Dispelled by the Light of Nature: A Physico-Theological Treatise*, London, 1652, p. 329.

106 Thomas Burnet, *The Sacred Theory of the Earth containing an account of the original of the Earth, and of the general changes which it had undergone, or is to undergo, till the consummation of all things*, London, 1684, pp. 106–107.

107 The distinction was considerably clarified by William of Ockham, see *opus Nonaginta Dierum*, chap. 95, Guillelmi de Ockham: *Opera Politica*, Manchester, 1940–63, II, 718 ff. Especially relevant is *Nota di duplici potentia dei*. For a classic study of voluntarism in natural law and legal theories see Otto von Gierke, *Political Theories of the Middle Ages*, Cambridge, 1927; also the informative article of Francis Oakley, "Medieval Theories of Natural Law: William of Ockham and the Significance of the Voluntarist Tradition", *Natural Law Forum*, **6** (1961). The distinction is found in Suarez, *Metaphysicarum disputationum*, Salmanca, 1597, disp. 7, sect. 2, arts. 7–8, a treatise widely read in England, and in *Commentarii in octo libros physicorum Aristotelis*, Conimbria, 1592, Lib. II, cap. 7, quaest. 16, art. I.

108 Robert Boyle, *The Excellency of Theology, Works*, III; *Notion of Nature, Works*, IV; *Of the High Veneration Man's Intellect owes to God, particularly for his Wisdom and Power, Works*, IV; *The Christian Virtuoso, Works*, V; John Locke, *The Reasonableness of Christianity, Works*, VI, London, 1824; Walter Charleton, *The Darkness of atheism Dispelled by the Light of Nature: A Physico-Theological Treatise*, London, 1652; Henry More, *op. cit.*; William Derham, *op. cit.*; John Keill, *An Examination of Dr. Burnet's Theory of the Earth together with some remarks on Mr. Whiston's New Theory of the Earth*, Oxford, 1698; Lady Anne Conway, *The Principles of the Most Ancient and Modern Philosophy, Concerning God, Christ and the Creatures*, London, 1692; George Cheyne, *Philosophical Principles of Religion*, London, 1715; John Ray, *Three Physico-Theological Discourses*, London, 1693. With respect to morality and natural law some of the Cambridge Platonists were against voluntarist theology. Edward Fowler, *The Principles and Practices of Certain Moderate Divines of the Church of England Abusively Called Latitudinarians*, London, 1671, and Ralph Cudworth, *A Treatise Concerning Immutable Morality*, London, 1731. The distinction can be traced through the vast Puritan and "Calvinist" literature of the century. For a treatment of some of these themes see Francis Oakley, "Christian-Theology and the Newtonian Science: The Rise of the Concept of the Laws of Nature", *Church History* **30**, 443–470, and Edgar Zilsel, "Genesis of the Concept of Physical Law", *Philosophical Review* **51** (1942). This distinction between the two modes of God's power and the doctrine of imposed Laws is found in F. Bacon's "A Confession of Faith", *Works*, vol. VII, pp. 220–221.

109 The Latin is published in Crauford Gregory's article, *op. cit.*, note 38. The original is in the David Gregory MS. 245, fol. 14a, Library of the Royal Society.

110 *Ibid.*, fol. 14a.

111 *Correspondence*, vol. I, p. 364. A similar conception of the process of creation is found in the writings of More and Boyle. In the general preface to the *Collection* (note 94), p. xv, More says of the spirit of nature subordinating matter: "This therefore being a mute copy of

the eternal *Word* (that is, of that Divine Wisdom that is entirely everywhere), is in every part naturally appointed to do all the best services that matter is capable of, according to such or such modifications, and according to that *Platform* of which it is a *Transcript*, I mean, according to the Comprehension and Purpose of these *Ideas* of things which are in the external Intellect of God". In one of his Commonplace Books, 185, Library of the Royal Society, fol. 35$^{v}$, Boyle tells us in a fascinating passage: "I conceive yt ye omniscient as well as almighty Creator of ye universe & al ye substances it contains has himself a most intimate & perfect knowledge of the whole system they make up & of every one of them in particular they being al his own creatures & works & their particular natures being but transcripts of those eternal ideas of ye divine mind according to wich they were originally framed . . .". From internal evidence this was probably written in the 1680's.

112 Newton would seem to have accepted the idea of creation *ex materia praeexistente*, or secondary creation, as did Whiston and Burnet in their analysis of Chaos, *op.cit.*, note 108. See also Newton's letter to Burnet, *Correspondence*, vol. II, 321–329. None of these thinkers, however, denies that God created formless matter *ex nihilo*.

113 Hall and Hall, *op. cit.*, note 1, pp. 138–139.

114 *Ibid.*, p. 139.

115 *Ibid.*, p. 139.

116 McGuire, *op. cit.*, note 1.

117 More, *op. cit.*, note 94, p. 60.

118 For a general discussion of this see Samuel I. Mintz, *The Hunting of Leviathan*, Cambridge, 1962, and McGuire, *op. cit.*, note 1.

119 Kubrin, *op. cit.*, note 1.

120 A. Koyré, *Newtonian Studies*, London, 1965, p. 92. Koyré quotes Pierre Coste's report in his translation of Locke's *Essay* of a discussion which he had with Newton in the Latter's old age. Newton's account of the creation of matter by God is the same as that of *De Gravitatione*. Coste recorded: On pourrait, dit-il [Newton], se former en quelque manière une idée de la création de la matière en supposant que Dieu ent empêché par sa puissance que rien neput entrer dans une certaine portion de l'espace pur, qui de sa nature est pénétrable, éternel, nécessaire, infini, car des de la cette portion d'espace aurait l'impénétrabilité l'une des qualités essentielles à la matière . . .". I have suggested that Newton may have adopted this conception from Henry More, see Chapter 3.

121 *Opticks*, *op. cit.*, note 46, pp. 403–404.

122 *Ibid.*, p. 339.

123 *Ibid.*, p. 402.

124 McGuire and Rattansi, *op. cit.*, note 1.

125 These differences were first noticed by Alexander in his edition of the *Leibniz-Clark Correspondence*, pp. 178–179. He did not, however, attempt to judge the significance.

126 U.L.C. Add. 3965.12, fol. 269$^{r}$: (italics mine) Epicurei naturam totam in corpus et inane distinguentes Deum penegebant, at absurde nimius. Nam Planetae duo ab invicem longo vacuo intervallo distantes non petent se mutuo vi aliquo gravitatis neque ullo modo agent in se invicem nisi mediante principio aliquo activo quod utrumque intercedat, et per quod vis ab utrogue in alterum propagetur. . . . Ideoque Veteres qui mysticam Philosophiam rectius tenuere ut Thales et Stoici docabant spiritum quidam infinitum spatia omnia *in infinitum* pervadere & totam universum continere & vivificare; et hicce spiritus supremum numen, juxta Poetam ab Apostola citatum, In eo vivimus et movemur et sumus. It is interesting to notice that Newton uses the phrase "spiritum quidam", 20 years later in the *General Scholium* of 1713, to refer to his speculations on an electric agent in nature.

127 U.L.C. Add. 3970, fol. 619$^{r}$. This passage is from the same document as the third draft for Query 23 quoted in Section II above.

128 Alexander, *op. cit.*, note 84, p. 12. In one of its fundamental aspects the theological issue

between Leibniz and the Newtonians has to do with the nature of Divine Providence. Leibniz did not deny that the world "goes without God's interposition" (p. 18) or that "the creation wants to be continually influenc'd by its creator" but he "maintains it to be a watch, that does without wanting to be mended by him . . ." (p. 18). For if the world needed to be mended it "would indeed show the *power* of God; but it would not sufficiently show his *wisdom*" (p. 18). This is the issue: the Newtonians emphasize the attribute of *omnipotence* which is central to any coluntarist theology, and Leibniz emphasizes *omniscience* which is central to his view that God had created a world which is intelligible to human reason in terms of the logical necessities which led Him to create it. Thus creation can be understood by the exercise of reason, using the principle of non-contradiction and that of sufficient reason as tools in the analysis. But if, on the other hand creation is a *contingent* artifact of God's arbitrary fiat with laws imposed by Divine will, experimental methods will have to be employed to supplement our knowledge of Providence as gathered from nature; since in this case there are no intrinsic connexions among things.

[129] C. Adam and P. Tannery, *Oeuvres de Descartes*, I, pp. 145–146.

[130] *Ibid.*, VII, p. 380. Passages where Descartes explicitly uses the phrase "Puissance ordinaire et puissance extraordinaire de Dieu" are VII, 435, and VIII, 2ª, 167.

[131] E. Anscombe and P. Geach (trans.). *Descartes Philosophical Writing*, London, 1964, p. 88. Many of the Arabic philosophers held views similar to Descartes. See Moses Maimonides, *The Guide for the Perplexed*, trans. M. Friedlander, 1956, Part I, chap. 71–74.

[132] For the sources of Descartes' use of these terms, see e. Gilson, *Index Scholastico-Cartesian*, Paris, 1913, pp. 62–63.

[133] C. Adam and P. Tannery, *op. cit.*, VII, p. 369.

[134] The distinction between the two modes of God's *potentia* is coherent enough at a certain level; however, on analysis it reduces to the power of God *simpliciter*. To the seventeenth century, however, God's scriptural dispensation to work through the established orders of nature and grace was a serious matter. The world was also a contingent creation which, though depending on God, could be known by the finite mind independently of the creator.

[135] Comment by T. A. Lacey, *Nature, Miracle and Sin*, 1916, pp. 31–32.

[136] Koyré and Cohen, *op. cit.*, note 1, p. 101.

[137] U.L.C. Add. 3965.13, fol. 541ᵛ which is a draft of ideas found in the *Scholium* on space and time in the *Principia*, and ideas on the mode of God's existence found later in the *General Scholium*. There are a number of manuscript sheets on which similar ideas are developed, fols. 541ᵛ, 524ʳ⁻ᵛ and 545ʳ⁻ᵛ. These were probably written in the late 1690's.

[138] Koyré and Cohen, *op. cit.*, p. 101. The other place where the term *Makom* appears is in the manuscript cited in note 141 below. Newton's phraseology is similar to the passage quoted in the text above.

[139] Gregory MS. 245, fol. 14ᵇ.

[140] *Ibid.*, fol. 3. Newton had copied verbatim the passage in which this sentence is found – from *Macrobius' Commentary on the Dream of Scipio*. See W. H. Stahl's translation, New York, 1952, chap. XIV, p. 142.

[141] U.L.C. Add. 3965.13, 541ᵛ, translated from the Latin. See my 'Newton on Place, Time, and God: An Unpublished Source,' *Brit. J. Hist. of Sci.* **II** (1978), 114–129.

[142] *Ibid.*, translated from the Latin.

[143] Koyré and Cohen, *op. cit.*, note 1, p. 101. The idea that Providence is continually and actively guiding Nature abounds in the literature of the second half of the seventeenth century as it did in the first. Whiston uses Donne's metaphor of the "Vicegerent": we "shall attempt to consider and Remark his Vicegerent Nature in her Mechanical Operations therein", *op. cit.*, note 102, p. 296. Whiston elaborates in a fashion characteristic of the period: "Tis from him everything is ultimately deriv'd: He preserves the Natures, and continues the Powers of every Creature: He not only at first produc'd but perpetually disposes and makes use of the whole Creation,

and every part thereof, as the Instruments of his Providence . . .", *ibid.*, p. 294. Charleton criticizes the rhetorical traditional which likens God to a "*Pilot, Emperour*, and *General of Nature*" because there is "still this *difference* to be allowed, that though a *Pilot* is not *ubquitary* in all parts of his ship, nor an *Emperour* actually *omnipresent* in all places of his dominions, nor a *General locally present* in all quarters and stations of his Army; yet God is *intimately omnipresent* in every particle of the World: so that what is uncertainly said of the *Soul, Tota* in *Toto & tota in qualibet parte*: may most certainly be said of *God*", *op. cit.*, note 105, p. 114. For an important discussion of this theme in English though see C. A. Patrides, *Milton and the Christian Tradition*, Oxford, 1966.

144 Hall and Hall, *op. cit.*, note 1, p. 363.

145 Peter Sterry, *The Kingdom of God in the Soul of Man*, London, 1683, p. 29. Charleton firmly adheres to the same position. Speaking of "the Creation of the World *ex nihilo*, and the constant *conservation* of the same in its primitive order . . .", he says: "I doe not intend that those are *Acts really distinct* each from other (for in the demonstration of the existence of God its plainly, though succinctly evinced, that *the conservation of the Universe is nothing but the Act of Creation prolonged or continued*), *op. cit.*, note 105, p. 94.

146 Clarke, *op. cit.*, note 84, p. 22.

147 *Ibid.*, p. 24.

148 U.L.C. Add. 3965.17, fol. $257^{r-v}$ unpublished. The correspondence between N. Hartsocker and Leibniz appeared in the *Memoires de Trevoux* from 1706 to 1711.

149 Clarke, *op. cit.*, note 84, p. 114. Clarke's view of miracles is a faithful reflection of Newton's. The latter's clearest expressions of his doctrine occurs in a theological common-place book, see H. McLachlan, *Sir Isaac Newton: Theological Manuscripts*, Liverpool, 1950, pp. 17–18. Newton tells us: ". . . miracles are so called not because they are the works of God, but because they happen seldom, and for that reason crete wonder. If they should happen constantly according to certain laws impressed upon the nature of things, they would be no longer wonders or miracles, but might be considered in philosophy as a part of the phenomena of nature [notwithstanding their being the effects of the laws impressed upon nature by the powers of God] notwithstanding that the cause of their causes might be unknown to us". Newton sees clearly that voluntarist theology implies that laws of nature are *imposed* upon contingent phenomena, rather than being inherent in the nature of things. His view of miracles is not orthodox. Moreover, it entails the difficulty that in order to know a miracle had occurred it would be necessary to know that we knew all nature's law: since the doctrine involves God in using existing laws in an extraordinary way.

150 Ralph Cudworth (notes by J. L. Mosheim, trans. by John Harrison), *The True Intellectual System of the Universe*, I, London, 1845, p. 55. Professor Guerlac has conjectured that Cudworth influenced Newton, but he has not offered any evidence, see *op. cit.*, note 1. Newton made four folio pages of notes on Cudworth's *System*. The manuscript is in the William Andrews Clark Memorial Library of the University of California at Los Angeles. Cudworth's use of the Aristotelian terms *To paschon* and *To poioun* is interesting. In Aristotle they really function as adverbial qualifiers in the category of substance. Cudworth, however, uses them as terms which describe two basic categories of reality. This may indicate a Stoic influence; for they used these terms to distinguish two sorts of entities. See james S. Reid's *M. TVLLI CICERONIS ACADEMICA*, London, 1885, p. 124. Stillingfleet considers these two principles to be of Stoic origin. In his discussion of the production of the world he speaks of the active principle (God) working upon the passive principle (matter) *Origines Sacrae*, 1680, pp. 415–416. The distinction is attributed to the Stoics on the authority of Plutarch, Diogenes, Laertius, Seneca, Cicero and Stobaeus, p. 399 and pp. 415–416. A late edition of Stillingfleet's work was in Newton's library.

151 Charleton uses the term active principle when arguing that atomism is insufficient "unless it be supported by this additional base; that there be some first *active principle*, which by its

infinite power first created out of nothing . . .", *op. cit.*, note 108. More uses both terms: "Wherefore it is most rational to conclude, that no *Matter* whatsoever of its own Nature by any active Principle of *Motion* . . ." and "It is sufficiently manifest from sense and experience, that Matter is a principle *purely passive*", *op. cit.*, note 94. *The Immortality of the Soul*, pp. 43, 57. Newton certainly read the latter treatise. See Richard S. Westfall, "The Foundations of Newton's Philosophy of Nature", *Brit. J. for the Hist. of Sci.*, 1962, **I**, 172–182.

[152] *Cudworth, op. cit.*, p. 90.

[153] U.L.C. Add. 3790, fol. 252$^v$. This version is similar in content to the third draft passage quoted in Section II.

[154] In this respect Newton's doctrine of causation is similar to that found in Berkeley's *The Principles of Human Knowledge, Works*, II (ed. T. E. Jessop), Edinburgh, 1949, p. 63, Sections 50–53 and 25–32. At Section 27 Berkeley says: "A spirit is one single, individed, active being: as it perceives ideas, it is called *understanding*, and as it produces or otherwise *operates* about them it is called the *will*. Hence there can be no *idea* formed of a soul or spirit: for all ideas whatever, being passive and inert *vide Sect.* 25, they cannot represent unto us, by way of image or likeness, that which acts. A little attention will make it plain to anyone, that to have an idea which shall be like that active principle of motion and change of ideas, is absolutely impossible". The similarity between the thought of Newton and Berkely with respect to their use of the terms *Will* and *active principle* is striking. Newton's earliest use of these terms, in a context similar to Berkeley's, occurs in 1705 (see Section II above). Berkeley's *Principles* was published in 1710.

[155] Cudworth, *The True Intellectual System of the Universe*, London, 1678, p. 672: "Nevertheless we acknowledge, that *God* and *Nature* do things every where, in the most *frugal* and *compendious way*, and with the least *Operosseness* and therefore that the *Mechanik Powers* are not rejected, but taken in, so far as they could comply serviceably with the *Intellectual Model and Platform*. But still so, as that all is supervised by the *Understanding* and *Intending Cause*, and nothing passes without his *approbation: . . .*".

[156] *Correspondence, op. cit.*, II, Newton to Burnet, 1680–81, p. 334.

[157] Here again Newton is stressing the Divine attribute of omnipotence.

[158] *Opticks, op. cit.*, New York, note 46, p. 402.

[159] For an opposing opinion see Kubrin, *op. cit.*, note 1, p. 344.

[160] *Op. cit.*, note 84, p. 23.

[161] Koyré and Cohen, *op. cit.*, note 1, p. 110. Newton's point is that Leibniz's position leads to "very narrô Ideas of God's wisdom and power". *Ibid.*, p. 114.

CHAPTER 6

# THE ORIGIN OF NEWTON'S DOCTRINE OF ESSENTIAL QUALITIES

In the second edition of the *Principia* in 1713, there appeared for the first time the third 'Rule of Philosophising'. This is the most important and significant of the Rules in the prefatory section of the Third Book. Its origin and its implications for Newton's philosophy of nature have not been sufficiently analysed or appreciated. To this end lies the purpose of the present study.

I shall concentrate on a central problem arising from two early manuscripts of the Third Rule. This is Newton's treatment of a distinction between two basic sorts of physical qualities: usually characterised in the seventeenth century by the terms primary and secondary. And I shall be concerned mainly to clarify his arguments and assumptions rather than to criticise his shortcomings. There are, of course, other problems connected with the Third Rule, such as the relation between Hypothesis III of the first edition, that bodies transform into one another, and Rule III of the second edition, that certain qualities of matter are immutable. And related to this is the question of the different status accorded to hypotheses, axioms, and rules within the framework of the *Principia*. These questions raise rather large issues with respect to Newton's philosophy of nature, and will be the subject of another study.

## I

In the early 1690's, Newton set about strengthening the loosely organised third book of the *Principia*. He intended to bring its sparse mathematical character into line with his deepest intuitions of the physical world. In the first edition, Corollary II of Proposition VI explicitly refers to Hypothesis III of the same edition to justify the view that bodies are able to transform into one another in virtue of their difference being merely one of form.[1] In the second edition[2] the same argument is repeated, and as in the first it is a bit of *reductio and absurdum* reasoning to show that the gravity of a body cannot change while its quantity of matter remains the same. There is, however, no mention of Hypothesis III: it had been discarded. There are many intended emendations to Corollary III in Newton's manuscripts which indicate the way in which this change took place.[3] One is specially relevant to the subject of the present paper. It is found in Newton's own copy of the first edition in the Cambridge Library. On an interleaved sheet opposite page 411 which contains Corollary II, Newton has written and then crossed out the following emendation:

Gravity is a quality of all bodies which are round the earth, and on which we are allowed to make experiments, and is proportional to the quantity of matter in each thing and cannot be intended and remitted and therefore according to Hypothesis III is a property of all bodies.[4]

Hypothesis III is no longer that of the first edition, but refers to a new formulation which was to be called Rule III, and which in Motte's edition of the *Principia* is rendered as: 'The qualities of bodies, which admit neither intension nor remission of degrees, and which are found to belong to all bodies within the reach of our experiments, are to be esteemed the universal qualities of all bodies whatsoever.'[5] That all bodies within experience have weight, and that this by experiment has been shown to be as their quantity of matter, is not sufficient grounds on which to attribute to all matter whatsoever the quality of gravity. It is necessary that the mutual gravitation of bodies 'cannot be intended and remitted': that is, in the same latitude it must be an invariant quality with respect to their total amount of matter. Since this criterion is used to distinguish the essential qualities of matter from its accidental ones, as it is formulated the emendation implies that gravity is an essential quality. Newton vigorously dissociated himself form such a conclusion. Indeed, because of considerable misunderstanding of his position, he added to the Third Rule in the Third edition of the *principia* an explicit disclaimer:

Not that I affirm gravity to be essential to bodies: by their *vis insita* I mean nothing but their *vis inertia*. This is immutable. Their gravity is diminished as they recede from the earth.[6]

Thus gravity is associated with matter but not essential to it. Newton probably discarded the above emendation for Corollary II because he saw that it led to unwelcome consequences.

It will be argued below that this criterion of the essential qualities of matter – those which do not intend and remit – was first developed in what I shall call Draft No. 2 of the Third Rule. Although such a distinction in qualities is implicit in some of Newton's earlier writings,[7] it seems that he only systematically analysed it when he embarked in the 1690's on his revision of the Third Book. And it was then, I think, while concentrating on his epistemological position and on the ontology of the *De Mundi Systemate*, the world of gross bodies, that he began to see more clearly the implications of his theory of sensible mater. This led initially to an attempt at unifying his earlier view of the mutual transformation of bodies, expressed so vividly in the 1670's when he spoke of matter in alchemical terms,[8] with the conception of the immutability of certain qualities that was hardening in his mind in the 1690's. The older view was still appealing, and the continuity of matter associated with this conception had been expressed in Hypothesis III of the First edition. At least by the time he was preparing the second edition in 1712, Newton probably came to the conclusion that the whole problem-area was too complex for an official solution which could be set forth briefly in the *Principia*. Hypothesis III of the First edition was discarded: the doctrine of transformation was not, however. It was relegated to parts of the Third Book, and it appears prominently in the later Queries to the *Opticks*.[9]

There are three extant draft versions of the Third Rule. I shall present two of them in translation. A third which will not be considered is found in the interleaved first edition. it is substantially the same as that published in 1713, containing only a few minor verbal differences.[10] At the top of the page Newton had first written 'Hypoth. III'. Later he changed it to 'Reg. III'. It is interesting that in this draft there is no reference to the 'force of inertia'. In its place, Newton uses the term 'impetus'. Since this version was written in the early 1690's, after the publication of the first edition, it is not clear why he should do so. In any event, we cannot regard this as a reversion to the theoretical connotations of the term as it was used in Medieval and Renaissance thought. Newton's 'vis inertiae' is not identical with the 'impetus' of the Paris Terminists or the 'vis impressa' of Galileo. According to these conceptions the 'impetus' of a projectile would be used up or be diminished, while Newton's 'vis inertiae' could not be. Draft No. 1 is found on an end sheet of another annotated first edition which was in Newton's possession. Since the Latin is printed in Volume X of the *Opera Omnia* of Huygens,[11] I shall not reproduce it.

*Draft No. 1* N. Q. 16.200 Trinity College Library

The laws (and properties) of all bodies on which it is possible to institute experiments, are laws (and properties) of all bodies whatsoever. For the properties of bodies become known only by means of experiments, and consequently all those are to be declared general which in general agree with experiments. Certainly in opposition to the tenor of experiments, imaginary ideas [literally 'dreams'] must not be rashly invented, nor must we withdraw from the analogy of nature, since that is commonly simple and always in harmony with itself. The extension of bodies becomes known only by means of the senses, but is not perceived in all cases, yet because it applies to all perceptible things it is affirmed of all things whatsoever. The greater number of bodies we find by experience to be hard or solid, and therefore the particles not of these only but also of all others, we with good reason conclude are hard. From these properties of perceptible bodies we conclude that all bodies are mobile and impenetrable, and that they by certain forces continue in motion or at rest. That the divided parts of bodies can be separated one from the other, we know from the phenomena; and that the indivisible parts can by reason be distinguished into small parts, is certain from mathematics. But whether the parts distinguished can be separated from each other by the forces of nature, is uncertain: yet if even by one single experiment it were established that by the fraction of some one undivided particle a hard body suffered division, we should by force of this hypothesis reach a general conclusion that not only were the divided parts separable but also that all the indivisible parts could be divided *ad infinitum*.

*Draft No. 2* U. L. C. Add 4005.15 Folio $81^{r-v}$

This major manuscript is considerably damaged. A section of about three inches extending into the sheet and the bottom margin has been torn away. I have refrained from conjecturing much about the missing Latin. The translation, therefore, has gaps in it. The sense, on the whole, is nevertheless tolerably clear. My conjectures are in square brackets. Newton's cancellations and interlineations are in round. This convention is uniform throughout the paper. Two other axioms are fully written with comments on this sheet.[12] It appears that Newton intended to include them together in his revised edition. The Latin transcription will be found in the appendix.

*Axiom 4*

The qualities of bodies which are perceptible and which cannot be intended and remitted, as far as can be [known] in experience belong to all bodies, must be regarded as the properties of all bodies, are the properties of all bodies.

Qualities which are intended (and remitted such as motion and rest,) heat and cold, wet and dry, light and darkness, colour and blackness (opacity and transparency, a good smell and a bad smell) acidity, bitterness and sweetness, volatility and immovability, (liveliness, health and sickness), these things do not come in for consideration here. The things which cannot be intended and remitted such (bulk) impenetrability (solidity) and motion [something scored out which is illegible] and that inertia which causes a resistance to motion and to changes of motion (solidity and extension) are usually considered to be the properties of all bodies. And the reason is because a quality which cannot be remitted cannot be taken away (from the whole); and on the other hand that which can be taken away, if it were to be taken away from some parts of the whole, it could be remitted in the whole. Impenetrability indeed is usually described as the essence of bodies and is hence attributed to all of them; but the essential properties of bodies do not become known by the light of nature. We gather (only) from the senses the fact that the bodies which we touch (all tangible things) are impenetrable and we (conclude that) because this quality (in these things) is not intended or remitted (conclude that) we attribute to all bodies alike, no less to the heavenly bodies and to bodies imperceptible to the senses than to those bodies which we touch: and that too with an argument which we think so strong (that nothing in the whole nature of things) that this quality is taken to be the essence of bodies and is therefore considered to be the firmest foundation of all philosophy. therefore the axiom on the strength of which we gather this property of bodies must not be repudiated. Of the same kind is the argument for the inertia and (mobility and force of resistance) of matter.

In (only) a few bodies we gather by experiments that this (force of inertia) is proportional [to the quantity] of matter and the mobility inversely proportional to the same without perceptible intension and remission and thence [we attribute it] to all [bodies]. Because . . . projectiles [move] in straight lines uniformly except in so far as they are impended by the air or gravity [there is a large tear in the manuscript] we attributed this law of motion to all bodies.

[At this point, because of the tear at the bottom of the page, nothing more can be conjectured with any degree of certainty. There is also a tear at the top of the sheet which affects the deciphering of the remarks on the verso side. It begins:]

. . . cannot be brought into a smaller space . . . but can . . . their least particles [yield] to compression. However from the compression of other bodies cannot be [rightly] (necessarily) inferred the compression of the least particles . . . we gather from [a word missing] do not yield to compression because the [least] particles of all bodies (be compressed) cannot be reduced to a smaller space by compression. (The foundation therefore) Therefore this axiom is the foundation of all philosophy. At all events nature would not be sufficiently simple and consistent with itself [unless] the qualities of bodies on which it is allowed to make experiments, as many of them as are found to be immutable, apply in common to all bodies.

I now turn to the dating of these documents. Fortunately this is a comparatively easy matter with the first. On the 8th/18th September 1691 Fatio de Duillier wrote to Huygens to say:

Vous m'obligeriez sensiblement Monsieur de m'envoier les Errata de son (Newton) livre que je Vous ai laissez et desquels malheureusement je n'ai point gardé de Copie. . . .[13]

Fatio had left the errata with Huygens on a visit to Holland in April 1690. It includes exactly as it was written by Newton a copy of the First Draft. We can conclude that this version of what was in part to become Rule III of the 1713 Edition was certainly written, at the latest, early in 1690. After taking a copy for himself, Huygens returned the errata on the 15 September. The Huygens copy of this was published in 1701 by Groening in the appendix to his *Historia Cycloeidis*.[14] The same draft was copied by David Gregory in *Notae in Newtoni, 'principia'*,[15] and it was also paraphrased by John Keill in his lectures on Newton's philosophy given at Oxford in the late 1690s.[16] Thus Newton's ideas were certainly widely known before the appearance of Rule III in 1713.

The date of composition of the second draft is not as certain. But it is clearly an early attempt by Newton to clarify his thoughts regarding the epistemology and ontology of the Third Book. The most striking difference between this Draft and the First is that it includes the 'intendi et remitti' criterion. And the body of the remarks is mainly concerned with working out the implications of this criterion. On the other hand the first draft is much closer in substance to the version published in 1713. This in itself, however, does not afford sufficient grounds for supposing that it is later than the second. For it can plausibly be argued that the second is probably later than the first along the following lines.

It is plain that the second draft is more 'philosophical' in character and much less like the printed version. Might not this be a result of Newton's close relationship with Locke in the 1690's? The two men probably met at Lord Pembroke's in 1689, after Locke's return form Holland.[17] Newton paid Locke a visit at Lady Masham's in 1691, and Locke was in Cambridge for a few days in May of that year.[18] Again, in 1693, the period during which Newton was suffering from acute mental exhaustion, he wrote to Locke: "I beg your pardon for . . . representing that you struck at ye root of morality in a principle you laid down in your book of Ideas & designed to pursue in another book & that I took you for a Hobbist".[19] Again, in a letter to Locke on February 23, 1691/92 we find Fatio declining an invitation to the country on the grounds that: "I find our meeting would be but lame, since Mr. Newton hath told me he could not possibly join with us".[20] It is interesting to speculate that there were previous meetings at which all three were present. Here is clear evidence that, some time after 1689, Newton had examined Locke's *Essay*, at least in part, and that he may well have discussed various topics in the presence of Locke and Fatio. Although the surviving correspondence with Locke reveals a predominantly theological interest, there is no reason to suppose that philosophical matters were not discussed, especially when we recognise Newton's intense interest in the epistemology of his *De Mundi Systemate*, an interest which coincides with Locke's preparations for the second edition of his *Essay*, published in 1694.

For these reasons, and in view of the basic similarity in the empiricism which the two thinkers shared it is probable that Newton and Locke discussed the primary and secondary quality distinction, and that the latter stimulated Newton to re-consider his manner of introducing it in the first Draft. Thus Locke may well have pointed out that the Hypothesis which states in this Draft 'The laws (and properties) of all bodies on which it is possible to institute experiments, are laws (and properties) of all bodies whatsoever', is not only comparatively vacuous as it stands, but that it gives no reason why certain qualities, such as extension, are essential to bodies and others are not. It is precisely this deficiency which the second Draft attempts to remedy, the arguments in the remarks to the 'axiom' being more elaborate formulations of those which appear as outcrops in the published version. I am not suggesting that Newton was in any way indebted to Locke for his method of introducing and defending this distinction in qualities. As will be indicated below, his discussion is quite different in character from anything to be found int he *Essay*. But, if the above arguments are plausible, the second draft was probably written later than the first, being the result of the philosophical stimulation of Newton's association with Locke.

The third Draft would seem to be later that the other two. It combines the main ideas of each. From the second Draft comes the criterion for distinguishing qualities, and a compressed reference to its more elaborate supporting arguments. And from the first comes an emphasis on an experimental determination of matter, and many of the amplificatory remarks that Newton develops there. Thus it seems that it grew out of these earlier attempts. The second Draft affords a greater insight into Newton's ontology than has been hitherto available, and it will therefore serve as a basic source for the following discussion.

## II

As we have indicated, in Axiom 4 of the second Draft Newton is drawing a distinction between those qualities which can be said to be essential to matter and those which cannot be considered as such. After some false starts the Axiom states that:

> The qualities of perceptible bodies which cannot be intended and remitted are the properties of all bodies.

Notice that Newton has interchanged the terms 'quality' and 'property'. Usually he preserved the distinction, using 'quality' to speak philosophically of matter, and 'property' to speak of characteristics determined by experiment. In the final statement of this criterion in Rule III, the term 'property' does not appear. He first decided to formulate the statement in a more elaborate way by adding after 'intended and remitted' the phrase 'as far as can be (known) by experience (*experiri*) belong to all bodies'. He then cancelled this and wrote in its place 'must be regarded as the properties of all bodies'. This is a stronger claim. Finally he discarded this, and formulated the statement as it is above. It has the appearance of a factual claim. But, as the use of the 'must' in the second formulation suggests, Newton at one point seemed to consider the axiom as necessary. He quickly recognised, however, that a contingent statement of this sort cannot in any sense be considered necessary. At all events, the statement is 'the foundation of all philosophy'. This makes plain the crucial role that he assumes it to have in his system of nature.

In view of the many cancellations evident in the formulation of Axiom 4, it would appear to be the first time that Newton attempted to put these ideas into the form of a written statement. To characterise the kind of qualities with which the axiom is concerned, he employs the verbs *intendere* and *remittere*. These terms are important for Newton's ontology. Thus, I shall first discuss their use by Newton and his contemporaries, next their meaning for Newton, and lastly the source in which he probably found them.

This is not, of course, the first time that he had used these terms. In an early treaties, *De Gravitatione et Aequipondio Fluidorum*, he employed them and their substantive cognates to characterise certain aspects of pressure, motion and gravity. Definition II defines the 'intensity' of these 'powers': "The intensity of any of the aforementioned powers is the degree of its quality".[21] Definition 12 defines them with respect to their extension: "Its extension is the quantity of space or time in which it acts'.[22] At this stage in his thought, Newton held that a 'power' was to be characterised with respect to both its intensity and its extension. He says of motion, for example:

Thus motion is more intense or remiss according as a greater or lesser space is traversed in the same time, for which reason a body is usually said to be moved faster or slower. Motion is greater or less extended in so far as the body moved is greater or smaller, that is in so far as it is spread through a greater or smaller body'.[23]

There is an unmistakable echo here of the scholastic distinction between qualities which can be diffused throughout an object or part of an object with greater or lesser intensity, and the extension of an object which becomes greater or lesser as the object becomes larger or smaller. The distinction is clearly illustrated in Definition 15 where inertia is taken as a primitive concept: 'Bodies are denser when their inertia is more intense, and rarer when it is more remiss'.[24] Thus density is defined in terms of the degrees of *intensity* of inertia, volume (extension) being constant. As we shall see this distinction is also reflected in the second Draft.

Again in the first edition of the *Principia*, before he had formulated the ideas that were to become Rule III, Newton uses the terms 'intension' and 'remission'. Two examples will suffice. In Corollary IV to Proposition VI, Book III, the forces of magnetism and gravity are distinguished. Magnetism is not as the quantity of matter attracted: "Estque vis magnetica longe major pro quantitate materiae quam vis gravitatis: sed & in eodem corpore intendi potest & remitti".[25] Motte renders the sense of the last phrase as, "The power of magnetism, in one and the same body, may be increased and diminished".[26] Also in Proposition XLI of the Third Book where the heads of comets are distinguished from their tails Newton says of gravity: "Cum autem gravitas corporum terrestrium sit ut materia in corporibus, ideoque servata quantitae materiae intendi & remitti nequeat'.[27] Motte translates the last phrase as, "Neither more or less in the same quantity of matter".[28]

Much more evidence could be cited to show that Motte's contemporaries used these terms in the same sense in similar contexts. For instance, John Keill in his *An Introduction to Natural Philosophy* says of scholastic terminology:

. . . we shall not be ashamed to use with the Peripateticks, the Terms *Quality, Faculty, Attraction*, and the like; not that by these words we pretend to define the true and physical Cause and Modus of Action, but as these Actions may be augmented and diminished, and therefore since they have the Properties of Qualities, the same Name may not unfitly be applied to them, so that we thereby only mean to express the Ratios of the Forces or their Augmentation and Diminution.[29]

Keill continues in characteristic Newtonian fashion to profess that "true Causes are hid from us", but he contends that we may speak of Gravity as a Quality. Accordingly "we may investigate the Intensions and Remissions of these Qualities which follow from some supposed Conditions".[30] By first considering gravity as an example Keill also shows that "the Intensions of Light, Heat, Cold, Perfumes, and the like Qualities, will be reciprocally as the Squares of their Distances from the Point whence they proceed".[31] Again John Clarke in his translation of the *Principia* rendered the phrase 'intensione virtutis' in Definition VI of the First Book, which refers to a magnet as the "Intenseness of its vertue".[32] This phrase was added to the second edition. Lastly, Pemberton in his *A View of Sir Isaac Newton's Philosophy* translated the first part of Rule III as, "Those qualities which in the same body can neither be lessened nor increased . . .".[33]

All the examples of the use of these terms in the period from 1685 to 1730 characterise qualities of bodies which can or cannot change through degrees, Motte's translation of the phrase 'intendi et remitti nequeunt' as 'admitt neither intention nor remission of degrees' is sanctioned by Newton's own practice in English. Writing about various kinds of 'virtues' in Query 29 of the *Opticks*, Newton says: "And as Magnetism may be intended and remitted, and is found only in the Magnet and in Iron: So this Virtue of refracting the perpendicular Rays is greater in Island-Crystal, less in crystal of the Rock, and is not found in other Bodies".[34] From a draft variant to part of Query 31 written about 1716 Newton wrote of gravity:

All bodies here below are heavy towards ye earth in proportion to the quantity of matter in each of them. Their gravity in proportion to their matter is not intended or remitted in the same region of the earth by any variety of Force, and therefore cannot be taken away.[35]

This passage expresses in English the substance of Axiom 4 and Rule III as it pertains to gravitational interaction among bodies in the *same* region. Although other processes and forces many change in the region, gravity itself does not.

From the two passages cited it is clear that Newton uses the English phrase 'intended and remitted' to express the Latin 'intendi et remitti'. Like his contemporaries who also spoke of 'increasing and lessening', 'augmenting and diminishing' and 'more or less', Newton thought that 'virtues', 'powers', 'forces' as well as 'qualities' could be thus characterised. The use of the terms 'intension' and 'remission' in this period would seem to indicate the residual influence of the scholastic doctrine of forms. It is clear, however, that late post-Restoration thinkers did not accept the conceptual structure which lay behind the earlier use of these terms. For such language had become a commonplace associated with the methods of instruction in seventeenth century teaching traditions. And these were, for the most part, Neo-Scholastic in character, representing merely the spirit of doctrines like the latitude of forms. So in the main, influence was restricted to vocabulary and idiom. As were many Cambridge men of his day, Newton was exposed to this atmosphere of thought. Its phraseology became a part of his intellectual equipment.

We have Newton's own practice in English as a sanction for rendering the Latin phrase in the axiom and the later Rule III as 'intended and remitted'. It must still be asked, however, what he wished to convey by this phrase. Although these verbs in classical Latin can literally mean 'to tighten' and 'to slacken', on the basis of the above discussion it is almost certain that they connoted something quite different for Newton. Taken together in their positive sense they characterise for him a certain kind of quality, a quality which can manifest continuous and successive degrees of intensity: as can the pitch of a sound or the depth of a colour. Thus, in his translation of Rule III, Motte's addition of the phrase 'of degrees' – it was not in the Latin – admirably catches Newton's meaning. Of this interpretation more will be said later.

As has been said, Newton is explicitly using 'intension and remission' as a criterion by which to distinguish qualities. Is there any *specific* source for this to be found in other seventeenth-century writers? We have seen that it was a commonplace to characterise certain qualities in these and equivalent terms. This in itself gives no precedent for Newton's use of them to express a fundamental criterion. We know, however, from one of his early commonplace books that he read Kenelm Digby's *Two Treatises*.[36] In Chapter XVI which is devoted to theories concerned with the 'operations and qualities of bodies' Digby says:

A fourth position among Philosophers is, that some notions do admit the denominations of Intension and Remission, but that others do not. The reason of which we shall clearly see, if we but consider how these terms of *intension and remission* do but express *more* or *less* of the thing that is said to be intended or remitted: for the nature of more of less doth imply a latitude and divisibility; and therefore cannot agree with the nature of such things as consist in an indivisible being. As for example to be a *whole*, or to be an *equall* cannot be sometimes more, sometimes less; for they consist in such a rigourous indivisible being, that if the least part imaginable be wanting it is no longer a *whole*, and if there be the least excesse between two thinges, they are no longer *equall*, but are in some other proportion of equality in regard of one another.

And hence it is that Aristotle teaches us that *Substance* and the Species of Quantity do not admit of intension and remission; but that Quality doth. For first, in *Substance*, we know that the signification of this word is, that which maketh a thing be what it is, as is evident by our giving it for an answer to the question what a thing is, and therefore, if there were any divisibility in Substance, it would be in *what* the thing is; and consequently, every division following that divisibility, would make the thing another *what*, that is another thing. And so the Substance that is pretended to be changed by intension or remission, would not be divided, as is supposed, but would cease to be, and another substance would succeed in the room of it. Whereby you see that every mutation in Substance, maketh a new thing: and that more or less in Quiddity can not be pronounced of the same thing.[37]

Digby goes on to point out that what is true of substance is also true of 'Quantity' and 'Species', and "so likewise in continued extension, a spanne, an elle, an ounce, or any other measure whatever, ceaseth to be a spanne, and the rest, if you adde to it or diminish from it the least quantity imaginable. And peradventure, the same is also of figures; as of a Sphere, a Cube, a Circle, a Square, &c. though they be in the rank of Qualities.[38] He continues:

But, if we consider such Qualities, as heat, cold, moysture, drynesse, softnesse, hardnesse, weight, lightness, and the like; we shall find that they may be in any body sometimes more, sometimes lesse (according as the excesse of any element or mixture is greater in it, att one time than att another) and yet the body in which these qualities are intended or remitted, remaine still with the same denomination. As when durt continueth still softe, though sometimes it be lesse softe, other whiles softer. . . .[39]

Digby's examples of things which do not admit of more or less are different from those of Newton, but in these very Aristotelian passages he does contrast the immutability of substance, 'what a thing is', and geometrical qualities, with qualities such as 'softnesse' which do admit of degrees. Indeed the discussion is very like that to be found in the *Categories*. For example, Aristotle says:

Substance, it seems, does not admit of a more and a less. I do not mean that one substance is not more a substance than another (we have said that it is), but that any given substance is not called more, or less, that which it is. . . . Again, a thing is called more, or less, such-and-such than itself; for example, the body that is pale is called more pale now than before, and the one that is hot is called more, or less, hot. Substance however, is not spoken of thus. . . . Thus substance does not admit of a more and a less.[40]

It is significant that Digby unlike Aristotle, who in this passage is simply concerned to distinguish substance from Qualities, extends the distinction to cover qualities: some admit of more or less, whereas others do not, "though they be in the rank of Qualities". It is precisely this distinction in qualities that Newton makes in his axiom. And like Digby he uses the terms 'intension' and 'remission' for this purpose: not Aristotle's contrariety criterion.[41] Nowhere else, to my knowledge, are these ideas thus combined in seventeenth century writings, not even in the Neo-Aristotelian teaching texts used at Cambridge in Newton's student days. Nor is such a discussion to be found in other possible sources for this sort of distinction, in the writings of Boyle, Barrow and Charleton. As I suggested earlier, Newton was squarely faced with an ontological problem, namely, how to express and to justify a fundamental distinction in the qualities of bodies. Digby's discussion would have provided him with what he needed for this task. For the passages in the first *Treatise* distinguishes qualities in a way which preserves the sense in which Newton thought of certain qualities as having the potential of transforming into one another: those that are said to intend and remit. In contrast to this the essential qualities are immutable. Thus it is not so much that Newton was influenced by Digby, as that he probably borrowed from him certain conceptual tools which seemed to be consistent with the ideas on matter which he was developing at the time. Therefore Newton may have been indebted to Digby at least for the distinction, though probably not for the language in which he expresses it. And when he was pondering the nature of matter in the 1690's, he may well have consulted Digby's treatise again.

It is time to consider the amplificatory remarks to the second Draft. From the contrasting list of qualities at the beginning of the passage it is clear that Newton is concerned with a distinction which Boyle and Locke called primary and secondary qualities. In both manuscripts as well as the published Rule, Newton's list of essential qualities is much the same. In effect they are, extension, being either in a state of moving or of resting, impenetrability, hardness, and inertia. The list includes spatial qualities such as extension and qualities such as hardness and inertia which are connected with weight and density. Both weight and density are metrical qualities which can be straightforwardly measured. By means of pendulum experiments in Proposition VI of Book III,[42] Newton accurately estimated the proportionality of the quantity of matter and gravity of various bodies in terms of weight and density. By this means he also measured their *vis inertiae*. Extension, on the other hand, is the quality of occupying space in virtue of the fact that bodies have a volume and are thus three-dimensional solids. This spatial quality entails that bodies have some size and shape. Since they stand in some relation to all other occupants of space, this also entails that they either persist or alter their position in time: thus they will be either at rest or in motion. Thus matter, for Newton, is essentially a configuration of spatial and metrical qualities.

But there are some difficulties about his list. What does he mean by hardness? And how is it to be distinguished from impenetrability and inertia? By a hard body Newton means something that is solid. He declares in Draft No. 2 that "the least particles of all bodies . . . cannot be reduced to a smaller space by compression". For Newton this is so because they are devoid of internal pores, and are consequently non-deformable and unalterable. In this sense, then, they are solid or hard. This line of reasoning Newton takes to be appropriate only for the 'least particles' or atoms.[43] Ordinary material objects are seen and felt to be impenetrable in varying degrees. Not being absolutely rigid and solid, they are more or less deformable, and are more or less penetrable by other things. In experience, however, material objects cannot *physically* get into the same place. Newton therefore considers impenetrability to be essential to bodies, and also to the least particles since solidity entails impenetrability. Inertia, on the other hand, is different from either of these qualities. For Newton it is intrinsic to bodies, being something that they have inherently, innately and essentially. This conception he inherited from Descartes and Gassendi, especially from the former.[44] But it was developed and transformed in his thought. Elsewhere I have argued that it was conceived by him, unlike Descartes, to be dispositional property which preserves bodies in a state and which *acts* to maintain that state from being changed by an external force.[45] Thus inertia *qua* resistance is distinguishable from the resistance due to a body's physically occupying a place, because it explains the resistance of that body to change whether it is in a state of resting or of moving. And this is so because of the inherent dispositional property *vis insita* which, manifesting itself by an inertia state, acts simultaneously as resistance internal to bodies and as impulse away from them.[46] It is only 'exercised' when they come into mutual contact, or into contact with a force.

In these drafts and in the published Rule, Newton repeatedly states that we know the essential qualities of matter only through sense-experience. In each case they are those qualities which satisfy the intension and remission criterion. But Newton further declares that bodies have to be amenable to experimental determination. Three interpretations suggest themselves. It might be said that he has another and independent criterion for determining essential qualities. Thus, as well as not intending and remitting, and essential quality must be such that can be determined experimentally. A second view is that the only qualities which can be thus determined are to be considered essential. This interpretation has some support in the first Draft and in the published Rule. Lastly, Newton might be interpreted as holding that the intension and remission criterion is fundamental as a means of according to qualities the status of being essential, and that for this purpose experimental amenability plays no part. The last interpretation seems correct to me and, as I shall further argue, it has the merit of indicating that Newton's criterion is essentially philosophical in character.

Despite the reference to experiment in these documents, Newton is primarily concerned to show that knowledge of matter is rooted in sense-experience. For example, we 'gather from the senses' that bodies are extended and impenetrable. This is not knowledge that needs to be justified by an inductive procedure, or that needs to be established inductively as Newton was later to claim for his laws of motion.[47] "We find by dayly experience," he tells us in a passage written about the same time as the Drafts, "that bodies resist one another as often as they come together & cannot by any force be made to penetrate one another's dimensions. And this holding true without exception in all bodies here below in wch we can come to make observations we conclude it to be the property of all bodies whatsoever. And such observations occurring every day to every man this property of bodies is acknowledged by all men without any dispute . . .".[48] Again in another document written about 1715 Newton defines body as "everything which can be moved and touched, in which there is resistance to tangible things, and its resistance, if it is great enough, can be perceived".[49] He immediately adds that it is "in this sense that the common people always accept the word body".[50] This definition and conception of sensible matter, Newton therefore takes to be in accord with the opinions and experience of the common man. The Definition was intended as part of a revision scheme of the beginning of the Third Book for the Third edition of the *Principia*. Thus the physics of that Book is not concerned with the abstract generalities of mathematical philosophy, but with gross bodies which exist in the real world.

If this analysis is correct, the intension and remission criterion is a means of categorising the essential qualities of bodies in the light of experience. "Nature would not be sufficiently simple," he tells us in the second draft, "unless the qualities of bodies on which it is allowed to make experiments, *as many of them as are found to be immutable*, apply in common to all bodies." Thus, like Locke and other empiricists, Newton is engaged in classifying certain features of the visible world, in this case the qualities of bodies: and he is doing so from a consciously held epistemological position which is conceptually prior to experimentation. What can now be said of the first two interpretations? Neither seems at all plausible. In the first place there is no suggestion in the draft or the published Rule that certain configurations of qualities are essential because they can be treated experimentally. What Newton does assert is that we must be able to experiment with at least *some* bodies, for others, such as those in the heavens, are beyond our reach. Furthermore, it is because terrestrial bodies have essential qualities that they can be determined by experiment, and not the converse. This is clearly the sense of the Rule as it appears in 1713. Secondly, we may be sure that Newton's reference to experiment involves the procedure of measurement. Was his distinction between essential and non-essential qualities in any way connected with the use of measuring techniques? Did he suppose essential qualities to be those which are exclusively measurable? I do not think that this contention is defensible. From what has been said it is clear that for Newton knowledge of essential qualities is independent of their possible measurement. So too is the intension and remission criterion, for it has its origin in the philosophical tradition discussed above.

Apart from such general considerations, there are some particular reasons for rejecting this interpretation of the justification of essential qualities in the *Principia*. What Newton is mainly concerned to define in terms of measurement is the 'quantity of matter' and the 'quantity of motion' of bodies. These are the most important magnitudes connected with the theory of force developed in the *Principia*, the consequences of which are so brilliantly displayed in the Third Book. Such basic measurements involve only the metrical qualities, weight, density, inertia, and motion in the sense of momentum. For this purpose the spatial qualities of being extended, having size and shape, and being either at rest or in motion are irrelevant. From this experimental point of view, extension and size, as such, are trivial measurements. And 'being either at rest or in motion' cannot be measured, for unlike size, weight, density, inertia and momentum, it is not an additive quality. It makes sense, for example, to say of something that it is twice as heavy as something else, but it does not make strict sense to apply this notion to perceived colours, auditory sounds, or to 'being either at rest or in motion'. This last quality, however, along with extension and metrical qualities, is held to satisfy the criterion for essential qualities by not intending and remitting; but, according to Newton, this is not so with colours and sounds. For both these general and particular reasons it is therefore somewhat misleading to suppose that Newton's criterion can be straightforwardly characterised in terms of a modern distinction between intensive and extensive qualities. The second sort of quality, thought of as a continuous magnitude, is said to be directly measurable and additive. An intensive quality, on the other hand, such as the degree of hardness of a surface, is said to be treated quantitatively only as a derived measurement in terms of a correlated scale.[51]

Newton's attempt to introduce and justify the primary and secondary quality distinction is novel. I know of no other late seventeenth-century thinker who approaches the problem in this way. What Newton does, in effect, is to appropriate the traditional distinction between qualities and substance, express it in the language of the latitude of forms, and then apply it to the general conception of matter more or less acceptable in his time. As we have seen, his conclusion is that only certain qualities are intrinsic to it. He refers to these as being 'such as are not liable to diminution (and) can never be quite taken away'.[52] Although it is related to the intension and remission criterion, the full implication of this phrase is not apparent as it stands. We are now in a position to see that it is an outcrop of an argument in the second Draft whereby Newton attempts to justify his distinction in qualities and which states:

> And the Reason is because a quality which cannot be remitted cannot be taken away (from the whole); and on the other hand that which can be taken away, if it were to be taken away from some parts of the whole, it could be remitted in the whole.

Although this is cumbersomely expressed, it can reasonably be interpreted in a way which relates it to the Digby passage. Thus secondary qualities can exhibit continuous and successive degrees of change in intensity. At zero degree qualities which manifest more and less in this sense would no longer be in the body. "And yet the body in which these qualities are intended and remitted," declares Digby, "would remain still with the same denomination".[53] This is probably the significance that Newton wishes to convey when he says of such qualities that in the limit they "could be remitted in the whole". And since they seem, therefore, to sit more loosely on the world, he supposes this to be adequate grounds for considering them inessential to matter. In contrast, the primary qualities are immutable in the sense that they do not intend and remit in and of themselves. For example, extension cannot be more or less *qua* the extension of some particular body, because it "cannot be taken away from the whole."

Enough has been said to establish that the chief contention of Newton's argument is this. He is not merely using his criterion as a means of categorising primary and secondary qualities – this in itself would not imply that the primary are the essential qualities – but he is also arguing that it is because certain qualities *cannot* be held to intend and remit that they are essential. He is *not* arguing that they fall under the criterion because they are essential. Thus the two lists of qualities, in connection with Newton's argument, purport to distinguish those which a body *must* have from those which it *might* have under certain conditions. In this respect he goes explicitly beyond Digby.

This purely philosophical argument presents us with what will be seen to be connected problems. Why did Newton think that qualities which do not "intend and remit" are essential? This in itself is surely not a sufficient condition. Since all our ideas are rooted in sense-experience, Newton's criterion, as was pointed out above, merely demarcates the qualities of matter into two categories. Moreover, the argument in the second Draft, as it stands, does not seem to be sufficient justification for considering one sort of quality as essential. Secondly, Newton also claims to have knowledge of matter which goes beyond the data of experience. For the main burden of the argument of the drafts and the Third Rule is to show that all physical objects possess these essential qualities, including those which are too small to be sensible, and even those which are in principle non-observable. Essential qualities are those which "we attribute to all bodies alike, no less to the heavenly bodies and to bodies imperceptible to the senses, than to those bodies which we touch . . ." But how can the claim be justified if essential qualities can only be known in experience? It cannot be justified by the intension and remission criterion since it merely characterises our direct experience of bodies. Moreover, is not this claim inconsistent with his professed experimental empiricism, the view that physical knowledge is uniquely based on inductive generalisation?

Enough has been said to show that Newton did not come to his belief in the nature of the essential qualities of bodies by inductive reasoning. In so far as he gives an answer to *how* he knows these qualities to be essential to all bodies whatsoever, he purports to rely on induction. But his belief in the uniformity of nature, an unquestioned and primary commitment, is certainly fundamental to his whole philosophy of nature. This is clearly illustrated in the case of matter. Newton never doubted that the least particles, which were beyond possible experience, had the same essential qualities as ordinary material objects. Nor did he doubt that we could infer their existence, even though such inferences would have only indirect inductive backing. Nature is uniform. Thus what is true of the intrinsic nature of matter in experience is also true of particles, the existence of which could never be given in experience. This is the basic presupposition of the Third Rule. Thus unlike induction nature's uniformity is not merely a methodological principle of justification.

This interpretation can be strengthened if some important features underlying Newton's empiricism are made clear. Although he held that knowledge of matter was to be established in experience, Newton did not suppose that the senses were in themselves a criterion of truth. In common with Locke, however, he did tend to assimilate the sensory to the intellectual. This meant that he held a tight relationship to obtain between the content of the sensory – the given of experience – and concepts which are applicable to it. This is illustrated in a passage from one of the drafts of the *General Scholium* written in 1712:

. . . but we do not have any idea of substances. We see but the shapes and colours of bodies, we hear but sounds, we touch but external surfaces, we smell odours, we taste flavours; but we know the substances or essences themselves by no sense, by no reflex action, and therefore have no more idea of them than a blind man has of colours . . .[54]

There are three things to notice in this passage. The term *idea* is employed in a technical sense; in fact in a sense close to that in which Berkeley was later to employ it. Each sensory organ has its proper object of awareness, as sight has shape and colour. On this basis Newton seems to suppose that the mind forms a concept of the material. Thus, for Newton, mere sensing is not knowing. However much he may have limited the basis of knowledge to ideas of immediate awareness, he clearly recognised the role of the 'reflex action' of the mind. We know the existence of material objects in sensing, but this does not entail that their 'real' qualities are also given in direct experience. Having had experience of bodies is only a necessary condition for knowledge of them; it is not sufficient. Thus, even knowledge of the intrinsic qualities of perceptible bodies must go beyond the evidence of the senses.

Newton also states that we can have no idea of either material or spiritual substance, and this seems to suggest that we cannot know the essence of matter or of God. It must be realised, however, that substance is used in this passage in the metaphysical sense of the schools. Newton tells us that in no way can we "know the substances or essences themselves". He is clearly using the term in a traditional sense, such that substance means the *essence* of something. The empiricism of the *Principia* makes no such pretensions to knowledge: whether matter can think, for example, is not a question that is to be decided by Newton's treatise. On the other hand, Newton never doubted that we could know something about the essential qualities of material objects. These are open to experience, have a separate and individual existence in space and time, and can be systematically investigated. Hence the nature of material objects must not be confused with substance used in the sense of the schools. Nevertheless the passage appears to express a sceptical position. Earlier, a similar attitude towards knowledge of God and matter was expressed in *De Gravitatione*.[55]

Although it is difficult to appraise the seriousness of this attitude, it is certain that Newton, like Glanvill and Boyle, would refuse to be identified with the extreme sceptical position that truth is unknowable. While rejecting the dogmatic principles of 'Hobbism', at the same time Newton did not wish to discourage the philosophical quest for knowledge. We must, however, be aware of the weakness and possible deception of our faculties and senses. In the *General Scholium* he seems to suggest that we cannot know the essence of matter, but in the Third Rule and the Queries, he states that *all* matter is extended, and so forth. Still, we must not too hastily conclude that these statements are contradictory. In the passage under discussion, Newton is insisting on caution with respect to our pretensions to knowledge. But if we could not know at least some of the essential qualities of matter, the whole argument of the Third rule would seem to be pointless. That we can in principle know such qualities is borne out by a statement written in the 1690's. It is in the last of five 'hypotheses' which were intended to form a conclusion to a projected Book IV of the *Opticks*. It reads, "The essential properties of bodies are not yet fully known to us".[56] But some are: and the Third Rule unmistakably characterises them.

In the light of the foregoing analysis it seems clear that Newton is appealing to certain ideas and assumptions not mentioned in any of the drafts or the published Rule itself. Although these go some way towards showing that the intension and remission criterion is not sufficient as a means of determining essential qualities, more important is another unstated commitment of Newton's philosophy. Like many late seventeenth-century thinkers, he accepted the view that primary qualities were the basic sorts of concepts necessary for a coherent natural philosophy. And we now know that he was early committed to an atomistic conception of reality, the main characteristics of which had been inherited from Gassendi, Descartes, Charleton and Boyle.[57] His acceptance of it, however, eventually transformed it into something distinctively Newtonian. For Newton this commitment meant that a quantitative approach to phenomena must theorise in terms of solidity and inertia. Thus these are among the qualities which are conceived as belonging to matter. But in the 1690s a fundamental idea seems more clearly to have shaped itself in his mind: being extended, being impenetrable, being hard, being endowed with inertia and being either at rest or in motion are integral elements of our conception of matter. And thus it became impossible for Newton that bodies, existing as phenomena independent of us, could be otherwise conceived. And these qualities themselves he thought of as determinables not as determinates. Newton was concerned, for example, with extension in general, not with its exemplification in any particular body.

This is doubtless a philosophical conception of matter, which is clearly presupposed by the intension and remission criterion. Does this mean that the criterion is superfluous? On the contrary: it enables Newton to connect – though he does not do it explicitly – relevant empirical facts and experimental evidence with his conception of matter and the assumptions which lie behind it. This was also Locke's procedure by means of his various arguments for the alleged differences in experience between primary and secondary qualities.[58] Both were concerned to connect empirical facts with the conceptual relationship which they held to obtain between primary-quality concepts and their concept of the material. Newton, however, restricted his attempt at connection to the intention and remission criterion which he took to be universally exemplified by certain qualities of matter. Thus he was attempting to clarify what was for him a deep conceptual fact, not on a par with the definition of such terms as 'unmarried man' and 'bachelor'. Thus to be a body *is* to be extended, impenetrable, and so forth. These are the necessary and sufficient conditions for something's being material. Thus in such terms something true is stated about the structure of matter. And so Newton regarded his criterion as having a clear empirical meaning in the light of the natural philosophy of the *Principia* and in that of general experience.

But what were these empirical connections which Newton took to establish this substantial truth about matter, and which were characterised by his criterion? Here we are on less certain ground. What follows is partly conjectural. I do not think, however, that it is incompatible with Newton's conception of matter. In company with the definition of 'body' given above, the following passage will serve as a helpful point of departure:

> For although philosophers do not define substance as an entity that can act upon things, yet all tacitly understand this of substances, as follows from the fact that they would readily allow extension to be substance in the manner of body if only it were capable of motion and of sharing in the actions of body. And on the contrary they would hardly allow that body is substance, if it could not move nor excite in the mind any sensation or perception whatsoever.[59]

As the passage suggest, there are many general observations of bodies which Newton could have taken to support his distinction between qualities. After all, he wrote in an early commonplace book that "The nature of things is more securely & Naturally deduced from their operacions one upon another yn uponn or senses".[60] What are the general facts about bodies which indicate that they 'act' in the way expressed in the quotation and in the definition of the term 'body'? It is possible that Newton had in mind not only the individual essential qualities of matter but also their inter-relations. In contrast to the secondary qualities, the primary seem to exhibit countless and close relations among themselves. Thus if a body were bereft of its extension, so too would it be of its figure, size, and quality of being at rest or in motion. If its solidity were removed, it would become a three-dimensional spatial occupant, and consequently would be devoid of inertia. In neither case could such a body be said to interact with another body, nor with the sensory organs. For Newton such consequences would be meaningless, indeed absurd: this would mean a change in the 'quiddity' of matter. Thus unlike the secondary qualities, the primary stand or fall together, and "can never be quite taken away".

The foregoing enables us to contrast further the differences in the *way* the two sorts of qualities can change. A chunk of matter could be divided into many pieces, but however long this process of division continued, each bit would be extended, impenetrable and so forth. Thus the essential qualities can only change – become more or less – as a consequence of the total *amount* of matter in body being changed. These qualities change altogether or they do not change at all. The exception, of course, is impenetrability. The degree of impenetrability of a body does not necessarily change when the body's bulk is altered. A body could be made or become larger or smaller and its impenetrability remain the same. This is because impenetrability, like the secondary qualities, does change in terms of degrees of itself. Although it may be an essential quality of matter, it is not so on Newton's grounds; for it does not, as he thought, fall under his criterion for essential qualities. It changes like a coloured surface which can be more or less pale at one time (all else being equal) than at another; whereas a body cannot be more or less solid (all else being equal) at one time than at another. If its solidity changes, so too do its other primary qualities, and it becomes a different body.

Newton may also have had in mind the difference in operations necessary in order to change respectively the primary and secondary qualities of bodies. In his *Essay* Locke says:

> Pound an almond, and the clear white colour will be altered into a dirty one, and the sweet taste into an oily one. What real alteration can the beating of the pestle make on any body, but an alteration of the texture of it?[61]

Locke's point would[62] seem to be, that not only are the primary qualities changed, but *through* them so also are the secondary. Similarly, Newton implies in the *Opticks* that the manifest colour of a body which can intend and remit, is capable of being changed by altering the size and arrangement of the body's constituent particles.[63] There is no analogue with secondary qualities; no change in the primary ones can be affected by altering them.

I have emphasised the philosophical character of Newton's discussion of essential qualities. In many respects his conception of matter is similar to the views of contemporaries such as Boyle and Locke: his intellectual indebtedness, however, is from a different source. I have referred many times to the deep-rooted classical formation of Newton's mind, but it was a mind capable of transforming the philosophical conceptions of the past so that they could apply to current problems. And prominent among these in the 1690's was the nature of physical qualities.

APPENDIX

Axioma 4. Qualitates corporum sensibilium quae intendi et remitti neque sunt proprietates corporum omnium (quatenus experiri licet corporibus omnibus competunt, pro corporum omnium proprietatibus haberi debent).

Qualitates quae intenduntur (et remittuntur ut motus et quies) calor et frigus, humiditas et siccetas, lux et tenebrae, color et nigredo, (opacitas et perspicuitas, odor bonus et malus) aciditas, amaritudo et dulcedo, volatilitas et fixitas, (vivacitas salus et agritudo) hic considerandae non veniunt. Quae vero intendi et remitti negueunt (magnitudo) uti impenitrabilitas, (soliditas) et mobilitas [seven words are quite illegible] & inertia (illa) qua motibus et motus mutationibus resistentur (soliditas, extensio) pro corporum omnium proprietatibus haberi solent. Et ratio est quia qualitas quae non potest remitti non potest (omnino) auferri, (et contra) quae auferri potest is auferatur ex totius partibus aliquibus, remittetur in toto. Impenitrabilitas quidem ad corporum essentiam referri solet et inde omnibus attribui sed corporum proprietates essentiales non innotescunt lumine naturae. Colligimus (tantum) ex sensu quod corpora quae tangimus (res omnes tangibiles) sunt impenitrabilia et (concludimus qua) quoniam haec qualitas (in estis) non intenditur nec remittitur (concludimus quod) tribuimus corporibus universis tam caelestibus et insensibilibus quam hisce quae tangimus: Idque argumento, uti putamus, tam valido, ut (nihil in tota rerum natura) haec qualitas ad corporum essentiam referatur & habeatur philosophiae totius fundamentum firmissimum. Axioma igitur cujus vi colligimus hanc corporum proprietatem repudiari non debet. Simile est argumentum ab inertia (et mobilite et vi rsistenti) materiae. In paucis (tantum) corporibus per experimenta colligimus quod (vis inertiae) sit proportiona [lis quant] itati materiae & mobilitas eidem reciproce proportionalis [tear in the manuscript] sine intensione et remissione et inde [tribuimus corporibus] universis. Quoniam projectilia in rectis uniformibus [tear in the manuscript] nisi quatenus ab aere impediuntur vel gravitate [a large tear in the manuscript] hinc motus legem tribuimus corporibus unive[rsis] (There follows five liens about which it is impossible to conjecture anything because of the large tear. The verso side begins:) in spatium minus redigi non [possunt] autem posssent eorum particulae minimae compressione [cedent] ex aliorum vero corporum compressione (necessario) non [recte concludimus] concluditur compressio (particularum) minimarum (concludimus) colligimus ex [a word missing] (compressio non cedunt) quod corporum omnium paticulae [minimae] (compressi) in spatium minus per compressionem reduci non possunt. (Fundamentum igitur) Philosophiae igitur totius fundamentum est hoc Axioma. Natura utique satis simplex sibique conformis non essset [nisi] qualitates corporum in quibus experimenta facere concessum est, (there is an eight word phrase crossed out which, owing to lack of syntax, makes no sense) quotquot immutabiles reperiuntur, competerent corporibus universis.

## NOTES

[1] Isaac Newton, *Philosophiae Naturalis Principia Mathematica*, London, 1686, p. 411.
[2] Isaac Newton, *Philosophiae Naturalis Principia Mathematica*, Cambridge, 1713, p. 368.
[3] These are fully discussed in Chapter 7.
[4] U.L.C. Adv. b. 39. I, The Latin reads: "Gravitas (in Terra) est qualitas corporum omnium (quae circa *Terram* sunt &) in quibus experimenta instituere licet & quantitati materiae in singulis proportionalis existens non potest intendi et remitti & propterea per hypoth III proprietes est corporum universorum".
[5] Isaac Newton, *The Mathematical Principles of Natural Philosophy* Tr. Benjamin Motte, Vol. II, London, 1729, p. 203.
[6] Sir Isaac Newton, *Mathematical Principles of Natural Philosophy* ed. Florian Cajori, Berkeley, 1962, p. 400 Vol. II.
[7] For instance in his reply to P. Pardies, *Isaac Newton's Papers & Letters on Natural Philosophy*, ed. I. Bernard Cohen, Cambridge, 1958, p. 108, and also in the manuscript published by Cohen in an article entitled "Versions of Isaac Newton's first published paper" *Archives internationales d'histoire des Sciences* 11, 1958, pp. 360–361. In using the term 'essential' quality of matter, I do not attribute 'essentialism' to Newton. The term denotes that certain qualities are invariably associated with matter.
[8] This view is especially evident in his 1675 letter to Henry Oldenbury in *The Correspondence of Isaac Newton* Vol. I, p. 366. It is discussed in the article cited in Chapter 7.
[9] Cajori, *Principia*, p. 522, pp. 526–530 and p. 542 which was only added in the second edition. In the *Opticks* these ideas are found in Query 30.
[10] U.L.C. Adv. b. 39. I. p. 402a.
[11] C. Huygens, *Oeuvres Completes*, The Hague, Vol. X, pp. 47–55.
[12] One is 'Hypothesis III' of the First Edition, and the other is a statement containing ideas similar to those in Rule IV of the Third Edition.
[13] Newton, *Correspondence* Vol. III, p. 168.
[14] Johannes Groening, *Historia Cycloeidis*, Hamburg, 1701. In 1705 David Gregory criticised this work as being anti-Newtonian. It was also known to Newton, viz. *Correspondence*, p. 169.
[15] Gregory MS. 210 at the Royal Society.
[16] John Keill, *An Introduction to Natural Philosophy: or, Philosophical Lectures Read in the University of Oxford.* Anno Dom. 1700 London 1726, p. 100.
[17] Maurice Cranston, *John Locke: a biography*, London 1957, p. 337.
[18] *Ibid.*, pp. 357–358.
[19] *Correspondence*. Vol. III, p. 280.
[20] Bodleian Library, MS. Locke, c. 18.
[21] A. Rupert Hall & Marie Boas Hall, *Unpublished Scientific Papers of Isaac Newton*. Cambridge 1962, p. 115.
[22] *Ibid.*, p. 115.
[23] *Ibid.*, p. 115.
[24] *Ibid.*, p. 115.
[25] *Principia Mathematica*. London, 1686, p. 411.
[26] Motte, *op. cit.*, p. 225. Motte's translation is based on the text of the Second and Third Editions in which Corollary IV is Corollary V.
[27] *Op. cit. Principia*, p. 505.
[28] Motte, *op. cit.*, p. 369.
[29] Keill, *op. cit.*, p. 4.
[30] *Ibid.*, p. 4.
[31] *Ibid.*, p. 7.
[32] John Clarke, *A Demonstration of some of the Principal Sections of Sir Isaac Newton's principles of Natural Philosophy*, London 1730.

33 Henry Pemberton, *A View of Sir Isaac Newton's philosophy*, London 1728, pp. 24–25.
34 Sir Isaac Newton, *Opticks, or a Treatise of the Reflections, Refractions, Inflections and Colours of Light*. Based on the Fourth Edition, London 1730, p. 373.
35 U.L.C. Add. 3970 f. 243$^v$.
36 *Quaestiones Quaedam Philosophicae* F 107, U.L.C. Add. 3996. Newton had taken notes from the first edition published in Paris. *Two Treatises. In the one of which, the Nature of Bodies; in the other, The Nature of man's Soule, is looked into: in way of discovery, of the immortality of Reasonable Soules*. Paris 1644.
37 *Ibid.*, pp. 141–142.
38 *Ibid.*, p. 142.
39 *Ibid.*, p. 142.
40 J. L. Ackrill, *Aristotle's Categories and the Interpretatione*, Oxford, 1966, Chap. 5, 3$^b$, 33, p. 10. The same distinction is found in Descartes', *Discours de la Methode*, e. Gilson, Paris 1947, p. 2. Gilson has traced this to the influence of Eustachio a Sancto-Paulo's *Summa Philosophiac.*
41 *Ibid.*, p. 10. "Another characteristic of substances is that there is nothing contrary to them".
42 Cajori, *Principia*, pp. 411–412.
43 For a more thorough discussion of these questions, see J. E. McGuire, "Body and Void and Newton's De Mundi Systemate:Some New Sources", *Archive for History of Exact Sciences*. Vol. 3. Number 3, 1966, p. 206–248.
44 See John Herivel *The Background to Newton's Principia*, Oxford 1965. Chapter 2, pp. 44–45.
45 "The 'State of *Inertiae*' and *vis Insita*". To appear shortly in the 'Texas Quarterly' as a contribution to a conference held at the University of Texas in November 1966 celebrating Newton's 'Annus Mirabilis' of 1666.
46 Cajori, *Principia*, Vol. I, p. 2, Definition III.
47 Joseph Edleston, *Correspondence of Sir Isaac Newton and Professor Cotes*, London 1850, p. 155.
48 U.L.C. Add. 3970, f. 479$^r$. The autograph entitled *Principles of Philosophy* was probably intended to be a preface to the Optics.
49 See Chapter 3.
50 *Ibid.*, p. 220.
51 For a discussion of this see M. R. Cohen, and Ernest Nage, *An Introduction to Logic and Scientific Method*. New York, 1934, pp. 293–297.
52 Cajori, *Principia*, Vol. II, p. 398.
53 Digby, *Two Treatises*, p. 142. Kant's position with respect to physical qualities is quite the opposite to Newton's. He argues that in sensory experience "intensive" qualities, admitting of degrees, are more fundamental, and more "real" in the content of perception that "extensive" magnitudes. A discussion of this interesting differences is beyond the scope of the present paper. Cfr. N. K. Smith: *The Critique of Pure Reason*, London, 1929, 201–208.
54 Hall & Hall, *op. cit.*, p. 361.
55 *Ibid.*, p. 138.
56 U.L.C. Add. 3970, f. 338$^v$.
57 Richard S. Westfall, "The Foundations of Newton's Philosophy of Nature". *British Journal for the History of Science*. Vol. I. 1962, pp. 171–182.
58 John Locke. *An Essay Concerning Human Understanding*. Ed. A. C. Fraser, Oxford, 1984, Vol. I, Book II, Chapter VIII, pp. 166–179.
59 Hall & Hall, *op. cit.*, p. 132.
60 *Quaestiones*. U.L.C. Add. 3996, f. 101$^v$.
61 Locke, *op. cit.*, Bk. II, Chapter VIII, p. 176.
62 *Op. cit.*, Part III, pp. 245–269.

CHAPTER 7

# TRANSMUTATION AND IMMUTABILITY: NEWTON'S DOCTRINE OF PHYSICAL QUALITIES

## INTRODUCTION

In the 1687 edition of *Principia* there appeared an interesting statement. Labelled Hypothesis III, it asserted that all bodies can transform into one another.[1] There was no commentary added, and as the Hypothesis was formulated it seemed to indicate that Newton accepted a doctrine which allowed for unrestricted transmutation. Very little has been written about this statement or about its relation to the body of Newton's natural philosophy[2] and it has remained imperfectly understood. Hypothesis III did not appear in the 1713 edition of the *Principia*. In its place is a statement, labelled Rule III, about the essential qualities of matter.[3] How this change took place has been, because of a lack of evidence, a matter for conjecture.

However, in the early 1690's Newton was engaged in revising the *Principia*. And from this period there are a number of fragmentary documents which throw light on a possible connection between these two statements; and which give the historian a clearer idea of what led to the final rejection of Hypothesis III of the first edition. These and related documents show that Newton was concerned with the nature of matter and, in particular, that he explicitly attributed to it a particulate structure. In the documentary precursors of Rule III he worked out, probably for the first time, a distinction in the physical qualities of bodies which was usually expressed by the terms "primary" and "secondary".[4] This led Newton, as I shall argue, to a systematic reconsideration of his ideas on transmutation, and to distinguish his doctrine from that of the Cartesians and the Peripatetics.

I shall also be concerned to trace the development and changes in Newton's conception of transmutation. For one of the main contentions of this paper is that he formulated, in the early 1690's, a new doctrine which restricted transmutation to a certain type of quality. Thus the eventual discarding of Hypothesis III is connected, as I hope to show, with this new conception and also with Newton's general reconsideration of the nature of matter. As a consequence of this the tougher line on hypotheses, which also marks at this time a transition in his thought, is really a secondary factor in explaining the final rejection of Hypothesis III.

## I. TRANSFORMATION AND IMMUTABILITY: THE EVIDENCE

On a manuscript sheet which I shall call A, and which forms part of the revision of the *Principia* undertaken in the early 1690's, Newton made a number of changes in the Corollaries to Proposition VI, Book Three. The difference between the forces of gravity and magnetism was to be stated in Corollary 1; it is Corollary 4 in the first edition, and Corollary 5 in the last two editions. The idea that the weight of bodies are independent of their mutable "shapes and textures" was to be Corollary 2; it is Corollary 1 in all editions. Corollary 3 (it is Corollary II in all editions) states:

All bodies whatsoever which are found about the Earth have gravity towards the Earth, and the weight of all which (at least in the same latitude) are equally distant from the centre of the Earth are as the quantities of matter in them. This is clear by experience in all perceptible ones, and consequently is to be understood of imperceptible ones, by Hypoth III. If the aether or any other body whatsoever [there are a series of dashes in the manuscript indicating that Corollary II was to be supplied as it appears in the First Edition] – – – (by Hypoth IV) . . . in the previous corollary.[5]

Newton mentions two hypotheses: "Hypoth III" is a new statement which was eventually to become Rule III of the second edition of the *Principia*. In this edition the Hypothesis reads in Motte's translation: "The qualities of bodies, which admit neither intension nor remission of degrees, and which are found to belong to all bodies within the reach of our experiments, are to be esteemed the universal qualities of all bodies whatsoever".[6] "Hypoth IV" is the old Hypothesis III of the first edition which reads: "Every body can be transformed into body of any other kind, and can assume successively all intermediate degrees of quality".[7] Thus at this point, about 1690, Newton intended using both statements in this Corollary. "Hypoth III" (to become Rule III) provides grounds for affirming that all bodies gravitate, and for ascribing immutable qualities, such as extension, to all matter. And "Hypoth IV" asserts that bodies can change into one another and, because of continuity in nature, can manifest all "intermediate degrees of quality".

Hypothesis III of the first edition is only appealed to once in the entire Third Book. However, it was probably connected, as will be shown later, with an elaborate theory of the origin of matter, and of "active principles" developed at the end of that Book. The single direct reference to the Hypothesis in the first edition occurs in Corollary II of Proposition VI. In order to clarify his conception that the gravity of bodies at equal distances is as their quantity of matter Newton says:

For if the aether, or any other body, were altogether void of gravity, or were to gravitate less in proportion to its quantity of matter, then because there is no difference between that and other bodies but in form of matter, by a successive change of form, it might be transmuted into a body of the same condition with those which gravitate most in proportion to their quantity of matter (by Hypoth III) – – –[8]

It is this passage that he intended should copied into the revision of Corollary II given above: as we have seen the reference to "Hypoth III" of the first edition was to be altered to "Hypoth IV". By the time of the publication of the second edition of the *Principia* in 1713, Newton had dropped Hypothesis III from the prefatory section of the Third Book. and in Corollary II the view that change of form could denote a change of gravity is treated as a *reductio ad absurdum* argument, the conclusion of which being contrary to experience. Thus, gravity is as the quantity of matter, and is not a property of a body's shape. In the second edition this particular conception of transformation is attributed to "Aristotle, Descartes and others".[9] The significance of this will be discussed later.

The revision of what was to be Corollary IV to Proposition VI in manuscript A is of considerable interest. In all three printed editions, it is Corollary III and is concerned with the existence of void space. After indicating that the justificatory remarks for the existence of the void in Corollary III of the first edition remain unaltered,[10] Newton added to them the following comment:

This demonstration is valid against those who admit the third or the fourth hypothesis. If anyone should deny these Hypotheses and have recourse to a third hypothesis, namely, that one admit some matter with no gravity by which the gravity of perceptible matter may be explained; it is necessary for him to assert two kinds of solid particles which cannot be transmuted into one another: the one [kind] of denser [particles] which are heavy (have gravity) in proportion to the quantity of matter, and out of which all matter with gravity and consequently the whole perceptible world is compounded, and the other [kind] of less dense particles which have to be the cause of the gravity of the denser ones but themselves have no gravity, lest their gravity might have to be explained by a third kind and that (again by a fourth and) so on to infinity. But these have to be very much less dense so as by their action to shake apart and mutually scatter the denser ones: by which means all bodies composed of the denser one would be quickly dissolved. And since the action of the less dense upon the denser will have been proportional to the surfaces of the denser, while gravity arises from that action and is in proportion to the matter of which the denser ones consist, it is necessary that the surfaces of the denser ones must be in proportion to their solid content, and therefore that all those particles must be equally dense and that they can neither be broken nor worn away nor in any manner destroyed: or else the ratio of the surfaces to the solid content, and consequently the ratio of gravity to the quantity of the mater would be changed. Therefore one must altogether determine that the denser particles cannot be changed into the less dense ones, and thereupon that there are two kinds of particles, and that these cannot pass into one another.[11]

Newton is making some important points. The first is that a vacuum exists in nature. If anyone denies this, and attempts to explain the mutual gravitation of large bodies by postulating non-gravitational matter, he will encounter difficulties. In the latter part of the passage Newton claims that the notion of non-gravitational matter makes little sense, since the two sorts of "dense" particles, one of which is devoid of gravity cannot together account for the fact that gravity is proportional to the quantity of matter. Still, if anyone should wish to explain the cause of gravity in this way, one sort of particle must be void of gravity, otherwise its gravity would have to be explained by a third kind of matter and so forth *ad infinitum*. Another point is that both Hypothesis III and Hypothesis IV are held to be compatible with the existence of a void. The hypothesis of non-gravitational matter, however, is compatible with neither. For if a body is devoid of gravity if cannot transform into one that is, and this violates the transformation Hypothesis, namely, Hypothesis IV in this revision sequence. On the other hand, if there is such matter, then it violates Newton's firm conviction that all matter must have a "power" of gravitating which remains invariable in the same region. Thus it would seem at this stage in his revision of Corollaries II and III of the first edition, that Newton thought either Hypothesis III or Hypothesis IV could be maintained, and with it an acceptance of the void.

That Newton is not setting the doctrine of the vacuum against these hypotheses is more clearly seen in another related document written in the same period. It is also an intended revision of the Corollaries to proposition VI in the third book. In a passage which is connected with the vacuity of celestial spaces discussed in Corollary III of the First Edition, Newton refers to Fatio De Duillier's hypothesis concerning the cause of gravity. This was based on the condition that matter has a large number of void interstices: ". . . since the more tenuous particles must be borne in all directions by motions which are rectilinear and very rapid and uniformly continued and these particles must experience non-resistance unless they impinge upon denser particles. Therefore the vacuum is to be completely taken for granted since it follows from this hypothesis [Fatio's] and equally from the third and fourth".[12] As in the other document, the hypotheses referred to are the new statement which was to be Rule III, and Hypothesis III of the first edition. From this passage and from his comments on non-gravitating matter there can be no doubt that Newton held these Hypotheses to be compatible with the existence of the void. Thus, Newton's ambiguous phrase "this demonstration is valid against . . ." is unfortunate. What he probably wished to convey was that the doctrine of the void is compatible with either Hypothesis. This does not of course imply that, at this stage, newton was unwilling to accept both together. The sentence from manuscript A, "If anyone should deny these Hypotheses and have recourse to a third . . .", leaves this possibility open.

There is, however, another document from the same period connected with the revision of Book Three, in which Newton planned to present both the new and old "Hypoth III" with comments. Immediately after Hypothesis III of the first edition, he has directed that the following explanatory remarks be added:[13]

*Thus the Peripatetics, Epicureans and Cartesians lay down.* It is an Hypothesis of the Peripatetics, *Epicureans* & Cartesians and *is not absolutely laid down but* is directed against *the prejudices of those by whom it is laid down* their prejudices only. *If anyone should think otherwise against him . . . not* [the Fragment ends at this point].

In this passage which I shall call manuscript B, it is clear that Newton has decided to reject the old "Hypothesis III" as an expression of his own position. However, this does not mean that he also rejects the conception that certain qualities can transmute into one another so as to change the appearance of bodies. It does mean, as I shall argue later, that Newton has reconsidered his doctrine of the transformation of bodies in relation to the general transmutation of qualities. For the formulation of Hypothesis III, as it appears in the first edition, states that *every* body can change into any other body, and implies that *all* qualities can transmute successively into one another. This seems inconsistent with the new Hypothesis III, which states that certain qualities are immutable and essential to all matter whatsoever. Next the commentary on the old Hypothesis is followed by the new statement:

Hypoth III. The qualities of bodies which cannot be intended and remitted and which apply to all bodies upon which is possible to set up experiments are the qualities of all bodies whatsoever.[14]

Not only did this statement provide a justification for extending gravity to all matter whatsoever, but it also justified, in Newton's mind, inferences to the existence and qualities of particles beyond possible sensory experience. The explanatory remarks which are written after the statement reveal how important the justification of the latter was for Newton's philosophy:

The same is to be understood of the qualities of all bodies of the same kind. This seems to be the foundation of all Philosophy. By no other means may one derive the qualities of imperceptible bodies from the qualities of perceptible [ones].[15]

Next Newton has written in full the old Hypothesis and, as in the other documents, it is numbered IV. Immediately after it is written a shortened version of the commentary on the Peripatetics and Cartesians:

This is an Hypothesis of the Peripatetics and Cartesians and it is directed only against their prejudices.[16]

I wish to turn now to a major document, hereafter to be called C, in which Newton has written a more extended commentary on Hypothesis III of the first edition. This document contains two other statements with full commentary. One is perhaps the earliest version of what was to be Rule III, and the other is similar to what was to be Rule IV of the third edition. All of these statements are entitled axioms in this document. This almost certainly indicates that Newton considered them, at this stage,[17] to be solidly based in experience. Hypothesis III and its commentary read as follows:

Axiom: Every body can be transformed into body of any other kind and can assume successively all the intermediate degrees of quality. This is laid down by almost all philosophers. All teach that from some common matter taking on various shapes (& textures) all things arise and are dissolved back into the same matter by the privation of the shapes and textures; like the Cartesians who claim that matter is extension divided and modified in various ways: the

Peripatetics who imagine some kind of primal matter which is shapeless but yet capable of all shapes; (Moses, Thales, Epicurus) [then cancelled] and the rest, (Pythagoras, Democritus and the rest) who compose other things either from atoms combined in various ways or from the four elements which arise out of a common matter. The only exception is [Anaxagoras].[18]

But his theory of similar parts never gained acceptance. Now it is possible to conceive of the existence of some kind of imperceptible bodies from which perceptible bodies are never formed and (consequently) therefore that the properties of these cannot be gathered from the properties of perceptible bodies. But this hypothesis is not founded on the appearance of things and (against dreams I do not dispute). I do not consider dreams [write this at y$^e$ y$^r$ end of y$^e$ foregoing axiom *written in English by Newton and bracketed*]. This hypothesis sins against the axiom.[19]

In this interesting commentary, Newton clearly accepts the doctrine that bodies transform. Indeed, he emphasizes that many philosophers have held a similar position. His list explicitly mentions, "Moses, Thales, Pythagoras, Democritus and the rest". The exceptions are Anaxagoras and Epicurus whose name he has cancelled in the manuscript. I shall return to the significance of the cancellation in the fifth section below. As for Anaxagoras, it is clear that Newton rejects his views because they are not based on phenomena. The polemical tone of the last paragraph almost certainly indicates that Newton also had in mind certain contemporaries who proposed extravagant hypotheses to explain the properties of matter. From pejorative and gratuitous conjectures of this sort, Newton is concerned to distinguish his own views on transformation.

As in the first two manuscripts discussed above, it seems that Newton intended to use both Hypothesis III (now called an axiom) and the new statement concerning the immutability of certain qualities. Since both are supplied with full commentaries in the same document,[20] it seems plausible to conclude that these statements were connected in his mind in the early 1690's. And that they were considered for a time – a point to which I shall return – as being separate but complimentary expressions of a unified theory of matter.

## II. TRANSFORMATION AND TRANSMUTATION

In manuscript B, we noticed that Newton unmistakably rejected Hypothesis III, (= IV in this revision sequence) on the grounds that it was no longer an expression of his position. The first question I want to ask is this: what conclusions did Newton draw regarding his general theory of physical qualities in the light of his new statement that certain qualities cannot intend and remit? Before answering this, it is necessary to consider more carefully what he wishes to convey in the formulation of Hypothesis III, in the First edition.

The Hypothesis is expressed in two distinct parts. The first part refers to the transformation of bodies, and the second characterizes the nature of qualities. The connection between the two parts seems clear. If a body is capable of becoming any other body, its appearances can be entirely changed:

and therefore it will be able to "take on" all degrees of any given quality. Leaving transformation aside for the moment, what sort of change is ascribed to qualities in the Hypothesis? Newton uses Hypothesis III in Corollary II of Proposition VI to justify successive changes in the "shapes and textures" of bodies. There is little doubt, then, that in the first edition the second part of the Hypothesis refers to the transmutation of the forms and qualities of bodies. Thus the evidence that a body has transformed into another, is the transmutation of its qualities.

It must not be concluded, however, that Newton made a clear and general distinction between the terms "transformation' and "transmutation". for the most part, they are used interchangeably. However, in an intended "*Conclusion*" to the first edition of the *Principia*, he does use them separately. This occurs in a passage in which various changes in natural substances are discussed in terms of the operation of forces on their constituent particles. Concluding a section on the structure of composite bodies Newton writes:

Their particles coalesce in new ways and by means of the attractive forces of contiguous ones they come together more densely: for which reason that rare substance water can be transformed by continued fermentation into the more dense substances of animals, vegetables, salts, stones and various earths. And finally by the very long duration of the operation be coagulated into mineral and metallic substances. For the matter of all things is one and the same, which is transmuted into countless forms by the operations of nature, and more subtle and rare bodies are by fermentation and the processes of growth commonly made thicker and more condensed. By the same motion of fermentation bodies can expel certain particles, which thereupon by their repulsive forces are caused to recede from each other violently; if they are denser, they constitute vapours, exhalations and air; if on the other hand they are very small they are transformed into light.[21]

It would be difficult to find another passage in Newton's writings which better illustrates the theory presupposed by Hypothesis III. With respect to changes in the density of bodies, Newton claims that this results from changes which take place in the configurations of their constituent particles: "they coalesce into the form of highly regular structures".[22] This is so because the micro-attractive forces bring such particles together to form denser bodies. Thus in this passage Newton suggests that transformation is a change which occurs in the inner structure of a composite body, such that it can become another kind of thing. On the other hand, transmutation is used to characterize perceptible changes in the qualities of bodies since matter can be "transmuted into countless forms". Again in a passage from Query 31 of the 1717 edition of the *Opticks* these assumptions are evident:

The Particles of Fluids which do not cohere too strongly, and are of such a Smallness as renders them most susceptible of those agitations which keep Liquors in a Fluor, are most easily separated and rarified into Vapour, and in the Language of the Chymists, they are volatile, rarifying with an easy Heat, and condensing into Cold. But those which are grosser, and so less susceptible of Agitation, or cohere by a stronger Attraction, are not separated without a stronger Heat, or perhaps not without Fermentation. And these last are the Bodies which Chymists call fix'd, and being rarified by Fermentation, become true permanent Air; those Particles receding from one another with the greatest Force, and being more difficulty brought together, which upon Contact cohere more strongly.[23]

Here, as in the passage from the *Conclusion*, repulsive and attractive forces operate on the constituent particles of bodies only after agents such as heat, cold and Fermentation have affected them: though, of course,these two types of change can and do occur together. Thus, again Newton is suggesting that transformation is a physical change and that transmutation is a chemical one. In a well known passage from Query 30 of the 1717 *Opticks*, however, he uses the term transmutation in the sense in which transformation is used in Hypothesis III. As one of many illustrations that nature "seems delighted with Transmutations" Newton observes:

Water, which is a very fluid tasteless salt, she changes by Heat into Vapour, which is a sort of Air, and by cold into Ice, which is a hard, pellucid, brittle, fusible Stone; and this Stone returns into Water by Heat, and Vapour returns into Water by Cold. Earth by Heat becomes Fire, and by Cold returns into Earth. Dense Bodies by Fermentation rarify into several sorts of Air, and this Air by Fermentation, and sometimes without it, returns into dense Bodies.[24]

The Aristotelian character of this passage need hardly be pointed out. The four elements, water, air, earth and fire, in that order, are said to transmute successively into various stuffs, vapours and airs by the operation of such external agents as heat, cold and fermentation. In all cases, when Newton considers the natural elements, the chain of transmutations begins in Aristotelian fashion with water. Thus in this way the appearance of bodies can be changed by the transmutation of their qualities. As in the passage from the draft *Conclusion*, such changes in the appearance of bodies are ultimately explained in terms of quantitative changes in the configuration of their constituent particles. A distinctively Newtonian conception of these particles and their associated forces is developed at great length in the latter half of Query 31.

## III. THE PRIMORDIALS AND THE TRANSMUTATION OF QUALITIES

Enough has been said to show that from 1687 to the end of his life, Newton accepted the conception of transformation in nature, and that in the first edition of the *Principia*, Hypothesis III presupposed a particulate theory of matter. This brings us back to the question of his theory of physical qualities. What implications for qualities are to be drawn from the second part of Hypothesis III? If qualities can manifest all intermediate degrees, as continuous degrees of intensity can be manifested by the loudness of a sound, in the limit such qualities could disappear entirely from bodies. But the new hypothesis states that certain qualities are essential to matter and thus define its nature. These are qualities such as extension and inertia which Newton holds cannot be characterized in terms of degrees of intensity: in the language of the Hypothesis, they do not intend and remit. Thus Hypothesis III states that *all* qualities admit of degrees, whereas the new Hypothesis states that some do not. It seems, therefore, that there is an inconsistency between these two statements with respect to what they contend about qualities. I do not think that this is so: at least I do not think that transformation in itself is necessarily incompatible with a view which holds that certain qualities are immutable. In order to establish this, an interpretation of the relation of the old to the new Hypothesis III will be helpful. What follows is a reconstruction, but it constitutes a plausible connection between the Hypotheses, and one which probably occurred to Newton.

As we have seen he holds a particulate theory of matter, which allows that the constituent particles of bodies can transform into various configurations. Change in this sense is neutral with respect to certain sorts of qualities transmuting and other sorts remaining immutable. This can be seen from the new Hypothesis III. The various drafts which Newton produced, in the early 1690's, as precursors to Rule III, considerably affected his thinking on the nature of matter, and especially his conception of transmutation. For the purpose of this statement and its commentary is to introduce a *distinction* in physical qualities, and to argue that only those which are immutable – do not intend and remit – are universally essential to matter. Thus, even though certain qualities of sensible bodies can transmute, the essential and primary qualities which characterize both them and their component particles do not. And it is these particles, by changing their spatial configuration, that bring about transmutations in such manifest qualities of bodies as texture, smell and colour. There is no reason, then, why Newton could not have made such a relationship between these two hypotheses explicit in the light of his reconsideration of the nature of matter. It was only necessary to build into the formulation of the old Hypothesis III the restriction which is explicit in the new Hypothesis concerning the sorts of qualities that can transmute. Thus the inconsistency in the two Hypotheses – that qualities take on *all* intermediate degrees as opposed to some qualities not manifesting *any* degrees – cannot be a serious reason for the rejection of the old Hypothesis as Professor Cohen has argued.[25] Since the new Hypothesis clearly makes a distinction in qualities, it seems more reasonable that Newton should have reformulated the old Hypothesis, rather than have rejected it as he did: the more so since he still accepted transmutation in nature. Also, by restricting transmutation to certain manifest qualities of perceptible matter, the old Hypothesis III would clearly have become a theoretical consequence of the particulate theory of matter Newton was reconsidering in the early 1690's.

Newton, however, did not make any of these changes, and he eventually discarded the old Hypothesis III altogether. As a preliminary to my own analysis, I want to consider an explanation offered by Professor Koyré for its rejection. According to Koyré, the old Hypothesis III is a statement which "affirms the unity of matter and the possibility of transforming any given body into any other".[26] By the "unity of matter" Koyré seems to mean that the atomic structure which Newton attributed to matter entails that each atom is qualitatively identical. That is, each one would have the same size, shape and bulk. Koyré points out, however, that in the Latin *Opticks* of 1706 matter is no longer a unity. He says of Newton's new position "that God and in all probability created in the beginning atoms of different shapes and masses – which implies the non-transformability of one kind into another".[27] Koyré then concludes that at least by 1706 Newton was "obliged to suppress his Hypothesis III which affirms the contrary".[28]

It seems to me that Professor Koyré's argument is open to objection; but in order to make this clear, an interpretation of Newton's conception of primordial particles is necessary. In the first place, there is abundant evidence to show that he considered them to be immutable. "And therefore that Nature may be lasting", he says in the *Opticks*, "the Changes of corporeal Things are to be placed only in the various Separations and new Associations and Motions of these *permanent* particles".[29] In all editions of the *Opticks* in Book II, Part III, Proposition VIII, he states that the least particles are without pores, and that, in structure, gross bodies are probably exceedingly porous. Newton argues that the latter doctrine is important for giving an account of colours in terms of the various internal reflections of light resulting from the different spatial arrangements of the component particles of bodies. ". . . in like manner these smaller Particles", Newton concludes, "are again composed of others much smaller, all which together are equal to all the Pores or empty Spaces between them, and so on perpetually till you come to solid Particles, such as have no Pores or empty Spaces within them".[30] Clearly implied in this passage is a difference in status between the least particles and gross matter. Since the primordials are absolutely solid, unlike gross bodies, they are neither deformable nor dividable. Such changes in Newton's sense, can only occur if a particle has internal vacuities. Thus the primordials, being without pores, are immutable and permanent.

Again from the reference to the ancients in the remarks to the Axiom (Hypothesis III of the first edition), it is clear that when they were written Newton considered the Axiom as stating that all things arise out of some common matter. Thus in this sense it is a statement of the unity of matter; and Newton is willing to ascribe *this* conception to "almost all philosophers". Although, as we have seen, his own articulation of it is closer to that of the ancient atomists than it is to the others in the list. Clearly an important presupposition of this conception is that the common matter, whether it is conceived atomistically or otherwise, itself remains unchanged. This view is also implied in the following passage from Newton's intended *Conclusion* to the first edition of the *Principia*. With reference to the primordials he says:

For through the forces by which contiguous bodies cohere, and distant ones seek to separate from one another, these particles will not collect together in the composition of natural bodies like a heap of stones, but they coalesce into the form of highly regular structures almost like those made by art, as happens in the formation of snow and salts. Undoubtedly, following the laws of geometry, they can be formed into very long and elastic rods, and by the connexion of the rods into rectiform particles, and by the composition of these into greater particles, and so at length into perceptible bodies.[31]

In Query 31 of the *Opticks* the same view is stated in rather more prosaic language:

Now the smallest Particles of Matter may cohere by the strongest Attractions, and compose bigger Particles of weaker Virtue; and many of these may cohere and compose bigger Particles whose Virtue is still weaker, and so on for divers Successions, until the Progression end in the biggest Particles on which the Operations in Chymistry, and the Colours of natural Bodies depend, and which by cohering compose Bodies of a sensible Magnitude.[32]

Here we have evidence of a deeply held conviction; that the order and intelligibility of nature presupposes the existence of "simples" which remain qualitatively invariant throughout successive changes in their spatial ordering. And the old Hypothesis III is really an expression of this conviction. For, as we have, it assumes that the primordials can be arranged into various structures, resulting in emerging levels of substructural composition which eventually produce the various sensible qualities of gross bodies. By the early 1690's, these primordials were the common matter of Newton's system of nature, and were characterized uniquely by primary qualities. Thus in Rule III Newton considers qualities such as size, shape and solidity as determinables, and is not concerned with their exemplification in any particular thing. Since every particle and every body has these essential attributes, Newton's conception of the unity of matter is unmistakenly clear. This conception is quite different from Koyré's interpretation of matter as a unity in Newton's philosophy before 1706, namely, that he held that any material thing, particle or gross body, could transform into any other.

This can be clarified if we now consider Koyré's explanation of why Hypothesis III was rejected. There is no evidence that Newton ever thought that the primordial particles could mutually transform, or transform into a gross body. They are permanent and changeless "even so very hard, as never to wear or break in pieces; no ordinary Power being able to divide what God himself made one in the First Creation. While the Particles continue entire, they may compose Bodies of one and the same Nature and Texture in all Ages: But should they wear away or break in pieces, the Nature of Things depending on them would be changed".[33] And in a scholium written in the early 1690's and intended for Proposition VII, Book III of the *Principia*, which states that all matter has gravity proportional to its quantity of matter, Newton says of the primordials of matter that they can never be "decreased or ever assume any increase, but abide eternally as the immutable seeds of things: and this is the reason why the visible forms of objects are conserved in perpetuity".[34] Hence, for Newton, the very principle of transmutation presupposes the changelessness of the ultimate particles, which are thus in a different category from other sorts of material objects, and are the common matter from which phenomena arise.

It might be objected that this analysis does not discredit Koyré's explanation of why Hypothesis III was discarded. According to Koyré, the characterization of the ultimate particles in the 1706 edition of the *Opticks* as being "of such Sizes and Figures, and with such other Properties, and in such Proportion to Space, as most conduced to the End for which he form'd them"[35] can only be reasonably interpreted in the following terms. Bodies of different sorts are composed of different sorts of primordials and hence they are not transformable – but the old Hypothesis III contradicts this.

Such, however, is not Newton's position. Whether or not the primordials have the same or different sizes and shapes is quite independent of the doctrine of transformation. As we have seen, Newton never contended that such particles could transform. In any even, as the above quotations testify, he did not rely on the geometrical qualities of the primordials in his explanation of the manifest qualities of bodies. This was the position of Descartes, Gassendi and Boyle. Newton, on the contrary, based an explanation of such qualities on the operation of forces. "I therefore propose the inquiry" he tells us in a *Draft Preface* to the first edition of the *Principia*, "whether or not there be many forces of this kind, never yet perceived, by which the particles of bodies agitate one another and coalesce into various structures".[36] Thus he stresses the *way* in which various forces arrange the constituents of bodies into increasingly complex structures, and not the geometrical structure of the particles themselves. Also, the earlier passage from the *Opticks* clearly indicates that the level of composition which accounts for chemical and optical properties is far removed from that of the primordials: which, without changing intrinsically, provide the ultimate basis of those particles which are immediately responsible for the nature of observable qualities.

## IV. NEWTON'S ALCHEMICAL COSMOGENY

Thus far there is no compelling reason to suppose that Newton thought the conception of matter expressed by the two Hypotheses had to be decided in terms of the one or the other. Indeed, I have indicated that he could have easily removed any inconsistencies in their formulation. This interpretation will, for course, have to be made compatible with the third manuscript in which Newton categorically states that the old Hypothesis III is an expression of the Peripatetic and Cartesian position. Before this can be done adequately, however, Newton's ideas on transmutation must be clarified further, and then contrasted with those of the Peripatetics and Cartesians. But these ideas developed in the period from 1670 to 1690. So if we are fully to understand the problems confronting Newton in the 1690's, these changes must be analysed.

In the 1670's Newton's cosmogeny was certainly alchemical in character. In a letter of 1675 to Henry Oldenburg, he suggested that nature seemed to have its origin in the transmutations of certain aetherial spirits into different forms.

Perhaps the whole frame of Nature (may be nothing but aether condensed by a fermental principle) [may be nothing but various Contextures of some certain aethereall Spirits or vapours condens'd as it were by precipitation, much after the manner that vapours are condensed into water or exhalations into grosser Substances, though not so easily condensible; and after condensation wrought into various formes, at first by the immediate hand of the creator, and ever since by the power of Nature, wch by vertue of the command Increase & Multiply, became a complete Imitator of the copies sett her by the Protoplast.][37]

Newton clearly has in mind a sort of spirituous matter which is convertible into many different forms, and which is the primitive *arche* out of which all things originate. the existence of this aether he regards as probable because of the behaviour of such phenomena as electric effluvia. They "instruct us, that there is something of an aetherall Nature condens'd in bodies".[38] Newton then describes the operations of this aether in unmistakable alchemical terms:

For nature is a perpetual circulatory worker, generating Fluids out of solids, and solids out of Fluids, Fixed things out of volatile & volatile out of fixed, subtile out of gross & gross out of subtile. Some things to ascend & make the upper terrestrial juices, Rivers and the Atmosphere; & by consequence others to descend for a Requital to the former. And as the Earth, so perhaps may the Sun imbibe this Spirit copiously to conserve his Shineing, & keep the Planets from receding further from him. And they that will, may also suppose, that this Spirit affords or carries with it thither the solary Fewell & material Principle of Light; And that the vast aethereall Spaces between us, & the stars are for a sufficient repository for this food of the Sunn & Planets.[39]

There is little doubt that the aether is conceived as a source of activity for the maintenance of the cosmos. And as it perpetually circulates, it changes into countless forms of things, such that possibilities for transmutation are limitless. Without such incessant activity the whole system of the stars and planets would soon run down; but it is constantly being reformed by the aethereall matter, a prototype of what Newton would later call in 1706 "active principles".

Later in his letter to Boyle of February 1678/79 Newton simplified his aethereal hypothesis. Most of the operations in nature were now accounted for by the "maine flegmatic body of aether", and the "aethereall spirits", with which it had been compounded, are not rejected. However, Newton still stresses that there is evidence for the transmutation of qualities in nature. And nature herself still seems to be a "perpetual circulatory worker".

If in the 1670's the interplanetary system was regarded by Newton as "a gigantic alembic or distillation apparatus",[40] within which natural objects were able to transform from state to state, this was not so in 1687. The first edition of the *Principia* contains no mention of an aethereall matter diffused into every part of the cosmos. This is so because in this period and in the 1690's, Newton seems to have been less convinced of the existence of the aether. But in particular, he was also becoming interested in the operations of comets as a means of maintaining activity in nature.[41] The Third Book in all editions of the *Principia* contains a fascinating theory centering on the "replenishing" nature of the vapours which make up the tails of comets. Newton discusses their role and nature in analogy to water-cycles on and near the earth. Just as water must be continually supplied for the nourishment of vegetables and for replenishing rivers: . . .

so for the conservation of the seas, and the fluids of the planets, comets seem to be required, that, from their exhalations and vapours condensed, the waste of the planetary fluids spent upon vegetation and putrefaction and converted into dry earth, may be continually supplied and made up; for all vegetables entirely derive their growth from fluids, and afterwards in greater measure, are turned into dry earth by putrefaction . . . and hence it is that the bulk of the solid earth is continually increased; and the fluids if they are not supplied from without, must be in a continual decrease, and quite fail at last. I suspect, moreover, that it is chiefly from the Comets that spirit comes, which is indeed the smallest but most subtle and useful part of our air, and so much required to sustain the life of all things with us.[42]

In the second edition Newton developed this theory further. Added for the first time in that edition, and preserved in the third, is the following passage which ends Book Three:

The vapours which arise form the sun, the fixed stars, and the tails of the comets, may meet at last with, and fall into, the atmospheres of the planets by their gravity, and there be condensed and turned into water and humid spirits; and from thence, by a slow heat, pass gradually into the form of salts and sulphurs, and tinctures, and mud, and clay, and sand, and stones, and corals, and other terrestrial substances.[43]

Here the vapours are not restricted to the tails of comets, but arise from the sun and the fixed stars. And they supply not only the earth, but all of the planets. Both these passages are concerned with the nature of physical objects, knowledge of which is based on sensory experience and on astronomical observations. Taken together they constitute a clear expression of Newton's theory of transmutation as it applies to interactions between celestial "vapours" and terrestrial objects. Unlike the aetherial matter of the 1675 hypothesis, which provided fuel for the sun, stars and planets, the circulation of the tails of comets only provides for the maintenance of the plants. Also, the fact that the second passage extends the theory to include the atmospheres of the sun and stars as active agents, shows how different this conception is from the aetherial spirit of the 1670's. And the balance between the matter lost and gained which is evident in the earlier hypothesis is no longer true in the *Principia* of 1687, for "The bulk of the solid earth is continually increased . . .". Comets, however, continually supply a "vapourous spirit" which considerably rarefies and dilates in the heavens. By the power of gravity this vapour is drawn slowly to the earth and, condensing into water, augments the dwindling supply of nourishing fluids necessary to sustain life: and thereupon the various transmuting agents produce physical stuffs such as mud, clay, sand, etc. Unless the stock of terrestrial water was thus sustained, it could not maintain the perpetual condensing transmutations demanded of it.

Gregory reminds us, however, that the vapour of a comet's tail is not an ordinary fluid such as water, "so great a part of which is yearly turned into solids"[44]. He adds ". . . that subtle spirit that does turn fluids into solids, a very small Aura or particle of this may be able to do the business".[45] Accordingly, we have a full fledged theory which attempts to account for the supply and origin of the nutritive factor in water, namely, the "humid spirits". And it attempts also to account for the active agent in the air which supports life and fire. Thus we have developed within the *Principia* itself a complete theory of transmutation, which is clearly related to similar passages in the *Draft preface* and *Conclusion* to the *Principia* of 1687, and to those in the 31st Query. Against this background a statement along the lines of Hypothesis III would seem to have a place at the head of Book Three.

This interpretation is strengthened by the fact that Newton intended to say more about the subject in a revision sequence of the second edition of the *Principia*. Begun about 1716, the plan was to introduce before the *Regulae Philosophandi* of Book Three, a set of definitions. These were to include such key terms as "body", "void", "phenomena", "rule", and "hypothesis".[46] A body is said by Newton to be anything that can be moved and touched and which offers resistance to being touched. he observes that,

> . . . in this sense . . . the common people always accept the word body. And of this sort are metals, stones, sand, clay, mud, earth, salts, wood, bones, flesh, water, oil, milk, blood, air, wind, smoke, exhalations, flames, and whatever can be included under the four elements or flow from them by exhalation and return into them by condensation.[47]

The list of material objects is similar to those mentioned in the passage from the Third Book of the *Principia* cited above. Thus this intended revision is connected with the theory of the Comet's tail. As will become more evident later, Newton's conception of transmutation is in certain respects quite different from the various contemporary articulations of the doctrine from an Aristotelian point of view. However, from the reference to the "four elements", the examples of material objects cited, and from the idiom in which the ideas are expressed, it is not too much to conclude that the passage is an expression of latent Aristotelianism. This is especially striking as it comes in the wake of a widespread rejection of the four element theory towards the close of the seventeenth century. Many chemists followed Van Helmont's theory of water as a basic principle. Others developed from the Paracelsians a five element theory, and some followed J. J. Becher in combining Van Helmont with a theory of three earth elements. In any event, the passage is undoubtedly a compressed way of referring to the contents of the perceptible world: to the theory which explains their transmutations, the origin of which lies in the comet's tail: and to the role of such intermediary states as vapours and exhalations.

Newton's ideas on transmutation have certainly changed since the 1670's. After the 1680's he seems no longer to have believed that everything originated from the aether. By 1690 transmutation is restricted to interactions among ordinary terrestrial bodies, and between these and the vaporous spirit emanating form the tails of comets. Since the earlier aether conception is discarded by Newton, it is thus no longer a source of activity and motion in nature. Thereafter this function is taken over by cometary mechanisms, partly by a return to a type of aether in the 1717 *Opticks*, and partly by a doctrine of "active principles" developed in that edition. Still, throughout this long period a doctrine of transmutation was never abandoned, though in character it lost some of the overt alchemical aspects that were so evident in the formulation of the earlier hypotheses.

## V. THE NEW CONCEPTION OF TRANSMUTATION

Yet the question still remains: why did Newton discard the old Hypothesis III? There are a number of factors to be considered; but first I want to make two general points. On considering the content of the above manuscripts, it is evident that in the early 1690's Newton became clearer about the assumptions and implications of the old and new Hypothesis III. This is mainly because in clarifying through many revisions what was to be rule III, he was almost certainly led to see more clearly what sorts of qualities could transmute. As was suggested earlier Newton became explicitly committed, in this period, to deriving the secondary qualities of bodies from the relative positions and motions of their qualitatively neutral constituent particles. Since immutable qualities are held to be essential to bodies, and the sole attributes of imperceptible particles, it follows that transmutation is a characteristic of only the secondary qualities of sensible matter. This contention is borne out by any passage after 1690 in which Newton speaks of transmutation in nature: and the secondary qualities which can intend and remit, provide evidence that bodies do undergo such change. Thus not only did his views on transmutation change, but his general conception of matter became more complex and rigorous.

Why did Newton not avail himself of the relationship which, as I have shown, could be maintained to hold between Hypothesis III and what was to be Rule III? It was only necessary to make the sort of modifications in their formulation that have been discussed. The old Hypothesis III as it stood did not embody any distinction between two sorts of physical qualities. Thus it could have been interpreted by Newton's contemporaries as stating that *all* qualities would eventually vanish leaving matter qualityless. For, if there are "intermediary degrees" there is, by continuity, a last or zero degree of any given quality. And thus it would have seemed that even the essential qualities of matter could have vanished in the limit. As we have seen, this conclusion is incompatible with the sense of the new hypothesis. Still, to clarify these conceptions in the detail they demanded would have involved a considerable and major revision of the whole prefatory section. Newton may have perceived that had he completed this task an unwelcome consequence might have occurred. The doctrine of transmutation would have been "canonized" by being placed in the company of the basic and general principles of the Third Book. Such a frank avowal of his position might have invited criticism which would have detracted from the central purpose of the Rules and the Third Book – the justification of the doctrine of universal gravitation. Thus, the essential tactical purpose of the prefatory section might have been misunderstood. And after all the argument for gravitation and its consequences, not transmutation, is the main concern of the last book.

Professor Cohen has argued that Newton was willing to include Hypothesis III in the *Principia* only so long as Book Three began with hypotheses. Since it was "Hypothetical" it did not matter that he did not personally believe it, or that he took it to represent a doctrine of the Cartesians and Periptetics which, as manuscript B indicates, he was now in fact repudiating. "But once he hardened his philosophical line, some time after 1692 and prior to 1706", writes Professor Cohen, "this was no longer possible; and the *Regulae Philosophandi* stated at the beginning of Book Three in the second edition . . . were a statement of what Newton believed – and were in no sense hypothetical".[48] At this stage in his thought, Cohen concludes, Hypothesis III no longer had any rightful place in the prefatory section to Book Three.

There is no doubt that in the 1690's Newton began officially to eschew hypotheses, no matter in what sense they were employed. We must, however, be clear about an important distinction that he is making. As we have seen from manuscript B, Newton labels Hypothesis III an assumption of the Cartesian and Peripatetic position. Hence as formulated it is relegated to the category of the hypothetical. It would be a mistake to suppose, however, that the manuscript also supports an outright rejection of the doctrine of physical transmutation. The evidence presented in section four of this paper is clearly contrary to such a conclusion. But such a conclusion seems tacitly to be accepted by professor Cohen:[49] at least his analysis of the manuscript strongly suggests this. But these manuscript remarks can be interpreted in a way which is consistent with Newton's continuing to accept a doctrine of transmutation in a restricted sense.

The first obvious point about these remarks is that they are unfinished: in fact the last sentence suddenly breaks off. It is therefore reasonable to suppose that Newton intended to say more about why Hypothesis III was an assumption of the Cartesians and the schools, and perhaps to contrast this with his own revised views. This interpretation makes sense of the phrase: "If anyone should feel otherwise, against him . . . not . . .". That is, the unwritten arguments would probably have been aimed at those who supported the doctrine as formulated by the Hypothesis, not at those who accepted views similar to the Newtonian position. Light is also thrown on the polemical intention of the remarks by this interpretation. Newton was contrasting his own views with those of his opponents. The latter were mere hypotheses; whereas, his own views on transmutation, as developed in the Third Book, were considered to be based on phenomena. The unfinished remarks might well have stated this explicitly. Newton would not have lost such an opportunity had this intended revision found its way into the second edition of the *Principia.*

In manuscript C, Newton cites Descartes, the Peripatetics, Moses, Thales, Epicurus, Phythagoras and Democritus as thinkers who held a doctrine of transmutation. It is clear that the theories of Epicurus and Democritus, and those of Moses, Thales and the Pythagoreans – all of whom were interpreted by Newton and the Cambridge Platonists to be atomists[50] – are closer to Newton's theory of matter than to that of the Cartesians and Peripatetics. Descartes held that matter is infinitely divisible.[51] This provided the possibility of an indefinite transformation of particle into particle, which is opposed to Newton's position that the primordials are indivisible. The Aristotelians too – at least they could be so interpreted – held that the four qualities could pass into one another in a cycle, so that any one of the four elements is potentially any other. But Aristotle also held that informed matter was continuously divisible.[52] And his doctrine that form is essential to perceptible matter may well have become connected in Newton's mind with the Cartesian view that matter can assume indefinite shapes by divisibility. At this point, as indicated by manuscript B, Newton probably concluded that the Peripatetic position was equivalent to the Cartesian, and that it was thus incompatible with his own. This line of reasoning is consistent with the new reference to the "mind of Aristotle, Descartes and others" which occurs in Corollary II, Proposition VI, Book Three in the second and third editions. Newton attributes to these schools the view that all differences between bodies is merely that of shape or form. Because of the innumerable possibilities, anything could become anything. Since this doctrine makes weight, inertia and quantity of matter a function of shape or form, it is inconsistent with universal gravitation. Thus the reference to Aristotle and Descartes in such a context is clearly related to manuscript B where, in the 1690's, Newton had clarified his assumptions about transmutation.

In the light of this, the cancellation of the reference to the Epicureans in the second manuscript is significant. It further supports the view that Newton intended to repudiate Hypothesis III, in order to contrast it with his own revised views on transmutation. As we have seen, he was now clearly restricting transmutation to the secondary qualities of matter. And Newton would almost certainly have interpreted Epicurus as also being committed to such a restriction: since in the system of the ancient spatial changes in the position of the immutable atoms also entailed changes in manifest qualities of things. In his *Notae in (Isaaci) Newtoni Principia* written in the early 1690's David Gregory, a close friend of Newton's, also concluded that Hypothesis III was a Cartesian in spirit. He comments:

> The Cartesians readily agree to this, the Peripatetics not in the same way, who makes a specific difference between celestial and terrestrial matter, nor the followers of the Epicurean philosophy, who make atoms the immutable seeds of things.[53]

The Hypothesis is thus for Gregory a clear formulation of the Cartesian position. It only reflects the Aristotelian position with the restriction that transmutation applies to terrestrial bodies, since for the schools the quintessence is immutable. But Gregory is quite definite that the Hypothesis is not an expression of the Epicurean position. In fact it is to be contrasted with that philosophy. His interpretation seems to be that the "immutable seeds" are terrestrial objects; but since they are changeless they cannot transform into perceptible bodies. This is clearly contrary to the Cartesian position, and Gregory thereby seems to imply that the Epicurean conception is similar to Newton's own particulate theory of matter. This is the conclusion that Newton also reached in the 1690's. And it thus explains his cancellation of the reference to Epicurus in manuscript B.

In manuscript C Newton attributes a doctrine of transmutation to a large number of ancient philosophers. The notable exception is Anaxagorus whose views he holds are not related to phenomena. The doctrine of ὁμοιομἑρεια was to Newton's mind an extreme expression of unrestricted transmutation. Thus at this stage, though he did not think that a reference to Descartes and the Peripatetics would be misleading, Newton did not wish his own views to be confused with those of Anaxagoras. Later, as we have seen, he rejected the opinions of the Cartesians on transmutation and those of the Peripatetics interpreted in an unrestricted sense. I have suggested that this was probably the result of Newton's becoming clearer about the relation between the old and new Hypothesis III: but there is another reason. In manuscript A, he argues that both Hypotheses are compatible with the acceptance of the void. This makes implicit the fact that he does not think that a Cartesian or Peripatetic plenum is necessary for his conception of physical transmutation.[54] Mere change of shape is not a change in the nature of a body as a plenist position seems to imply. Nor can essential qualities change in a way that allows for transmutation, since they do not intend and remit. This further supports my interpretation that Newton's theory became restricted sometime in the 1690's, to changes in the secondary qualities. Indeed, his acceptance of the void plainly indicates that it is an integral part of his conception of transmutation, since it is necessary for thus explaining such a change.

In the manuscript Newton tells us that this doctrine is "laid down by almost all philosophers". What is the significance of this? The fragment must be seen in connection with another manuscript which Newton probably sent to David Gregory some time in the 1690's. This document, which Newton intended to include in the same revised edition as the other manuscripts discussed in this paper, contains a number of lengthy scholia on the philosophy of the ancients. These were to be inserted after the first six propositions of the Third Book, which state the argument for universal gravitation. They discuss, among other things, the anticipation by the ancients of the atomic philosophy and the doctrine of gravitation itself. With reference to this document Rigaud says: "Harassed as he was by opposition to his philosophy his object was evidently to show that many portions of it had occurred to the philosophers of antiquity".[55] Rigaud then maintains that when physical truths are found int he writings of those "untutored to legitimate philosophising (they) become good proof of there being nothing unnatural in the opinions they illustrate".[56]

It is clear that Rigaud regards Newton as simply selecting certain passages from the ancient writers, which superficially seem to parallel his own doctrines, and hoping thereby to put the Newtonian philosophy in a venerable tradition. But why should Newton do this? He could not possibly have hoped to convince any continental critics by this tactic. Given that Newton continually stated that his philosophy stood or fell by its experimental foundations, does it not seem odd that he should have bothered to construe his natural philosophy in this manner? Especially when the task must have cost him considerable labour. It would seem that a plausible answer can be given for these questions, only on the assumption that Newton considered his philosophy to be a *prisca sapientia*. Such a conception involves more than merely placing a body of doctrine in a venerable tradition; it also involves the notion that there is one true philosophy of nature to which certain other thinkers have given a partial expression. This true philosophy of nature Newton considered himself to be expounding in these "classical" scholia to the Third Book.[57] Since many of the same ancients are mentioned in these scholia as in the remarks to Hypothesis III in manuscript C, it would seem that the latter should be seen in the same light. Thus, in the spirit of these scholia Newton is claiming that the Hypothesis is an expression of a universal truth, in a perennial philosophy of nature, to which other thinkers have given different or partial expression. Furthermore this sort of historical approach to truth in theology and natural philosophy allows for the selection or rejection of doctrines expressed by other thinkers in terms of the criteria of the *prisca* employed: or, since no thinker has been able to express the complete truth about phenomena, the nature of the doctrines being defended. Thus, Newton could at one time refer to the views of Descartes and Aristotle on transformation, as another expression of his own position. But on reflection, he could later reject them as being in opposition to the "true" doctrine, which finally included, for Newton, only the philosophy of Epicurus and the atomists.

Thus the rejection of Hypothesis III is intimately connected with Newton's changing views on the character and scope of transmutation in nature. this in turn is a result of fully working out the implications of his new Hypothesis III, namely, that certain qualities are immutable and essential to matter. The old Hypothesis III was not dropped because it was incompatible with Newton's atomism or because there was a contradiction between it and the new Hypothesis regarding qualities: these apparent difficulties could have been clarified. Although it would have involved a considerable re-organization of the prefatory section, and would have obscured, as I have suggested, the main tactical purpose of Rule III. But another reason suggests itself. As Professor Cohen has stated, Newton came to regard Hypothesis III as just that: a hypothetical doctrine which went beyond its evidence. But Newton's new conception of transmutation restricts such changes to the secondary qualities, which are a data of experience, and which according to Rule III admit of degrees of intensity. Every example of such change that Newton discusses in his published work is an ordinary fact which any contemporary could have accepted. Apart from the theory of the tails of comets, there is no reference, after 1690, to any of the sort of phenomena discussed in alchemical literature. Thus, transmutation probably became for Newton a phenomena of nature, abundantly supported by a wide range of examples.[58] Such that any statement which formulated this new conception of transmutation would be more akin to an axiom: that is a statement "deduced" from phenomena and explained by the theory of atomism. Thus not only was the old Hypothesis III incompatible with the category of Rule, but had Newton formulated his new conception of transmutation it, too, would have been in a different category – the category of an axiom which is incompatible with being either an hypothesis or a Rule. So again there would have been no room for a statement on transmutation at the beginning of Book Three.

## CONCLUSION

I have tried in this study to present the evolution of Newton's ideas on transmutation as they appear in the *Principia* and the *Opticks*, and to characterize the style of his thought. A great deal more could be said. In particular, a definitive view of Newton's ideas on transmutation will be impossible in the absence of a thorough analysis of his alchemical manuscripts. The new evidence that has been considered is fragmentary, and is, indeed, difficult to interpret. Nevertheless it is of first importance. It gives us a glimpse of the turns and complexities of Newton's thought, as he faced, in the 1690's, new problems in the theory of matter. It is clear that Newton underwent an intellectual struggle of some intensity, and that eventually he decided – there is no way of knowing precisely when – not to present a formal definition of his new ideas regarding transmutation. This, as I have argued, was for a prudential rather than an intellectual motive; for there is no evidence that Newton thought transmutation in itself was either an hypothesis or a misconceived doctrine.

## NOTES

[1] Is. Newton, *Philosophiae Naturalis Principia Mathematica*, London, 1687, p. 402. The Latin reads: "Corpus omne in alterius cujuscunque generis corpus transformari posse, & qualitatum gradus omnes intermedios successive induere".

[2] Only two scholars have been explicitly concerned with this problem, Professor Koyré and Professor Cohen. *Vide* I, Bernard Cohen, "Hypotheses in Newton's Philosophy", *Physis rivista Internazionale Di Storia Della Scienza*, vil. VIII, Fasc. 2, 1966, pp. 163–184, and Alexandre Koyré, *Newtonian Studies*; "Newton's Regulae Philosophandi", London, 1965, p. 263; see also Koyré's, "Pour Une Edition Critique Des Oeuvres De Newton", *Revue D'Histoire Des Sciences*, Huitième Année, tome VIII, 1955, p. 32.

[3] Isaac Newton, *Philosophiae Naturalis Principia Mathematica*, Cambridge, 1713, pp. 357–358.

[4] See Chapter 6. It attempts to throw light on the development of Newton's ideas regarding essential qualities, and on his individual method of introducing and justifying the primary/secondary quality distinction.

[5] MS. U.L.C. Add. 3965.6, folio 267$^r$. Coro 3. Corpora universa quae circa Terram sunt gravia sunt in Terram, et pondera ommium quae (in eodem saltem climata) aequaliter distant a centro Terrae sunt ut quantitates materiae in ÿsdem. Patet in sensibilibus omnibus per experientiam ideoque de insensibilibus etiam intelligendum est per Hypoth III. Is aether aut corpus aliud quodeunque – – – (per Hypoth IV) – – – in Corollario superiore.

Newton's interpolations and cancellations will be given in round brackets, and my glosses will be given in square brackets.

[6] Benjamin Motte, *The Mathematical Principles of Natural Philosophy*, vol. II, London, 1729, p. 203.

[7] *Vide* note 1, *Principia*, p. 402.

[8] *Vide* note 1, *Principia*, p. 411, The translation is mine.

[9] *Vide*, note 3, p. 368.

[10] "Therefore a vacuum is necessarily granted. For if all spaces were full, the specific gravity of the fluid which fills the region of the air on account of the extreme density of the matter, would fall nothing short of the specific gravity of quicksilver or gold or any other of the most dense bodies. And on that account neither gold, nor any other body, could descend in air; for bodies do not descend in fluids, unless they are specifically heavier than the fluids". The translation is mine.

[11] MS. U.L.C. Add. 3965.6, folio 267$^r$. Valet haec demosntratio contra eos qui Hypothesin vel tertiam vel quartam admittunt. Siquis Hypothesibus hisce repudiatis ad Hypothesin tertiam recurrat nempe materiam aliquam non gravem dari per quam gravitas materiae sensibilis explicetur; necesse est ut duo statuat particularum solidarum genera quae in se mutuo transmutari nequent: alterum crassiorum quae graves sint pro quantitate materiae et ex quibus materia omnis gravis, totusque adeo mundus sensibilis conflatur, & alterum tenuiorum quae sint causa

gravitatis crassiorum sed ipsae non sint graves ne gravitas earum per tertiam genus explicanda sit (& ea hujus per quartum et) sic deinceps in infinitum. Haec autem debent esse longe tenuiores sic per actionem suam crassiores discutiant & ab invicem dissipent: qua ratione corpora omnia ex crassionibus composita cito dissolverentur. Et cum actio tenuiorum in crassiores proportionalis fuerit crassiorum superficiebus, gravitas autem ab actione illa oriatur et proportionalis sit materiae ex qua crassiores constant: necesse est ut superficies crassiorum proportionales sint earum contentis solidis, et propterea ut particulae illae omnes sint aequaliter crassae utque nec frangi possint nec alteri vel ratione quacunque communui, ne proportio superficierum ad contenta solida et inde proportio gravitatis ad quantitatem materiae mutatur. Igitur particulas crassiores in tenuiores mutari no posse et propterea duo esse particularum genera quae in se mutuo transire nequeunt omn [ino] statuendum est.

12 A. Rupert Hall and Marie Boas Hall, *Unpublished Scientific Papers of Isaac Newton*, Cambridge, 1962, p. 312.

13 MS. U.L.C. Add. 3965.6, folio 266$^r$. Newton's cancellations are italiced. The Latin reads: (sic statuunt *deleted*). Peripateticorum [*changed from* Peripatetici] [Epicureorum *deleted*] & Cartesianorum [*changed from* Cartesiani] est Hypothesis et (non absolute statuitur sed. *deleted*) contra eorum praejudicia solummodo dirigitur (a quibus statuitur. Siquis aliter rentiat contra hunc non. *deleted*).

14 *Ibid.*, folio 266$^r$.

15 *Ibid.*, folio 266$^r$.

16 *Ibid.*, folio 266$^r$.

17 The other two statements are fully discussed in Chapter 6.

18 I have added the name Anaxagoras.

19 MS. U.L.C. Add. 4005, folio 81$^r$. It is difficult to interpret the instruction which Newton has written in English. It seems, however, that he full well intended to contrast the "hypotheses" outlined in the second paragraph with the axiom itself. The passage is translated from the Latin:

> Axiom. Corpus omne in alterius cujuscunque generis corpus transformari posse et qualitatum gradus omnes intermedios successive induere. Hoc a Philosophis (prope) omnibus statuitur. Ex materia quadam communi formas (& texturas) varias induente res omnes oriri et in eandem per privationem formarum et texturarum resolvi docent omnes: ut Cartesiani qui materiam illam extensione diversimode divisam et modificatam esse volunt; Peripatetici qui materiam (quandam) primam informen (& formarum omnium capacem) fingunt; Epicurus [*deleted*] (Moschus, Thales) caeterique (Pythagoras Democritus caeterique) qui vel ex atomis diversimode connexis vel quatuor elementis a communi materia orirendis res alias componunt. Solus excipiendus est [Anaxaguras] Sed hujus homoeomeria nunquam obtinuit. Fingi quidem potest corporum (insensibilium) genus aliquod existere ex quibus corpora sensibilia nunquam formentur et propterea horum proprietates ex corporum sensibilum proprietatibus colligi non posse. Sed (hujus) Hypothesis est quae in rerum phaenomenis non fundatur et (contra somnia non disputo) somnia non expendo [write this at y$^e$ y$^r$ end of ye foregoing Axiom] Pecal haec Hypothesis in Axioma.

20 MS. U.L.C. Add. 4005(2), folios 81$^{r-v}$.

21 Hall and Hall, *op. cit.*, pp. 341 and 342.

22 *Ibid.*, p. 341.

23 Sir Isaac Newton, *Opticks or a Treatise of the Reflections, Refractions, Inflections & Colours of Light*, Based on the Fourth Edition London. 1730. Dover Publications, Inc., 1952, p. 396.

24 *Ibid.*, pp. 374–375.

25 I. Bernard Cohen, *op. cit., vide* note 2, p. 177.

26 Alexandre Koyré, *Newtonian Studies*, London, 1965, p. 263.

27 *Ibid.*, p. 263.

[28] *Ibid.*, p. 263.
[29] *Opticks, op. cit.*, p. 400. My italics.
[30] *Ibid.*, p. 268.
[31] Hall and Hall, *op. cit.*, p. 341.
[32] *Opticks, op. cit.*, p. 394.
[33] *Ibid.*, p. 400.
[34] MS. U.L.C. Add. 3965.6 folio 270$^r$. Translated from the Latin.
[35] *Opticks, op. cit.*, p. 400.
[36] Hall and Hall, *op. cit.*, p. 307.
[37] *The Correspondence of Isaac Newton*, ed. H. W. Turnbull, vol. I, Cambridge, 1959, p. 364. the brackets are in the printed text.
[38] *Ibid.*, p. 364.
[39] *Ibid.*, p. 366.
[40] Johs. Lohne. "Newton's Proof" of the sine Law and his Mathematical Principles of Colours", *Arch. for the History of Ex. Sc.*, I, p. 401 (1961). Newton imagined the earth to be like a sponge, sucking up spirituous matter from the heavens to quench its thirst. And as this stream of aethereal matter fell to the earth it propelled bodies above the earth to descend by its impact. Once in the bowels of the earth this matter sublimated, changing first into an aereal and then into an aethereal form as it rose again into the atmosphere and heavens. *Correspondence, op. cit.*, p. 366.
[41] Although in the 1670's, and early 1680's Newton had not been convinced of the periodic motions of the comets, by the time of the *Principia* he had shown that they move in orbits. See the correspondence with J. Flamsteed, February 28, 1680/81, *Correspondence, op. cit.*, pp. 340–347, and September 19, 1685, pp. 419–421. Since the tails of the comets when near the sun always pointed away from it, Newton concluded that they manifested a force which overcame gravity, and that their vapours could be pulled into the earth.
[42] *Principia*, Motte edition, vol. II, pp. 371–372.
[43] *Ibid.*, p. 387.
[44] W. G. Hiscock, Ed: *David Gregory, Isaac Newton & their Circle, extracts from David Gregory's Memoranda 1677–1708*, Oxford, 1937 [1705].
[45] *Ibid.*
[46] For a full account of these definitions see Chapter 3.
[47] *Ibid.*
[48] *Op. cit.*, note 2, p. 179.
[49] *Ibid.*, pp. 175, 177 and 178.
[50] J. E. McGuire and P. M. Rattansi, "Newton and the Pipes of Pan", *Notes and Records of the Royal Society*, Vol. 21, No. 2, 1966, pp. 108–143.
[51] E. Anscombe and P. T. Geach, *Descartes Philosophical Writings*, London, 1964, p. 186. "Again, a body cannot be divided into so many parts that we do not conceive of these parts as being divisible – – –."
[52] *De Caelo*, Book I, 268$^a$, 5–10. W. D. Ross ed. *The Works of Aristotle*, vol. II, Oxford, 1930. Newton knew Aristotle's *De Caelo, vide* Hall & Hall, *op. cit.*, p. 311.
[53] The Latin reads: "Hoc Cartesiani facile concedent. Peripatetici non item, qui specificam differentiam ponunt inter materiam caelestem ac terrestrem. Nec Epicureae philosophiae sectatores, qui atomos rerumque semina immutabilia ponunt. Royal Society MS. 210, folio 72.
[53] In a manuscript fragment connected with revisions to the old Rule III in the early 1690's, Newton makes the point explicit:

The Peripateticks placed the 4 Elements of earth water air & fire under y$^e$ orb of y$^e$ Moon & made all the heavens above them to be filled with a Quintessence, pretending y$^t$ Nature abhorred a vacuum, that is that matter was necessarily in all places (by the necessity of its nature *cancelled*) or that matter was self existing. The Cartesians tell us that matter is space itself & therefore must be everywhere. But what . . . . U.L.C. Add. 3965, folio $129^v$.

[55] S. P. Rigaud, *Historical Essay on the first publication of Sir Isaac Newton's Principia*, Oxford, 1838, p. 101.

[56] *Ibid.*, p. 101.

[57] For an account of these documents see J. E. McGuire and P. M. Rattansi, *op. cit.*

[58] I do not wish to imply that Newton became disinterested in alchemy after 1690. The problem, however, of relating his private alchemical ideas to his public utterances on transmutation is difficult, and remains to be done. I am happy to say that this has now been done. See B. J. T. Dobbs, *The Janus Faces of Genius: The Role of Alchemy in Newton's Thought* (Cambridge, University Press, 1991).

# INDEX

## The University of Western Ontario Series in Philosophy of Science

15. W. L. Harper, R. Stalmaker and G. Pearce (eds.): *Ifs. Conditionals, Belief, Decision, Chance, and Time.* 1981 ISBN 90-277-1184-4; Pb 90-277-1220-4
16. J. C. Pitt (ed.): *Philosophy in Economics.* 1981 ISBN 90-277-1210-7; Pb 90-277-1242-5
17. M. Ruse: *Is Science Sexist?* And Other Problems in the Biomedical Sciences. 1981 ISBN 90-277-1249-2; Pb 90-277-1250-6
18. N. Rescher: *Leibniz's Metaphysics of Nature.* A Group of Essays. 1981 ISBN 90-277-1252-2; Pb 277-1253-0
19. L. Laudan: *Science and Hypothesis.* Historical Essays on Scientific Methodology. 1981 ISBN 90-277-1315-4
20. W. R. Shea (ed.): *Nature Mathematized.* Historical and Philosophical Case Studies in Classical Modern Natural Philosophy. 1983 ISBN 90-277-1402-9
21. M. Ruse (ed.): *Nature Animated.* Historical and Philosophical Case Studies in Greek Medicine, 19th-Century and Recent Biology, Psychiatry, and Psychoanalysis. 1983 ISBN 90-277-1403-7
22. W. R. Shea (ed.): *Otto Hahn and the Rise of Nuclear Physics.* 1983 ISBN 90-277-1584-X
23. H. F. Cohen: *Quantifying Music.* The Science of Music at the First Stage of the Scientific Revolution (1580-1650). 1984 ISBN 90-277-1637-4
24. R. E. Butts: *Kant and the Double Government Methodology.* Supersensibility and Method in Kant's Philosophy of Science. 1984 ISBN 90-277-1760-5; Pb 90-277-2384-2
25. J. R. Brown (ed.): *Scientific Rationality.* The Sociological Turn. 1984 ISBN 90-277-1812-1
26. F. Wilson: *Explanation, Causation and Deduction.* 1985 ISBN 90-277-1856-3
27. J. C. Pitt (ed.): *Change and Progress in Modern Science.* 1985 ISBN 90-277-1898-9
28. H. B. Hollinger and M. J. Zenzen: *The Nature of Irreversibility.* A Study of Its Dynamics and Physical Origins. 1985 ISBN 90-277-2080-0
29. K. Okruhlik and J. R. Brown (eds.): *The Natural Philosophy of Leibniz.* 1985 ISBN 90-277-2145-9
30. G. Oddie: *Likeness to Truth.* 1986 ISBN 90-277-2238-2
31. F. Wilson: *Laws and Other Worlds.* A Human Account of Laws and Counterfactuals. 1986 ISBN 277-2232-3
32. J. Earman: *A Primer on Determinism.* 1986 ISBN 90-277-2240-4; Pb 90-277-2241-2
33. R. E. Butts (ed.): *Kant's Philosophy of Physical Science.* Metaphysische Anfangsgründe der Naturwissenschaft (1786-1986). 1986 ISBN 90-277-2309-5; Pb 90-277-2310-9
34. I. B. MacNeill and G. J. Umphrey (eds.): *Advances in the Statistical Sciences.* Festschrift in Honor of Professor V.M. Joshi's 70th Birthday (6 volumes). Vol. I: Applied Probability, Stochastic Processes, and Sampling Theory. 1987 ISBN 90-277-2393-1
35. I. B. MacNeill and G. J. Umphrey (eds.): *Advances in the Statistical Sciences.* Vol. II: Foundations of Statistical Inference. 1987 ISBN 90-277-2394-X

# The University of Western Ontario Series in Philosophy of Science

36. I. B. MacNeill and G. J. Umphrey (eds.): *Advances in the Statistical Sciences.* Vol. III: Time Series and Econometric Modelling. 1987 ISBN 90-277-2395-8
37. I. B. MacNeill and G. J. Umphrey (eds.): *Advances in the Statistical Sciences.* Vol. IV: Stochastic Hydrology. 1987 ISBN 90-277-2396-6
38. I. B. MacNeill and G. J. Umphrey (eds.): *Advances in the Statistical Sciences.* Vol. V: Biostatistics. 1987 ISBN 90-277-2397-4
39. I. B. MacNeill and G. J. Umphrey (eds.): *Advances in the Statistical Sciences.* Vol. VI: Actuarial Science. 1987 ISBN 90-277-2398-2
    Set ISBN (Vols 34-39) 90-277-2399-0
40. N. Rescher: *Scientific Realism.* A Critical Reappraisal. 1987
    ISBN 90-277-2442-3; Pb 90-277-2528-4
41. B. Skyrms and W. L. Harper (eds.): *Causation, Chance, and Credence.* 1988
    ISBN 90-277-2633-7
42. W. L. Harper and B. Skyrms (eds.): *Causation in Decision, Belief Change and Statistics.* 1988 ISBN 90-277-2634-5
43. R. S. Woolhouse (ed.): *Metaphysics and Philosophy of Science in the 17th and 18th Centuries.* Essays in Honor of Gerd Buchdahl. 1988
    ISBN 90-277-2743-0
44. R. E. Butts and J. R. Brown (eds.): *Constructivision and Science.* Essays in Recent German Philosophy. 1989 ISBN 0-7923-0251-6
45. A. D. Irvine (ed.): *Physicalism in Mathematics.* 1989 ISBN 0-7923-0513-2
46. J. van Cleve and R. E. Frederick (eds.): *The Philosophy of Right and Left.* Incongruent Counterparts and the Nature of Space. 1991 ISBN 0-7923-0844-1
47. F. Wilson: *Empiricism and Darwin's Science.* 1991 ISBN 0-7923-1019-5
48. G. G. Brittan, Jr. (ed.): *Causality, Method and Modality.* Essays in Honor of Jules Vuillemin. 1991 ISBN 0-7923-1045-4
49. W. Spohn, B. C. van Fraassen and B. Skyrms (eds.): *Existence and Explanation.* Essays in Honor of Karel Lambert. 1991 ISBN 0-7923-1252-X
50. J. C. Pitt: *Galileo, Human Knowledge, and the Book of Nature.* Method Replaces Metaphysics. 1992 ISBN 0-7923-1510-3
51. V. Coelho (ed.): *Music and Science in the Age of Galileo.* 1992
    ISBN 0-7923-2028-X
52. P. Janich: *Euclid's Heritage: Is Space Three-Dimensional?* 1992
    ISBN 0-7923-2025-5
53. M. Carrier: *The Completeness of Scientific Theories.* On the Derivation of Empirical Indicators within a Theoretical Framework: The Case of Physical Geometry. 1994 ISBN 0-7923-2475-7
54. P. Parrini (ed.): *Kant and Contemporary Epistemology.* 1994
    ISBN 0-7923-2681-4
55. J. Leplin (ed.): *The Creation of Ideas.* Studies for a Methodology of Theory Construction. 1995 ISBN 0-7923-3461-2
56. J.E. McGuire: *Tradition and Innovation.* Newton's Metaphysics of Nature. 1995 ISBN 0-7923-3617-8

KLUWER ACADEMIC PUBLISHERS – DORDRECHT / BOSTON / LONDON